RELIGION, EDUCATION, SCIENCE AND TECHNOLOGY TOWARDS A MORE INCLUSIVE AND SUSTAINABLE FUTURE

This book addresses topics relating to religion, education, science, and technology, and explore their role in developing a more inclusive and sustainable future. With discussions viewed through the lenses of religious and Islamic studies, education, psychology, social science, economics, and natural science, the book is interdisciplinary. It also brings together a range of diverse work by academics around the world including Indonesia, Malaysia, the United States, Australia, Kenya, Germany, and the Philippines.

The papers are derived from the 5th International Colloquium on Interdisciplinary Islamic Studies (ICIIS 2022), a prestigious event designed to provide a global forum for academicians, researchers, practitioners, and students to present their research findings to global experts. ICIIS was hosted by (State Islamic University/UIN) of Syarif Hidayatullah Jakarta's School of Graduate Studies, Indonesia in collaboration with UIN Sultan Thaha Saifuddin Jambi, Indonesia, UIN Mataram Nusa Tenggara Barat, Indonesia and Umma University Kajiado, Kenya.

PROCEEDINGS OF THE 5TH INTERNATIONAL COLLOQUIUM ON INTERDISCIPLINARY ISLAMIC STUDIES (ICIIS 2022), LOMBOK, INDONESIA, 19–20 OCTOBER 2022

Religion, Education, Science and Technology towards a More Inclusive and Sustainable Future

Edited by

Maila D.H. Rahiem
UIN Syarif Hidayatullah Jakarta, Indonesia

Routledge
Taylor & Francis Group
LONDON AND NEW YORK

First published 2024
by Routledge
4 Park Square, Milton Park, Abingdon, Oxon OX14 4RN

and by Routledge
605 Third Avenue, New York, NY 10158
e-mail: enquiries@taylorandfrancis.com
www.routledge.com – www.taylorandfrancis.com

Routledge is an imprint of the Taylor & Francis Group, an informa business

British Library Cataloguing-in-Publication Data
A catalogue record for this book is available from the British Library

Library of Congress Cataloging-in-Publication Data
A catalog record has been requested for this book

ISBN: 978-1-032-34416-4 (hbk)
ISBN: 978-1-032-34417-1 (pbk)
ISBN: 978-1-003-32205-4 (ebk)

DOI: 10.1201/9781003322054

Typeset in Times New Roman
by MPS Limited, Chennai, India

Religion, Education, Science and Technology towards a More Inclusive and Sustainable Future – Rahiem (Ed.)
© 2024 the Editor(s), ISBN: 978-1-032-56461-6
Open Access: www.taylorfrancis.com, CC BY-NC-ND 4.0 license

Table of Contents

Preface ix
Acknowledgment xi
Committee Members xiii
Organizing Committee xv

Religion, education & psychology

Obscuring the natural way: A philosophical and religious response to acquisition and growth in Enlightenment thought 3
C. Helsing

Harvesting the potential of diversity through peer collaboration in educational settings 13
L. Zander

Radicalization and proximal concerns for society: The case of the United Kingdom 24
P.M. Henry

Pakikipagkapwa (fellowship) as mode for an inclusive and sustainable future 33
F. Del Castillo

Muslim Nobel Laureates in science: The legacy of redefining the nexus of Islam and modernity 37
S. Naim & F.R. Widjayanto

Revitalization of uṣūl al-fiqh through iḥtiyāṭī principles 44
M. Rosyid, A.N. Kholiq, F. Bintarawati, M.K. Rofiq, L. Lutfi & M.N. Irfan

Seyyed Hossein Nashr's criticism of problems of modernity 50
B. Badarussyamsi, E. Ermawati & M. Ridwan

"Scenting the heart" and "two wings" of sufism 56
A.M. Usep

The term 'adl in the Qur'an perspective of tafsir Maqāṣidī Ibn 'Āsyūr 61
J. Azizy, Sihabussalam, Dasrizal, Asmawi, S. Jannah & G. Wasath

Contextualizing the transformations of Swahili Qur'anic translations within the contemporary religio-cultural-political dynamics of the world 66
H.I. Mohamed

The ecology of the Qur'an: Religion-based environmental preservation efforts 72
J. Azizy, B. Tamam, N.A. Febriani, Sihabussalam, H. Hasan & H.H. Ubaidah

Gotong royong, an indigenous value for a more inclusive and sustainable future 77
R. Latifa & N.F. Mahida

The implementation of multicultural education in Indonesia: Systematic literature review 83
M. Huda, A. Saeful, H. Rahim, D. Rosyada, M. Zuhdi & S. Muttaqin

Strengthening emotional intelligence, spiritual intelligence, and school culture on student character 90
K.I. Rosadi, A.A. Musyaffa, K. Anwar, M. El Widdah & M. Fadhil Idarianty

Problems facing digital learning in Jakarta (secondary schools as a case study) 97
A.A. Sihombing, M. Fatra, H. Rahim, F.D. Maigahoaku, L.R. Octaberlina & R.M. Jehaut

Muslim education in Kenya: Challenges and opportunities 103
A.A. Ali

Child-friendly schools in Indonesia: Validity and reliability of evaluation questionnaire 108
T. Wulandari, A. Mursalin, Sya'roni, F.K. Dewi, Atika & Baharudin

Challenges of mastering information and communication technology literacy competence for teachers in the age of digital learning 113
Reksiana, A. Zamhari, M. Huda, D. Rosyada & A. Nata

The effect of self-awareness and critical thinking on students' ability to do scientific activities 118
A. Syukri & Sukarno

Distance learning in the perspective of Sadd al-Żarāi'during the Covid-19 pandemic 123
A. Arifin, R. Yazid, I. Sujoko, H. Hasan & S. Muttaqin

Educational model in Tzu Chi school in terms of multicultural education dimensions 128
Nukhbatunisa, M.R. Lubis, A. Zamhari, A. Khoiri, W. Triana & I.Y. Palejwala

Innovative diffusion and social penetration in education: WhatsApp as a medium of communication for thesis guidance between lecturers and students 133
E. Hadiyana, Fahrurrozi & M. Fakhri

Integration and interconnection between Al-afidah and neurosciences in the human learning process 139
Syukri, E.A. Subagio & Zulyadain

Shifting education and technology through the lens of constructivism 144
A. Saeful, M. Huda, S.U. Masruroh, D. Khairani & S. Muttaqin

A scoping review: Nonsuicidal self-injury during emerging adulthood for a more empathetic understanding 149
Pihasniwati, H.L. Muslimah, R.R. Diana, A. Mujib, D. Saepudin & U. Kultsum

"The school does not belong to me:" Involving parents in an Islamic boarding school in Indonesia 156
D.E. Ginanto, Alfian, K. Anwar, Noprival, K. Putra, K. Yulianti & T. Mulyadin

The influence of Jambi Malay language use toward the implementation of Islamic *aqidah* educational values at Orang Rimba 161
D. Rozelin, D. Mustika, M. Arifullah, Musli, Mailinar & U. Azlan

Analysis of the utilization of instructional media and technology during instruction of Arabic language at Umma University 167
O.R. Omukaba

The mythology of Putri Mandalika in the Sasak Islamic tradition in Lombok 172
F. Muhtar, S.A. Acim, Ribahan & A. Fuadi

Society and humanity

Seloko adat as an identity and existence of the Jambi Malay society in Indonesia 179
M.I. Al Munir, M. Habibullah, P. Abbas, M. Rusydy, S.R. Jannah & Masiyan

Challenges facing Muslim converts in the Republic of Kenya: A case study of Mumias-Kakamega county 184
H.Y. Akasi

Participation of elderly workers in Jambi province 191
A.S. Prasaja, M. Soetomo, R. Ferawati, A. Halim & R.D. Ariani

The challenges faced by modern Muslim Indonesian society in the Industrial Revolution era 196
S. Ramadhan, A. Zamhari, M.I. Helmi, A.M. Albantani & I. Subchi

Digital participatory archive for justice: An inclusive future for vulnerable groups 202
T.Y. Sari & K.P. Silalahi

Can wives deradicalize their husbands? 207
E. Kurniati & A. Zamhari

Role of SUPKEM leadership in fighting against extremism and terrorism in Kenya 213
H.Y. Akasi

Consequeces of panai money in Siri and value of Maqasid Al-Syariah in tribe marriage Bugis in Indragiri Hilir Riau 218
B. Maani, Syukri & Aliyas

Knowledge sharing on local wisdom in the workplace: A systematic literature review 223
R.J. Fakhlina, Zulkifli, S. Basuki & A. Rifai

Grave pilgrimage as a transcendental communication media (Ethnographic study of Diamlewa and Oimbani grave pilgrimage) 230
Jamaludin & S.N. Yaqinah

The social movement of women in pesantren (Islamic boarding schools): From empowerment to resistance against patriarchal culture 235
K.N. Afiah, Kusmana, A.N.S. Rizal, D.A. Ningrum, A. Khoiri & A.J. Salsabila

The political Islam perspective of social movement theory: A case study of FPI in Indonesia 240
A. Khoiri, Kusmana, H. Hasan, Nukhbatunisa, D.A. Ningrum & J. Azizy

Economy and science

Sustainability challenges of community-based social security program in Indonesia: Effective collaboration and digital technology adoption 249
S. Hidayati, N. Hidayah & Kamarusdiana

Governance and outreach of Islamic microfinance (BMT) toward cash waqf 254
A.S. Jahar, R.A. Prasetyowati & I. Subchi

The concept of Dharurah: A review of its basic guidelines and application in shariah economic contracts 261
J.M. Muslimin, F.M. Thamsin & W. Munawar

Islamic feminism in Indonesia: The case of Fiqh an-Nisa program P3M/Rahima 267
N. Hidayah, S. Hidayati & K. Zada

Development of digital entrepreneurship programs in pesantren in Indonesia 273
Sarwenda, H. Rahim, D. Rosyada, A. Zamhari & A. Salim

Biodiversity of freshwater microalgae of Jangkok River, Lombok, Indonesia: An analysis for sustainable future 279
N. Purwati & E.T. Jayanti

Diversity of pollinators in cucumber plantation on organic and conventional farming in Jambi 285
B. Kurniawan, D. Putra, L. Anggriani, A.Q. Manurung, M. El Widah & Ramlah

Author index 293

Religion, Education, Science and Technology towards a More Inclusive and Sustainable Future – Rahiem (Ed.)
© 2024 the Editor(s), ISBN: 978-1-032-56461-6

Preface

Bringing together scholars from around the world, the 5th International Colloquium on Interdisciplinary Islamic Studies (ICIIS) provided an opportunity to share the most recent research and ideas on religion, education, science, and technology in an inclusive and sustainable world. This year's conference covered various topics, from religious philosophy, education issues, and health advancement to environmental concerns. ICIIS was hosted by (the State Islamic University/UIN) of Syarif Hidayatullah Jakarta's School of Graduate Studies, Indonesia, in collaboration with UIN Sultan Thaha Saifuddin Jambi, Indonesia, UIN Mataram Nusa Tenggara Barat, Indonesia and Umma University Kajiado, Kenya.

For the convenience of reading, this book is divided into three major sections: first, religious, educational, and psychological writings. The scope of the discussion extends from A philosophical and spiritual response to the acquisition and development of Enlightenment thought to Muslim Education in Kenya, multicultural education, mythology, and the Islamic tradition.

The second section consists of a collection of society and humanities-related essays on identity. The social movement of peasant women and society, Challenges confronting Muslim converts in the Republic of Kenya, the lives of the elderly, and the role of youth in developing peasant women's community and social initiative. Additionally, on Shifting education and technology through a constructivist lens.

Economics, various science and math-related themes, including the integration of Islam and science, science learning methods, and an overview of Indonesian students' high-order thinking, are all covered in the third section.

It is critical to understand that this book does not simply present research findings; it also makes recommendations to governments, educational institutions, and other stakeholders on improving the quality of education through technology-based educational programs.

Prof. Asep Saepudin Jahar, M.A., Ph.D.
Chairman of the 5th ICIIS
Director of School of Graduate Research 2019–2023
Rector of UIN Syarif Hidayatullah Jakarta 2023–2027

Religion, Education, Science and Technology towards a More Inclusive and Sustainable Future – Rahiem (Ed.)

Acknowledgements

This work would not have been possible without the strong collaboration of Syarif Hidayatullah Jakarta's School of Graduate Studies, Indonesia, UIN Sultan Thaha Saifuddin Jambi, Indonesia, UIN Mataram Nusa Tenggara Barat, Indonesia, and Umma University Kajiado, Kenya. I would like to explicitly thank the authors from Indonesia, Australia, the United States, Germany, Malaysia, Kenya, and the Philippines for contributing their ideas and research to this book. I would also like to express my deepest gratitude to everyone with whom I had the privilege of working on this volume, including the editorial support team, the chair, co-chairs, and members of the 5th ICIIS organising committee. Great appreciation for their efforts in organising the event, disseminating the invitation to contribute to the publication, organising writing seminars, and offering technical assistance to make this book a reality. In addition, my gratitude to the scientific committee's diligence in ensuring the quality of the papers published in this book is abundant and greatly appreciated, as are all the brilliant ideas that guided this conference from its planning phases to the publication of this book.

Maila D.H. Rahiem
Editor

Religion, Education, Science and Technology towards a More Inclusive and Sustainable Future – Rahiem (Ed.)
© 2024 the Editor(s), ISBN: 978-1-032-56461-6
Open Access: www.taylorfrancis.com, CC BY-NC-ND 4.0 license

Committee Members

Scientific Committee

Chair:

Dr. Minako Sakai	(*University of New South Wales, Canberra, Australia*)

Members:

Prof. Dr. Steven Eric Krauss	(*Universiti Putra Malaysia, Serdang, Malaysia*)
Prof. Dr. Melanie Nyhof	(*Carthage College, USA*)
Prof. Dr. Claire Edwina Smith	(*Flinders University*)
Prof. Dr. Zulkifli	(*UIN Syarif Hidayatullah, Jakarta, Indonesia*)
Prof. Dr. Kusmana	(*UIN Syarif Hidayatullah, Jakarta, Indonesia*)
Prof. Dr. Amelia Fauzia	(*UIN Syarif Hidayatullah, Jakarta, Indonesia*)
Prof. Dr. Ratna Sari Dewi	(*UIN Syarif Hidayatullah, Jakarta, Indonesia*)
Arif Zamhari Ph.D.	(*UIN Syarif Hidayatullah, Jakarta, Indonesia*)
Dr. Imam Subchi	(*UIN Syarif Hidayatullah, Jakarta, Indonesia*
Dr. Muhammad Zuhdi	(*UIN Syarif Hidayatullah, Jakarta, Indonesia*)
Dr. J.M. Muslimin	(*UIN Syarif Hidayatullah, Jakarta, Indonesia*)
Usep Abdul Matin, Ph.D.	(*UIN Syarif Hidayatullah, Jakarta, Indonesia*)
Dr. Rena Latifa	(*UIN Syarif Hidayatullah, Jakarta, Indonesia*)
Sabilil Muttaqin, Ph.D.	(*UIN Syarif Hidayatullah, Jakarta, Indonesia*)
Dr. Flori Ratna Sari, Ph.D.	(*UIN Syarif Hidayatullah, Jakarta, Indonesia*)
Prof. Siti Nurul Azkiyah, Ph.D.	(*UIN Syarif Hidayatullah, Jakarta, Indonesia*)
Prof. Mohamad Abdun Nasir	(*UIN Mataram, NTB, Indonesia*)
Dion Eprijum Ginanto, Ph.D.	(*UIN Sulthan Thaha Saifuddin, Jambi, Indonesia*)
Dr. Ahmed Abdi Ahmed	(*Umma University, Kajiado, Kenya*)
Dr. Ibrahim Hassan Ali	(*Umma University, Kajiado, Kenya*)
Dr. Ali Adan Ali	(*Umma University, Kajiado, Kenya*)
Dr. Fides del Castilo	(*De La Salle University, Manila, Philippines*)
Dr. Aria Nakissa	(*Washington University, St. Louis, USA*)
Dr. Nur Surayyah Madhubala Abdullah	(*Universiti Putra Malaysia, Serdang, Malaysia*)
Dr. Serafettin Gedik	(*Amasya Üniversitesi, Amasya, Turkey*)
Dr. Phil Henry	(*University Derby, Derby, UK*)
Dr. Andrew Bills	(*Flinders University, Adelaide, Australia*)
Dr. Eva Fakhrun Nisa	(*Australian National University, Australia*)
Fatwa Adikusuma, Ph.D.	(*University of Adelaide, Canberra, Australia*)
Dr. Mathew Piscioneri	(*Monash University, Melbourne, Australia*)

Religion, Education, Science and Technology towards a More Inclusive and Sustainable Future – Rahiem (Ed.)

Organizing Committee

Chair:
Prof. Asep Saepudin Jahar, M.A., Ph.D (*UIN Syarif Hidayatullah, Jakarta, Indonesia*)

Members:
Prof. Dr. H. Ahmad Syukri (*UIN Sulthan Thaha Saifuddin, Jambi, Indonesia*)
Prof. Dr. Fahrurrozi, MA (*UIN Mataram, NTB, Indonesia*)
Dr. Hamka Hasan (*UIN Syarif Hidayatullah, Jakarta, Indonesia*)
Dr. Asmawi (*UIN Syarif Hidayatullah, Jakarta, Indonesia*)
Dr. Imam Sujoko (*UIN Syarif Hidayatullah, Jakarta, Indonesia*)
Dr. Lalu Nurul Wathan (*UIN Mataram, NTB, Indonesia*)
Dr. Abdul Aziz (*UIN Mataram, NTB, Indonesia*)
Dr. Badarus Syamsi (*UIN Sulthan Thaha Saifuddin, Jambi, Indonesia*)
Dr. Diana Rozelin, M.Hum (*UIN Sulthan Thaha Saifuddin, Jambi, Indonesia*)
Dr. Sukarno (*UIN Sulthan Thaha Saifuddin, Jambi, Indonesia*)
Dr. Manswab Mahseen Abdirahman (*Umma University, Kajiado, Kenya*)
Mr. Hassan Yusuf Akasi (*Umma University, Kajiado, Kenya*)

Secretariat & Workshop:
Siti Ummi Masruroh, M.Sc. (*UIN Syarif Hidayatullah, Jakarta, Indonesia*)
Dewi Khairani M.Sc. (*UIN Syarif Hidayatullah, Jakarta, Indonesia*)
Dewi Aprilia Nigrum, M.A. (*UIN Syarif Hidayatullah, Jakarta, Indonesia*)

Religion, education & psychology

Religion, Education, Science and Technology towards a More Inclusive and Sustainable Future – Rahiem (Ed.)

Obscuring the natural way: A philosophical and religious response to acquisition and growth in Enlightenment thought

C. Helsing*
High Point University, US

1 INTRODUCTION

A sustainable future for human existence can only be secured when sustainability becomes the central value that organizes human activity. Despite this truth, the idea of sustainability is strikingly absent from the Enlightenment philosophies that gave rise to contemporary models of political and economic organization. The absence of sustainability as a central value in political and economic thinking means political and economic models struggle to accommodate the conceptual relationships necessary for ensuring a sustainable future. This struggle is reflected in the theoretical and practical failures of contemporary models and the failure of these models to address the immanent crisis of ecological collapse.

In this address, I examine the role of acquisition and growth in Enlightenment thought, particularly as they guide the writings of John Locke and Adam Smith. These assumptions in their respective philosophies sever individuals and society from ecological thinking. In response to this schism between humanity and nature, I consider writings from Indigenous American authors Corbin Harney and Joseph Bruchac. Indigenous writings not only acknowledge the interdependence of humanity and nature but contain clear warnings against separating the human from the natural. Looking next to Chinese thought, I argue that social-economic stability and ecological stability are not exclusive. In contrast to Enlightenment thought, classical Chinese thought explicitly recognizes the dependence of social-political stability on ecological stability. Finally, in response to Locke's appeal to theological claims to justify acquisition, I examine the relationship between humanity and nature in the Abrahamic traditions of Judaism, Christianity, and Islam. Despite Abrahamic religions predicating a hierarchical relationship between God, humanity, and nature, Abrahamic scriptures also recognize nature as a divine creation and as a source of spiritual inspiration. This recognition entails rejecting selfish, destructive desires and recognizing acquisition and growth as the ends of human activity. In place of these destructive values and goals, I argue that global religious traditions share a widespread recognition for the ascetic virtues of self-control and self-restraint in the pursuit of a greater spiritual existence.

2 ACQUISITION AND GROWTH IN ENLIGHTENMENT THOUGHT

The philosophical basis for acquisition and growth as a primary mechanism of social organization is found in the writings of John Locke and Adam Smith. Locke and Smith, seeking security against the abuses of the European monarchy, base their respective political and economic theories on conceptions of the individual self. However, in looking at the

*Corresponding Author: chelsing@highpoint.edu

DOI: 10.1201/9781003322054-1

individual as the basis of social organization, both Locke and Smith conceptualize the self as separate from and superior to environmental concerns and ecological relations.

2.1 *Acquisition and property in Locke*

A product of the fractious religious wars and political turmoil of post-reformation Europe, Locke's writings provide the philosophical basis for the general concept of a Western liberal society. For Locke, society emerges from the association of autonomous individuals, persons capable of freely selecting their own ends, who agree to form mutual social compacts for the purpose of greater social benefit. Locke's vision of an individualist society includes principles still viewed as central to contemporary liberal democratic societies: natural rights, social contracts, the rule of law, and the separation of religious authority and civic authority. However, in seeking to secure the philosophical autonomy of the individual against unjust legal or political authority, Locke conceives of the individual as separate from any relation except by contractual choice.

This includes separation from and superiority to ecological relations. This can be seen most clearly in Locke's writings on property, which seek to establish the basis for individualist property claims over natural resources. For Locke, property begins with self-ownership; ownership of one's own body and ownership of one's own labor (Locke 1999). Locke argues that labor applied to resources produces property (116). Locke translates this principle of ownership into justification for a natural right to enclosure, to seclude land and resources as individual property (Locke 1999). Locke does present one caveat: an individual is only entitled to as much land as they can use without leading to waste (Locke 1999). This seems to provide a corrective against overconsumption, as waste results in forfeiting a claim to resources. However, Locke argues that currency provides a means of transforming resources into nonperishable wealth (Locke 1999).

From a theoretical standpoint, Locke's claims rely on (1) the idea of an individual existing separately from relations except through an intentional contract, and (2) that such individuals possess an intrinsic claim on property. Regarding the first point, human beings exist within and through relations in the world, not separately from relations. Human consciousness emerges from our embodied existence within the world and from our interactions with dynamically changing phenomena and relations. Buddhism rejects metaphysically independent existence and instead regards all existence as dependently originating from interdependent, impermanent conditions. The American philosopher John Dewey argues that "[p]ersonality, selfhood, subjectivity are eventual functions that emerge within complexly organized interactions, organic and social" (Dewey & Hook 1981). In a Confucian sense, we develop our moral agency through recognizing and nurturing the mutual relationships that structure our existence. Contrary to Locke, natural and social relations are not constraints on freedom, but the vehicles through which we develop our capacities for cultural proficiency, human compassion, and moral agency.

The recognition of the dependence of human existence within an already existing world of natural relations also calls into question a second point in Locke's argument: that human beings possess intrinsic claims to property. Locke's argument for intrinsic property incorrectly ignores the role of social relations in recognizing, respecting, and regulating property claims. As communitarian thinkers demonstrate, there are no rights without communities to recognize and protect those rights. This is particularly true for property claims: property claims have no meaning except within civic and cultural frameworks that recognize and respect such claims.

In one sense, Locke's general vision of a liberal society crystallizes the philosophical transition away from European feudalism. In another sense, however, Locke's vision of the individual removes the individual from the natural relations that provide the necessary conditions of life. In more practical terms, Locke's claims ironically underpin rationalizations for the colonial seizure of territory and hostile European expansion. Locke himself argued that indigenous tribal lands in the Americas could be rendered more productive if managed under European agrarian practices (Locke 1999). Ironically, Locke leveraged the very same argument designed to protect individual property rights against those who would become victims of colonial conquest.

2.2 *Growth and nature in Smith*

If the writings of Locke elevate the ideas of individualism, contracts, and property as premises of social organization, the writings of Adam Smith elevate the ideas of growth and profit to the central role of social, political, and economic organization. Adhering throughout his work to the principles of supply and demand, Smith argues that the material welfare of human society improves with increased production and greater, more affordable access to the commodities necessary for life. For Smith, these improvements occur through greater specialization in labor, technological advancement, reinvestment of capital, and the guiding hand of rational self-interest (Smith 2000). However, Smith's reduction of human flourishing to the increase of material commodities blinds him to the necessities of sustainability and interdependence. Like Locke before him, Smith's model of human society exists separately from natural relationships, except as resources to be consumed for economic growth.

In attempts to secure the philosophical autonomy of individuals, Enlightenment thought broadly predicates a mythological "state of nature" in which human beings exist prior to social organization and live independently of social contracts. In this model, the state of nature serves less to articulate relationships between humanity and ecological systems and more to create an artificial distinction between feudalism and modernity. Enlightenment thinkers regard the state of nature as a time prior to rationalism, commerce, and autonomous contracts, not as a system of natural relations that includes human society. Smith's version of this story has two notable effects. First, Smith, like Locke, privileges growth and acquisition above harmony or sustainability. Second, Smith's mythology rewrites the history of enclosure in Tudor England, in which the landed nobility mobilized efforts to eliminate agricultural commons and consolidate territory for private use. Smith obscures this history by claiming that then-present social distinctions (particularly that of lord or landowner versus tenant) were the result of individual virtues, as opposed to class struggle and the consolidation of economic and political power.

Ironically, although Smith recognizes that his vision of economic growth requires continual consumption of natural resources, he oddly never recognizes the necessity of sustainable consumption. Appealing to the idea of supply and demand, Smith notes that every species of animal multiplies in proportion to its means of subsistence (Smith 2000). Instead of serving as a source of concern, Smith, who seems oblivious to overconsumption, regards the increase as playing a positive role in the production of wealth. For Smith, the commodification of labor regulates the growth of the population: increasing the reward of labor increases the growth of the population (Smith 2000). Smith regards the "liberal" (generous) reward of labor as the "necessary effect and cause of the greatest public prosperity" (Smith 2000). The liberal reward of labor is useful for the increase in population (and thus productivity) and the conditions of the working poor.

Like Locke, Smith also implicitly justifies the colonial acquisition of land and resources. Smith states, "it is in the progressive state, while the society is advancing to the furthest acquisition," that commerce is most rewarding to landless labor (Smith 2000). The closest Smith comes to recognizing the problems of overconsumption are in the effects on landless labor: a "static" society, in which production and consumption are equivalent, is a "dull" state; a declining state, in which consumption outpaces production, is "melancholy" (Smith 2000).

In their pursuit of individual autonomy, liberal society, and material wealth, Enlightenment thinkers conceptually sever the intrinsic relationships between human beings and our natural environments. Enlightenment thought privileges acquisition and control over integration and harmony, consumption over sustainability and renewal. This reorganization of values has broad implications for social and economic theory, most notably in the complete absence of consideration for sustainability. In social theory sustainability and harmony become relegated to personal values, not intrinsic requirements for human society. In economic theory, resources are only regarded in terms of the cost of acquisition. Waste is only regarded as an "externality." Neither is considered in terms of their impact on social stability or even the stability of the economic system itself.

While social and economic theory have undergone several centuries of development and details have changed, sustainability and natural harmony remain completely missing from standard models. This omission is evident in every abandoned factory town, in every clear-cut forest, and in every jungle slashed and burned for progress. This omission is visible in the exploitation and abandonment of labor, in the erosion and erasure of local production, and in the despair and unrest that come from resource insecurity. Globally, the collective forgetting of harmony and sustainability is seen in the complex fragility of commercial networks, in the opacity and abuse of regulations, and in the increasingly deleterious role of commercial networks in political conflicts.

3 RELIGION AND SUSTAINABILITY

Religion is not merely a collection of belief statements and rules. Religious traditions, broadly speaking, constitute integrated visions of life. These visions are shaped by voices within each tradition – voices that articulate values that bind together the disparate moments of human experience into a meaningful, coherent whole. Furthermore, the human capacities for imagination and expression that give rise to symbolic form are the same capacities that give rise to all arts and sciences. The values that shape religious expression are values that can also shape expressions of scientific inquiry, social development, and economic activity.

3.1 *Indigenous traditions: animals remember what people forget*

Where Enlightenment thought severs humanity from natural relations, religious and intellectual traditions throughout the world recognize the interdependence of humanity and nature. This interdependence is clearly recognized in indigenous traditions around the world. The words of native American author Corbin Harney draw attention to the dialogical relationship between humanity and other living organisms:

> Sometimes we, the people, don't realize all the living things out there have got a voice, they've got a spirit like we do … All the life on this Earth today has to take care of each other. Those are the reasons why my people used to tell me, "whenever you take a life of all different things, make sure you tell them why you're taking their life. Even if it's a plant, it understands you, it can work with you if you ask. When you ask those things to take care of you, it can take care of you." (Harney & Purbrick 2009)

All living beings want to live, and all living beings have the ability to help others live. In Harney's words, the moral obligation of all living beings is mutual care. Mutual care arises from mutual understanding, and mutual understanding comes from mutual communication and respect. This is most clearly seen in the gesture of the request. Articulating human needs in the form of a request treats other living organisms as moral equals. This respects the desire for life found in all living beings. In this sense, respecting life on earth means not reducing life on earth to merely the means of satisfying human ends. Respecting life means not subsuming the totality of relations that constitute the natural world to exclusively anthropocentric desires. Human beings must respect the need for natural processes to exist for their own ends. Failing to respect the multilateral complexity of nature leads to the disruption of those very processes that support human life.

Beyond simply respecting nature as a moral equal, Indigenous thought also emphasizes the importance of learning from nature.

> Later that same day, as I walked under a low autumn sky the scarred color of slate, I heard a sound overhead. Just above the clouds a flight of geese was passing. They were talking to one another. That was what my Abenaki grandfather always said the geese did—they talked to one another. Their yelping calls were so clear, so sharp, so insistently communicating—not just with one another but with everything that breathed and listened, from the ancient blue of Cayuga Lake to the deer in the field that also lifted its head—that they pierced my heart. I would have much preferred to fly off with those birds than go into that classroom again.
>
> Seeing animals as of lesser value than humans has always been called foolishness in American Indian cultures. Not only traditional stories, but personal experience, taught the elders of all our nations that the Animal People care for their families, feel love and sympathy, anger and despair just as human beings do. By observing the animals, humans can learn many things. (Bruchac 2016)

Joseph Bruchac also recognizes the voice of nature. By acknowledging the voice of nature, Bruchac recognizes the moral status of other living beings. As members of a moral community other living beings deserve respect. Bruchac explains that "the animals remember what people forget," meaning that human society forgets its interdependence with the natural world (Bruchac 2016). We accomplish this forgetting through the anthropocentric centering of human desires in our thoughts, words, and deeds. Human philosophy forgets the human dependence on natural relations. Human discourse displaces listening with property claims and plans for increased profits. Human actions build systems of consumption that neglect sustainability.

3.2 *Confucianism: social and ecological equilibrium*

Apart from indigenous traditions, other philosophical and religious traditions recognize the importance of sustainable resources and the dangers of profit and growth as dominant guiding ideals. Chinese and Indian traditions both recognize the relationship between ecological stability and social stability. In India, the Dharma Shastras detail expectations of flood-plain management and agricultural stability attached to the *dharma* (sacred duty) of the ruler. In classical Chinese thought the heavens, earth, and humanity are all part of a system of natural relations. These relations are not static but undergo constant transformation and change. This includes social relations and social institutions, ranging in scale from human familial relations to the cycles of dynastic change. In this light, Confucianism and Daoism both describe strategies for developing social stability within a world of political upheaval and continual transformation.

In Confucianism, the philosopher Mencius clearly recognizes the material basis of social stability and the danger of elevating profit as the central social and political ideal:

> 1A1: Mencius met with King Hui of Liang. The King said, "Venerable sir, you have not considered a thousand *li* too far to come. Surely you have some means to profit our state?" Mencius replied: "Why must the king speak of profit? I have only [teachings concerning] humanness and rightness. If the king says, 'How can I profit my state?' the officers will say, "How can I profit my house?' and the gentleman will say, 'How can I profit myself?' Those above and below will compete with one another for profit, and the state will be imperiled ... when rightness is

> subordinated to profit the urge to lay claim to more becomes irresistible. (Mencius *et al.* 2009)

Mencius' writings keenly identify the role of organizational values. When "profit" (here seen as gain, advantage, or power) becomes the center of organizational activity, the system loses coherence as individual members seek selfish advantage above their peers. The system no longer functions for the general welfare of its constituent members. Rather, the system functions to enrich a privileged minority at the expense of a subservient majority. When this occurs, discontent incentivizes members against the system, and those in power risk total personal and social ruin for comparatively short-term privilege.

Mencius keenly notes the critical role of ecological equilibrium and material stability in meeting human needs, particularly the obligations of family care. This reflects the broader Confucian connection between the material welfare of a population and social stability. Mencius conveys the need for proper social regulation to prevent overconsumption and scarcity.

> 1A3: If the agricultural seasons are not interfered with, there will be more grain than can be eaten. If close-meshed nets are not allowed in the pools and ponds, there will be more fish and turtles than can be eaten. And if axes are allowed in the mountains and forests only in the appropriate seasons, there will be more timber than can be used. When grain, fish, and turtles are more than can be eaten, and timber is more than can be used, this will mean the people can nourish their lives, bury their dead, and be without rancor. Making it possible for them to nourish their lives, bury their dead, and be without rancor is the beginning of kingly government. (Mencius *et al.* 2009)

At the heart of Mencius' moral vision is the cultivation of compassion: when an individual witnesses the suffering of another and feels sympathetic feelings of distress, the feeling of distress becomes the impulse that leads to an expression of compassion. The Confucian project in general recognizes that human beings express moral sentiments of compassion most immediately through familial relations. These familial relations become the basis for learning to grow compassion by extending circles of benevolence.

What Mencius recognizes is the material requirement of benevolence: caring for others cannot be accomplished merely as a feat of emotional expression but requires material goods. Caring for others requires food, clothing, and other resources necessary for survival and comfort. The purpose of political society for Mencius is to ensure the stability of civil society within a natural environment. Tempering human desires for consumption results in ensuring sufficient material resources for all. This allows individuals to fulfill both individual needs and familial obligations. Mencius' argument is that if those above carefully sustain the material conditions that allow individuals to fulfill their relationships, then the people will sustain the political system.

3.3 *Abrahamic religious traditions: Divine creation and self-purification*

Abrahamic religions have a complicated relationship with natural resources. Abrahamic religions all indicate a hierarchical relationship descending from God to humanity and from humanity to nature. In Genesis, God states that nature exists for the purpose of sustaining humanity, and God tasks humanity with cultivating nature for survival. This command also suggests that the difficulty of labor is meant as punishment for defying the Edenic Covenant.

It is no mistake that Locke's writings justify a natural right to property and enclosure by appealing to biblical directions to cultivate the land for nourishment. However, it is a mistake to conflate Locke's arguments with a full understanding of the Abrahamic traditions. In addition to the external challenges to Locke's position, Abrahamic religious traditions also possess important correctives against the wasteful overconsumption of natural resources.

Theologically, Abrahamic traditions view nature as a divine creation. As a divine creation, the natural world deserves respect and care, serves as a source of knowledge, and inspires sacred experience. Writings in Abrahamic religions are diverse and diversely interpreted, but passages from Judeo-Christian writings and Islamic writings both convey the divine nature of the natural world. In Judaism and Christianity, the Book of Job states:

> 12:7 But ask now the beasts, and they shall teach thee; and the fowls of the air, and they shall tell thee: 12:8 Or speak to the earth, and it shall teach thee: and the fishes of the sea shall declare unto thee. 12:9 Who knoweth not in all these that the hand of the LORD hath wrought this? 12:10 In whose hand is the soul of every living thing, and the breath of all mankind. (Cambridge University Press 1611)

In the Holy Qur'an of Islam, Sura 6:28, states:

> There is not an animal that walks on earth and no bird that flies on its wings but are communities like your own. No single thing have We left out of the Book. Then to their Lord shall they all be gathered. (Salahi 2019)

These readings reverse the conception of a strictly hierarchical relationship between humanity and nature. Rather than viewing the natural world as a means of satisfying human desires, these passages view the natural world as a divine creation deserving of respect. Furthermore, both passages view the natural world as capable of inspiring spiritual reflection. If all creation is the result of a divine will, then this divine will can be seen in the manifest nature of all creation. One who listens can hear the presence of the divine in the voices of living beings and the earth itself. One who sees other living beings can see their existence as part of a divine community.

Spiritually, Abrahamic traditions are all concerned with the idea of ethical motivation or intention. Actions are formed in accordance with divine law, which demands selflessness and devotion. Actions performed in the service of selfish motivations are immoral. This can be seen in the Gospel of Matthew, when Jesus calls for rigorous self-examination of one's motives, denounces selfish motivations, and extols the ideal of universal love.

> 6:22 The light of the body is the eye: if therefore thine eye be single, thy whole body shall be full of light. 6:23 But if thine eye be evil, thy whole body shall be full of darkness. If therefore the light that is in thee be darkness, how great is that darkness! (Cambridge University Press 1611).

The metaphorical language is particularly effective in the context of the Gospel's explanation regarding the fulfillment of the commandments; it is not sufficient to simply adhere to the letter of the law, one must examine one's motives for sinful desires and "pluck out" those desires that lead to destructive behavior.

Abrahamic scriptures are particularly concerned with selfish desires related to money and profit, as these quickly provide rationalizations for dehumanizing others or ignoring human suffering. As the Gospel of Matthew observes, the totalizing effect of one's motivations results in an impossible situation: one cannot be devoted to two different visions of the Good. In response to the impossibility of material permanence, Abrahamic religions call instead for spiritual-moral awareness.

> 6:19 Lay not up for yourselves treasures upon earth, where moth and rust doth corrupt, and where thieves break through and steal: 6:20 But lay up for yourselves treasures in heaven, where neither moth nor rust doth corrupt, and where thieves do not break through nor steal: 6:21 For where your treasure is, there will your heart be also. (Cambridge University Press 1611)

Promises of divine rewards may have served the purpose of motivating the faithful to moral behavior, but visions of heavenly kingdoms may also serve as spiritual metaphors.

Attachment to "earthly" sensory pleasures results in selfish, destructive behavior, while commitment to moral behavior connects the individual to a greater spiritual community. As noted by Jewish, Christian, and Islamic mystics alike, this spiritual vision cannot be attained through the selfish pursuit of external reward but can only occur through devotion to the spiritual vision for its own sake. This purity of vision is captured clearly in the poetry of the Sufi mystic, Rabi'a of Basra:

> My Lord, if I worship you in fear of the fire,
> Burn me in hell. If I worship you in desire
> For paradise, deprive me of it. But if I
> Worship you in love of you, then deprive me not
> Of your eternal beauty. (Cambridge University Press 1611)

3.4 *Ascetic traditions and ascetic virtues*

Lastly, religious traditions around the world recognize the importance of ascetic virtues. The ability to abstain from selfish pleasure serves as the first step to experiencing a larger spiritual vision. For example, the Hindu Upanishads teach *damyata*, *datta*, and *damyata*: self-control, generosity, and compassion. In learning to sacrifice one's own pleasure, one turns to the needs of others; in turning to the needs of others, one learns generosity. In learning generosity, one feels compassion.

> The Children of Prajapati, the Creator – gods, human beings, and asuras, the godless – lived with the father as students. When they had completed the allotted periods the gods said, "Venerable One, please teach us." Prajapati answered with one syllable: "*Da*." "Have you understood?" he asked. "Yes," they said. "You have told us *damyata*, be self-controlled." "You have understood," he said.
>
> Then the human beings approached. "Venerable One, please teach us." Prajapati answered with one syllable: "*Da*." "Have you understood?" he asked. "Yes," they said. "You have told us *datta*, give." "You have understood," he said.
>
> Then the godless approached. "Venerable One, please teach us." Prajapati answered with the same syllable: "*Da*." "Have you understood?" he asked. "Yes," they said. "You have told us *dayadhvam*, be compassionate." "You have understood," he said.
>
> The heavenly voice of the thunder repeats this teaching. *Da-da-da!* Be self-controlled! Give! Be compassionate!" (Easwaran 2007)

In Judaism, Christianity, and Islam, this can be seen in the ritual fasting and purification practices related to fasting holidays. Yom Kippur, Passover, and Ramadan (and other holy times) all teach purification through the denial of pleasure. These fasting practices are not limited to food but admonish many forms of sensory pleasure: food, drink, intoxication, sexual intimacy, and other forms of sensory gratification. Abstaining from sensory gratification in these areas teaches the practitioner the self-restraint necessary to not act impulsively or simply for selfish gratification. In ascetic traditions broadly conceived, self-control is the first step to entering a larger existence in relation to other members of a community. Lay practices of fasting and purification serve to introduce the practitioner to an important process of spiritual development. The development of self-control thus complements spiritual rules and guidelines prohibiting socially destructive behavior.

The strongest examples of asceticism might be found in the examples of Jain writings and the great Jain hero, Mahavira. Jainism regards all living beings as possessing *jiva*, a spiritual awareness, and a capacity for life and motion. Violence of any form produces karma, a spiritual impurity that clings to the *jiva* and clouds the consciousness. Karma invades the *jiva* through the senses, trapping the *jiva* in cycles of destruction and rebirth.

> These three classes of living beings have been declared by the Jinas: (1) earth, water, fire, wind; (2) grass, trees, and plants; and (3) the moving beings, both the egg-bearing and those that bear live offspring, those generated from dirt and those generated in fluids. Know and understand that they all desire happiness. By hurting these beings, people do harm to their own souls and will repeatedly be born as one of them. (Kritanga Sutra 1.7.1-9)

Jainism is particularly cognizant of violence toward nature:

> Kritanga Sutra 1.7.5-9: He who lights a fire kills living beings; he who extinguishes it kills the fire. Therefore a wise man who well considers the Law should light now fire. Earth contains life, and water contains life; jumping or flying insects fall in the fire; dirt-born vermin and beings live in wood. All these beings are burned by lighting a fire. Plants are beings that have a natural development. Their bodies require nourishment, and they all have their individual life. Reckless men who cut them down for their own pleasure destroy many living beings. By destroying plants, whether the plants are young or grown up, a careless man harms his own soul. (Voorst & E 2017)

When the *jiva* becomes obscure, the soul becomes blind to the spiritual awareness of other living beings, and suffering is inflicted through one's selfish actions. Separation of karma occurs through purification and austerity: fasting, renunciation, simplicity, and equanimity.

> Uttaradhyayana Sutra 28.30: Austerities are twofold, external and internal. Both external and internal austerities are sixfold. By knowledge one knows things, by faith one believes in them, by conduct one gets freedom from karma, and by austerities one reaches purity. Having destroyed their karma by austerities, great sages go on to perfection and get rid of all misery. Thus I say. (Voorst & E 2017)

In teaching restraint and compassion, the ascetic virtues simultaneously promote individual and social stability while providing a sense of spiritual fulfillment. In working for the good of others, the individual discovers a capacity for fulfillment that goes unserved by simply working for one's own self. In liberating the self from selfish desires, the individual becomes connected to a greater spiritual vision. This greater spiritual vision is nothing less than beholding the beauty of creation and loving the beauty of creation for its own sake, separately from any motivation of acquisition or growth.

4 CONCLUSIONS

Locke's attempt to conceptually decouple individual autonomy from social and political frameworks results in extracting human life from both natural and social relations. However, we argue here that moral agency only truly emerges within the context of a living environment. These living environments include the ecological and social relations that shape, inform, and sustain human life. Moral agency is not a question of radically unconstrained choice: no such choice exists. The conditions of human existence are inextricably entangled with the rest of existence. Morality is not simply a case of deductive reasoning from a set of static premises. Moral agency requires developing a conscientious awareness of the world, an understanding of the relationships that constitute human existence, and the capacity to act responsibly in response to those relations. This responsibility begins with self-awareness and self-restraint. The elements of moral consciousness are necessary for developing a moral agency that can respond to the challenges of cultivating sustainable societies in balance with nature.

Although the writings of Adam Smith seek to improve human welfare through the production and consumption of economic goods, Smith's political economy fundamentally misunderstands the relationship between human society and its ecological foundations. The failure to include sustainability as a key element of human flourishing results in omitting sustainability from measures and standards of economic success. While economic theory has undergone extensive development since the time of Smith's writings, the fundamental assumptions of consumption and production remain unchanged. Questions of sustainability remain "external" to questions of profit. Instead of accounting for the necessary conditions of economic activity, economic theory relegates sustainability to a domain of individual value.

This discussion highlights a few of the many possible responses to Enlightenment models of social and economic organization. There are more voices and traditions than can be recognized in this short space. Intellectual and cultural traditions around the world provide possibilities for remembering human dependence on sustainable ecological conditions. Numerous forms of spiritual and ritual practices exist that develop self-awareness and self-control. The goal is not to find any one intellectual, cultural, or religious tradition that provides a singular response to the global crisis of climate change: the goal is to find many voices that each recognize the interdependence of humanity and nature. Just as humanity exists through irreducibly complex relations in the natural world, the transformation of human societies can occur through overlapping visions of a sustainable future.

REFERENCES

Bruchac, J. (2016) *Our Stories Remember: American Indian History, Culture, and Values through Storytelling*. Fulcrum Publishing.

Cambridge University Press (1611) *Holy Bible, King James Version*. Cambridge University Press.

Dewey, J. and Hook, S. (1981) *The Later Works, 1925–1953*. Edited by J. A. Boydston and J. Ratner. Carbondale: Southern Illinois University Press.

Easwaran, E. (2007) *The Upanishads*. Nilgiri Press.

Harney, J. and Purbrick, A. (2009) *The Nature Way*. The University of Nevada Press.

Locke, J. (1999) *Two Treatises of Government. Hamilton: McMaster University Archive of the History of Economic Thought*. ProQuest Ebook Central.

Mencius. *et al.* (2009) *Mencius. Translations from the Asian Classics*. New York: Columbia University Press.

Salahi, A. (2019) *The Qur'an: A Translation for the 21st Century*. Markfield, Leicestershire: the Islamic Foundation.

Smith, A. (2000) *Inquiry Into the Nature and Causes of the Wealth of Nations*. Infomotions, Inc.

Voorst, V. and E., R. (2017) *Anthology of World Scriptures*, Ninth Edition. Cengage Learning.

Religion, Education, Science and Technology towards a More Inclusive and Sustainable Future –
Rahiem (Ed.)

Harvesting the potential of diversity through peer collaboration in educational settings

L. Zander*
Leibniz University Hannover, Germany

ABSTRACT: The ability to collaborate in diverse work groups has been identified as a critical skill to participate and thrive in changing economies, and researchers have advocated for it to be taught in schools and universities. Somewhat in contrast to the normative rectilinearity of this recommendation, research on the benefits of diverse teams has yielded inconsistent results. Moreover, self-selected collaboration networks are frequently characterized by homophily, and collaboration in groups high in surface-level diversity does not appear to be a regular ingredient of classwork in schools and universities. How then can the ability to collaborate in diverse teams be practiced in educational settings? To substantiate the previous propositions, and answer this question, the essay is structured into four sections: I will first define diversity in collaborative learning, differentiating surface and deep-level diversity. Building on this definition, I will review empirical findings regarding the benefits and challenges of diversity in work groups, drawing social, educational, and organizational research. I will report findings regarding the prevalence of collaboration in diverse groups in schools and universities before summarizing recommendations on how to foster the skills needed for collaboration in diverse teams in schools and universities, underscoring the role of successive collaborative practice. These recommendations are associated with the necessity of conducting systematic, accompanying research to better understand the preconditions for developing the skills required for collaborating in diverse work groups and alleviating students' transitions in increasingly diversified working contexts to inform educational practice.

Keywords: diversity, collaboration, collaborative learning, group work, higher education

1 INTRODUCTION

Collaborating with others, more than just working side by side, implies that learners or members of a team jointly pool their knowledge and resources to reach a common goal that they identify with (Scoular et al. 2020). As such it differs from simple cooperation, which can be achieved through coordinated division of labor. Collaboration is a complex interactive process that is so pervasive and recurrently required in everyday academic and professional life that it should be taught in schools and universities (Hesse et al. 2015). In an increasingly globalized professional context, however, it is not only the ability to *collaborate* but also the ability to collaborate in *diverse* working groups that has been identified by educational experts as a critical skill to participate and thrive in changing economies (Scoular et al. 2020).

Interestingly, research in educational settings suggests that collaborative group work per se is less frequently part of regular class work than could be expected (European Schoolnet 2014; Pianta et al. 2007; Rybińska & Zub 2018). Moreover, when asked for their academic

*Corresponding Author: lysann.zander@iew.uni-hannover.de

DOI: 10.1201/9781003322054-2

collaboration and help networks, students tend to prefer peers who are similar to them in terms of social group memberships, that is, for example, their gender, migration background, and religious affiliation (Zander et al., 2014, 2019), indicating that students' opportunities to practice and rehearse the skills required to successfully collaborate in diverse working groups are rather limited. This provides an interesting backdrop for social psychological and organizational research, which to date has produced mixed empirical evidence for the benefits of diverse work teams. While some studies have found that more diverse working groups are more innovative and productive, particularly in tackling complex tasks, others have highlighted the challenges and performance losses in more diverse working groups caused by less harmonious group processes (Ely 2004; Van Dijk 2012). Similarly, empirical research in educational settings on collaborative learning in diversified small groups, such as that practiced, e.g., in the Jigsaw technique (Banyard and Grayson 1996), has yielded mixed results regarding the benefits (Stanczak et al. 2022). Thus, while students' competence to collaborate in diverse groups is normatively highly valued by educational experts (e.g., Binkley *et al.* 2012), it is still comparatively rare in everyday school and university life. Also, there is little research in education that systematically investigates the effects of successive collaborative practices in diverse groups.

To substantiate these propositions, the essay is structured into four questions: (1) How can we define diversity in collaborative learning? (2) What do we know about the benefits and challenges of diversity and working groups? (3) How common is collaboration in diverse groups in schools and universities? (4) How can the skills required to collaborate in diverse teams be fostered in schools and universities?

The first question pertains to the definition of diversity in collaborative (study) groups in educational settings. Drawing on the insights from social psychological research, I suggest that a better understanding of the dynamics, challenges, and outcomes of collaboration in diverse work teams in research *and* practice requires recognition of the surface-level or "high-visibility" characteristics, but more importantly, the recognition of deep-level diversity (Harrison et al. 2002) as they relate to the collaborative process. The second question pertains to the benefits and challenges of diversity in working groups, a flourishing field of research in social and organizational psychology that has greatly benefited from the refinement of the concept of diversity. Here, I will focus on pertinent findings from experimental and field research before addressing the third question, i.e., how selected findings are disseminated and presented in relation to the use of collaborative practices in schools and universities. In the final section, I will briefly review how teachers and lecturers may create opportunities allowing their students to develop their skills to collaborate in diverse groups and highlight the potential role of successive rehearsals of collaboration in varying student teams of surface- and deep-level diversity. Here, I will argue for educational research to systematically differentiate the introduced types of diversity and to use quasi-experimental research to better understand the preconditions for developing skills required to collaborate in diverse working groups and to facilitate students' transitions in increasingly diversified working contexts to inform educational practice.

1.1 *How can we define diversity in peer collaboration?*

To define diversity and its role in collaborative processes, it seems prudent to first define collaboration. According to the APA dictionary (APA 2022), *collaboration* is defined as "the act or process of two or more people working together to obtain an outcome desired by all" or "an interpersonal relationship in which the parties show cooperation and sensitivity to the others' needs." Collaboration "facilitates the exploration of new ideas, the discovery of solutions, and creation of knowledge unrestrained by a knowledgeable expert who might impose a solution of the problem addresses" (Hmelo-Silver et al. 2013). Implied in these definitions are the notion of status equality and interdependency. It is the symmetry, equality, and mutuality of the relationship that enables students to discuss divergent positions with equal rights and to experience the social and cognitive change arising out of the reconciliation of perspectives (Zander–Mušić 2010). Rather than working individually, side by side, toward a given goal as in cooperation, collaboration also implies that group members depend on each other in their outcomes and cannot fulfill the task independently (von Davier and Halpin 2013). More than that, responsibility for the group outcome is shared while the work necessary to obtain it is divided (Trilling and Fadel 2009; Scoular et al. 2020). In more abstract terms, then, collaboration requires a fine balance between divergence and synergy: while knowledge and expertise can be diverse, certain basic concepts and values need to be shared to pool and integrate the existing knowledge and perspectives. How can this be achieved? According to the framework of teachable collaborative problem-solving skills by Hesse and colleagues (2015), a set of social sub-skills are required to interact smoothly with others in collaborative tasks, which can be coarsely distinguished into three classes: *participation* (e.g., general activity in group work; interacting with, prompting, and responding to others' inputs; perseverance); *perspective taking* (e.g., readapting one's own behavior and explanations to suit others' needs), and *social regulation* (e.g., negotiating goals and strategies; metacognition and transactive memory: recognizing one's own and others' skills given the group's composition). It becomes clear that to engage in a collaborative process with others, students must be strongly committed to developing and maintaining a working alliance and be motivated to overcome and integrate differences in perspectives, knowledge levels, and viewpoints.

What is the role of diversity in collaboration? To answer this question, we must first define diversity. *Diversity* has been broadly defined as differences between individuals on any characteristic or attribute suitable to perceive dissimilarity to one's self (Harrison et al. 2002; Van Knippenberg & Schippers 2007). In research and educational practice, the term diversity often refers to *surface-level differences* in age, gender, ethnicity, and race – characteristics often inferable from appearance. One might add any other element of physical appearance potentially relevant in students' interactions. Rooted in the propositions of Social Identity Theory and Self-Categorization Theory (Tajfel 1978; Worchel & Austin 1986), this type of diversity has also been described as social category diversity (Harrison et al. 2002; Jehn et al. 1999). A critical point in inferring diversity from surface characteristics is the potential discrepancy between the ascribed and *actual* identities experienced by the categorized individual.

Deep-level diversity, on the other hand, has been described as differences in psychological attitudes and values. As shown in work and study groups, deep-level diversity becomes visible over time: in group members' verbal and nonverbal behaviors (Harrison et al. 2002). In academic contexts, other forms of deep-level diversity should be added, including neurodiversity (in terms of sociability, attention, learning, and other mental functions) or socioeconomic background (Clouder et al. 2020). Tasheva and Hillman (2019) describe these latter forms of diversity as human capital and add social capital as "resources and opportunities available through individuals' networks of relationships" (p. 747). The distinction between these types of diversity has significantly contributed to organizational researchers' understanding of the conditions under which particular types of diversity in working groups benefit or harm teams' productivity and personal achievement in organizational contexts (Stahl et al. 2010; van Knippenberg & Mell 2016). In the following section, I will discuss research that has examined selected findings on the effect of diversity in working teams on the outcomes of collaborative processes.

1.2 *What do we know about the benefits and challenges of collaboration in diverse working groups?*

To answer that question, it is worth looking beyond the disciplinary boundaries, as research on the impact of group composition on performance and other outcomes has a long history in the social sciences, with the study of diversity gaining popularity in the past decades

(DeLucia-Waack 2011). Interestingly, social psychological and organizational research is somewhat disconnected from educational research. I will refer to both, as they offer important complementary insights and perspectives.

Research on the outcomes of working group diversity in organizational settings has been guided by two traditions: the information and decision-making perspective and the social categorization and identity perspective (Van Knippenberg and Schippers 2007; Williams & O'Reilly III 1998). The basic assumption of the *information and decision-making perspective* is that groups high in diversity are more likely to possess a broader range of attributes that can be relevant for the collaborative task at hand, particularly when these are more complex nonroutine tasks (Van Knippenberg & Schippers 2007). Creative and new ideas are thought to develop out of the encounter of different perspectives when they, as described above, are integrated in the collaborative process. The basic assumption of the *social categorization and identity perspective* is the idea that varying (surface-level) characteristics of the members of a working group can be used to assert interpersonal similarity and difference and to differentiate ingroup and outgroup members within the group, given that they are salient in the respective context. Because cooperation, trust, and helping are more likely among ingroup members, this perspective often implies more harmonious collaboration in groups of high homogeneity (Van Dijk et al. 2017) According to this perspective, diversity would harm a team's performance and the collaborative process because intergroup biases would catalyze conflict, information exchange, and identification with the team (Guillaume et al. 2014; Stahl et al. 2010). The social category identity perspective offers some interesting ideas on how diversity can affect the collaborative process on the level of social sub-skills required for smooth collaboration (Hesse et al. 2015) in the team described above. Regarding the first facet, *participation* (including interacting with, prompting, and responding to others' inputs), research has found that members of teams with greater demographic diversity are less cooperative (Chatman & Flynn 2001). Additionally, stereotypic expectations of others' ability to contribute to solving collaborative tasks can forestall equal participation of the members in diverse teams, particularly when small groups are committed and oriented to reach a collective goal (as needed in collaborative learning). In their *Expectation States Theory*, Berger and colleagues (2018) suggested that the perceived obligation to reach the goal leads team members to make quick guesses about which members will most likely make a more valuable contribution. These performance expectations are often implicit and typically based on stereotypes associated with surface-level diversity attributes (e.g., gender and affinity to technological issues). They quickly shape the interactive behavior of group members in ways of a self-fulfilling prophecy, which manifests itself in behaviors such as overestimating the competencies of higher status group members or judging their contributions as more valuable (Correll & Ridgeway 2006). Interestingly, computer-mediated interaction has been found to be less shaped by intergroup attitudes, e.g., in terms of responding to communications by members of the same or other gender (Bhappu et al. 1997; Zander *et al.* no date). In this way, Expectation States Theory would predict that – particularly in newly formed teams – surface-level diversity will strongly affect the third class of social skills necessary in collaboration, i.e., *social regulation.* Additionally, participation, negotiation of goals and strategies as well as metacognition, including the knowledge of what others in the group know and are good at, can be shaped by stereotypical expectations (Van Dijk et al. 2017) in diverse teams when social categories and identities are salient. In summary, the positive effects of working group diversity (in terms of productivity) are often discussed in the context of the information and decision-making perspective, while challenges occurring at the interpersonal level in diverse work groups are often explained in reference to the social categorization and identity perspective. Yet, so far, research does not show a consistent pattern suggesting either perspective in its simplest form. Researchers have acknowledged that the effects of diversity and its underlying mechanisms cannot be accurately described by looking at simple main effects only. Instead, accounting for moderating variables, specific dimensions of diversity (surface-level and deep-level diversity) as well as considering interacting dimensions of diversity ("fault lines") contribute to our understanding of diversity effects. For example, in a recent meta-analysis of 44 studies, Wang and colleagues (2019) found that surface-level diversity, across all studies, was unrelated to creativity and innovation – outcomes that are often relevant to and expected from collaborative working groups – while deep-level diversity was positively related with creativity and innovation. The authors also examined several potential moderators including the role of digitality. They found that the relationship between deep-level diversity and creativity was positive in collocated, nonvirtually interacting teams, suggesting that the potential of deep-level diversity can only be harvested to face-to-face interaction. Yet, the authors acknowledge that there is currently too little systematic research including deep-level diversity attributes and moderating factors to offer robust conclusions.

In educational research, the role of group diversity has been extensively studied within the instructional paradigm of cooperative learning (Johnson & Johnson 1999). Johnson and Johnson (1999) have defined five key elements of cooperative learning, such as positive interdependence, individual accountability, face-to-face promotive interaction (including feedback), social and interpersonal skills (including trust and conflict management), as well as group-processing (including an orientation toward the goal).[1] When these elements are in place and instructions are well-prepared and executed to prevent negative group dynamics such as the "free rider effect," they have generally been shown to produce benefits in terms of achievement improvements or ameliorated relationships among members of different social groups (Kyndt et al. 2013). A cooperative learning technique developed in the early 1970s by Elliot Aronson and his students to harness the potential of diverse but segregated classrooms is the Jigsaw technique. Here, students collaborate in differently composed groups in various consecutive phases to discuss and integrate knowledge in other constellations (Banyard and Grayson 1996). The method is high in positive interdependence (Johnson & Johnson 2009); every team member's contribution is crucial for the success of the team. This precondition is also thought to be the mechanism bringing about more positive intergroup attitudes and higher levels of social integration in the classroom. Despite the enormous popularity of this technique, very few studies to date have tested the effects of Jigsaw collaboration on learning in schools (Stanczak et al. 2022). An early study by Moskowitz and colleagues (1985) found no positive effects on students (in terms of their perceptions of classroom climate, attitudes toward peers or school, locus of control, school attendance, or reading achievement – even when the quality of teacher implementation was considered). A Norwegian study by Bratt (2008) did not find positive effects on intergroup attitudes. A recent study by Stanczak and colleagues (2022) also did not find positive effects of a Jigsaw session on achievement outcomes compared with individual learning arrangements and teaching as usual.

It should be noted that while empirical studies rigorously investigating the effects of Jigsaw cooperation in diverse groups are surprisingly rare, studies systematically investigating the effects of time (repetition of collaborative practice, successive rehearsals in the same and varying teams) are, to the best of my knowledge, nonexistent. Given the intense preconditions and skills required for collaborative learning, it would be surprising if randomly composed, diverse teams of students were able to yield the benefits of diversity after one or two single sessions. A study by Harrison and colleagues (2002) indirectly supports this reasoning: Analyzing 144 student project teams over the course of a semester, they found that the effect of surface-level diversity on team outcomes tailed off over time, while the effects of deep-level (psychological) diversity increased.[2] Importantly, the authors noted that

---

[1]Strong overlaps with the "teachable collaborative problem-solving skills" defined by Hesse and colleagues (2015) are notable.

[2]They also included participant's perceptions of team diversity in their analysis. Interestingly, surface-level diversity was not related to team members' perceptions of psychological diversity. Some aspects of deep-level diversity in the team, for example, team members 'conscientiousness, were strongly related to perceptions.

"collaborating or getting together frequently to perform tasks can reduce the impact of demographic differences," that is, the impact of surface-level diversity on a team's outcomes, and that with repeated practice, after negotiation and conflict, deep-level diversity (or homogeneity) becomes the deciding factor.

In summary, results from organizational and academic contexts are far less conclusive than could be expected given the ubiquity of the demand for collaboration in diverse teams.

1.3 *How common is collaboration in diverse groups in classrooms of schools and universities?*

Given the benefits and challenges associated with collaboration in diverse study groups and working teams identified in educational, social psychological, and organizational research, it seems crucial to examine its prevalence in school and university contexts. Given the normative desirability of collaborative skills in diverse work and study groups, there is surprisingly little systematic research on the prevalence of collaborative work in schools and universities, even without considering diversity. Results from The Survey of Schools, commissioned by the European Commission based on data collected from 30 European countries in 2011/2012 show that on average only about 40% of grade 11 students engage in group work on a daily basis, with high variance across countries, meaning that this is not the case for more than half of the respondents (European Schoolnet 2014). Similar shares have been reported in the OECD TALIS study (OECD, 2014). In a recent analysis, Rybińska and Zub (2018) noted that the use of subjective labels, e.g., "frequently," likely leads to an overestimation of the actual use of collaborative practices in classrooms. To support this notion, the authors point to Polish case studies conducted by the Educational Research Institute (Karpiński *et al.* 2013; Muszyński *et al.* 2015, as cited in Rybińska & Zub 2018), finding that in observed lessons "working in pairs or groups constitutes less than 10% of the forms and methods used in the classroom" (p.10). Given the strong advocacy and desirability of collaborative group work voiced in academic publications, these results seem startling. Systematic indicators of the share of collaboration in diverse groups are hard to find. However, an analysis of self-selected peers for help and collaboration supports similar conclusions: Examining the academic networks of 1039 ninth graders, we found that students were more likely to ask a peer for help with learning and studying mathematics when they had the same gender and shared a background of migration and religious affiliation (Zander*et al.* 2019), suggesting that self-selection into diverse study groups is relatively rare. Similarly, in a recent micro-longitudinal case study in two classes, we found that help networks in mathematics and German 1 classes were strongly segregated by gender, a stable trend that only slightly diminished in the higher grades (Zander *et al.* no date).

What do we know about the frequency of collaborative work in university settings? Two decades ago, Sylvia Hurtado (2001) published the data from a faculty survey by the UCLA Higher Education Research Institute with encouraging results. On average, 71.5% of the surveyed faculty reported to require cooperative learning in their teaching. Still, there were significant gender differences, with the female faculty indicating that they would require cooperative learning in some or most courses. The findings, while somewhat dated, point to the importance of teachers as enablers of collaboration in university courses. The analysis does not include varying academic cultures in different academic fields. Yet, recent research suggests that there may be considerable variation across disciplines in the requirement and desirability of collaboration. Analyzing collaboration and help-seeking networks and strategies in computer science and education classes in a German university, we found that students overall were significantly more reluctant to share academic information with their peers in STEM fields (Zander *et al.* no date; Zander & Höhne 2021). This finding suggests an image problem of sharing academic information, help-seeking, and collaboration in computer science but not in educational science. Rather than an essential skill that needs to be taught, sharing academic information with peers could be seen as an indicator of an individual's inability to complete a given task on their own.

As learning has moved to the Internet during the COVID pandemic, school teachers and university instructors have likely decreased the amount of group and partner work given the additional challenges associated with collaborating in online learning environments. While university students may have found it easier to establish collaborative relations with peers during remote learning than primary and secondary students, current data suggests that collaboration is far from being a regular component of teaching arrangements in seminars and tutorials. In a survey conducted in a winter term (February 2022, during the pandemic, yet with classes returned to campus) at the Faculty of Humanities at a German university, we asked students to think of their learning times in seminars. We asked them to report the share of learning that, in their experience happened with other students. On a range from 0% (always alone) to 100% (always with others), 439 students indicated a percentage of 41.45% (SD = 28.56), suggesting the predominance of individual learning for most students. In the spring and summer of 2022, we also asked specifically about collaborative work with other students (excluding lectures as a different teaching format). We first explained collaborative work as group work in which members work intensively toward a goal rather than just distributing and dividing tasks, and gave specific examples. In the spring of 2022, 31.4% of respondents reported that they were not offered this opportunity in their seminars. In the summer of 2022, when classes were fully moved to face-to-face, 33% of students reported that they did not have the opportunity to collaborate in any of their seminars. While these data are not representative, they support the idea that collaboration is less prevalent than advocates of collaborative practice would hope. Systematic assessments of collaborative practices across campuses and countries, complementing students' and teachers' views, are needed to get a more reliable impression of the prevalence of collaborative practices. These assessments should also include student reports of opportunities to collaborate in diverse student groups by considering the basic diversity of the respective campus. While more research is needed to understand the extent to which students in schools and universities are provided with the opportunity to collaborate in diverse groups, it is worth examining what is known about the factors contributing to students' tendency to collaborate in diverse groups (rather than self-selecting into homogenous groups) and teachers' tendency to include collaborative activities in their courses.

1.4 *How can skills required to collaborate in diverse teams be fostered in schools and universities?*

A wealth of recommendations have been published on what can be done to minimize barriers to collaborative learning in classrooms. They range from specific hands-on suggestions for teachers to norm-oriented guidelines addressing principals and school cultures. Generally speaking, teachers are more willing to use collaborative learning when they are informed about or recognize the value of collaborative practices, are able to implement them correctly in their classrooms, have enough time to use them (given the additional time requirements of collaborative learning and extensive curricula), are supported by their principal and school management, and materials for collaborative learning practices are available and can be easily shared among colleagues (Rybińska & Zub 2018). These recommendations, however, address how to increase the prevalence of collaborative learning practices in general and often omit the question of how to manage the aspect of diversity in collaboration.

Given the scarcity of systematic empirical research addressing the processes involved when students collaborate in groups that differ in gender, ethnicity, age, sociability, or cognitive functioning, recommendations are based more on practical experience and conclusions drawn from the literature than on scientific evidence. To better inform and prepare educators, this research is needed, although it does not seem unreasonable to say that the role of teachers cannot be overestimated. Not only by learning about the method, deciding to include collaborative practice in their teaching portfolio, and selecting or developing appropriate material, but also by setting a supportive climate and communicating the relevance of collaborative learning, deciding about the composition of their collaborative groups in their class, moderating the process, and providing feedback, teachers become the critical contextual factor. Unlike researchers in laboratory settings, teachers can and have to consider a range of surface- and deep-level characteristics when composing groups, including gender, cultural backgrounds, bilinguality, neurodiversity, giftedness, and interests. Given these multiple facets, random group assignment may be an equitable option that is acceptable for students when appropriately contextualized by teachers.

In this final section, I will revisit several topics addressed in this essay that may help teachers and researchers alike to better understand the individual and contextual factors required for beneficial collaborative processes in diverse student groups.

From the perspective of the individual, Hesse and colleagues (2015) proposed three subsets of skills required for successful collaboration: participation, perspective-taking, and social regulation. Regarding participation, the selection of material is key. The nature of the task should prevent the applicability of performance expectations regarding surface-level status attributes (Berger & Zelditch 2018; Zander & Höhne 2021). Important insights regarding participation come from research by Cohen and colleagues (Cohen et al. 2002). Cohen and colleagues addressed the issue that neurodiverse students are often perceived as lower status groups in classrooms because they appear to lack social and academic competencies. In a series of studies, Cohen and colleagues (2006) underscored the importance of instruction. The participation of special needs students and the quality of group products increased when teachers underscored the variety of skills valuable and relevant in the classroom context. At the same time, tasks should be created in such a way that competent participation is possible for all participants. Responsiveness, engagement, and participation are likely to be higher when team members feel valued and their knowledge and potentially different perspectives are appreciated by others. Team members who belong to social groups that are underrepresented or even stereotyped in a given context can be particularly responsive to overt and subtle social cues (Inzlicht & Schmader 2012).

Yet Harrison and colleagues (2002) further suggest that when managing work group diversity, it seems prudent "to look beyond the presumed negative impact of surface-level diversity to its possible positive effects (Jehn et al. 1999), as well as to assess the deep-level differences that might erupt into negative affect and relationship conflict. (…) Maximizing differences in knowledge, skills, and abilities, while minimizing differences in job-related beliefs, attitudes, and values, might create especially effective teams (p. 1042)." Again, empirical research is needed to examine these propositions in educational settings.

A second group of subskills required for collaboration, as stated by Hesse and colleagues (2015), are skills involved in *perspective-taking*. Research has shown that perspective-taking can be particularly challenging in intergroup interactions where different social group memberships are salient (Dixon et al. 2005). If practiced, actively considering outgroup members' mental states can attenuate many intergroup biases, including more favorable intergroup appraisals, more positive non-verbal behaviors, increased helping, and less stereotypic evaluations (Todd & Galinsky 2014). Given these benefits, regular perspective-taking exercises, i.e., imagining oneself "in the shoes" of another student, or imagining one's own mental state as if one were in the situation of the other student may also be helpful in building the third set of subskills involved in *social regulation*.

Following this reasoning, occasional, scattered practice does not suffice to release the potential of diverse groups. It should be noted that teachers do not have to solve this challenge on their own. A variety of extracurricular activities and programs are available that provide students with opportunities to interact with peers who are not part of their regular cliques and engage in activities unrelated to class work. One example is the collaborative dance project. Evaluating a German initiative "Time to Dance," Zander and colleagues (2014) found that students who participated in a dance project with their class over the course of one school term had more diverse collaboration peer networks than students who did not participate. This effect was due to boys, who, after participating, were more likely to collaborate with girls than boys who did not participate in the dance project. Importantly, collaboration should be introduced playfully but repeatedly while giving students frequent chances to self-select their favorite collaboration partners to prevent reactance. As teachers, students will self-select into diverse collaborative groups if they recognize the value of collaboration, feel competent to interact in diverse groups, and are supported in doing so.

2 IMPLICATIONS FOR EDUCATIONAL RESEARCH

As noted earlier, a growing body of research shows that students' friendship and academic exchange networks are highly homophilic, suggesting a lack of opportunities for collaboration, even in educational settings characterized by high surface diversity.

At the same time, there is a lack of quantitative research (descriptive, correlational, and experimental) conducted in educational settings that systematically analyze (a) the prevalence and quality of learning and collaborating in surface- and deep-level diverse work groups, (b) the interplay of different dimensions of diversity in shaping the students' and study group outcomes, and (c) the role of contextual affordances in determining the benefits of or challenges associated with work group diversity. This research gap invites a systematic explanation of these issues, the development of theories about collaborative processes in diverse educational contexts, and the derivation of practical recommendations for teachers.

3 CONCLUSION

The way people work has changed rapidly in recent decades. It has become a truism that societies profit from a diverse workforce in terms of economic benefits (Ellemers & Rink 2016). Yet, merely assembling diverse workgroups – in educational or professional contexts – will not per se fulfill the promises of diversity. In the present essay, I aimed to substantiate the position that the ability to interact in teams with high surface and deep diversity, as any other skill, needs to be practiced and rehearsed more frequently by students, supported by their teachers and lecturers in primary, secondary, and tertiary education. At the same time, researchers are tasked with the responsibility of supporting educational practitioners by conducting research that helps teachers and instructors to better understand the conditions under which diversity catalyzes or hinders productive collaborative processes.

REFERENCES

APA (2022) *APA Dictionary of Psychology*. APA. Available at: https://dictionary.apa.org/.

Banyard, P. and Grayson, A. (1996) 'The jigsaw technique', in *Introducing Psychological Research*, pp. 65–68. doi: https://doi.org/10.1007/978-1-349-24483-6_10.

Berger, J. and Zelditch, M. (2018) 'Status cues, expectations, and behavior', in *Status, Power and Legitimacy*. Routledge, pp. 155–174.

Bhappu, A. D., Griffith, T. L. and Northcraft, G. B. (1997) 'Media effects and communication bias in diverse groups', *Organizational Behavior and Human Decision Processes*, 70(3), pp. 199–205. doi: 10.1006/obhd.1997.2704.

Chatman, J. A. and Flynn, F. J. (2001) 'The influence of demographic heterogeneity on the emergence and consequences of cooperative norms in work teams', *Academy of Management Journal*, 44(5), pp. 956–974. doi: 10.2307/3069440.

Clouder, L. *et al.* (2020) 'Neurodiversity in higher education: a narrative synthesis', *Higher Education*, 80(4), pp. 757–778. Available at: https://link.springer.com/article/10.1007/s10734-020-00513-6.

Cohen, E. G. *et al.* (2002) 'Can groups learn?', *Teachers College Record: The Voice of Scholarship in Education*, 104(6), pp. 1045–1068. doi: https://doi.org/10.1111/1467-9620.00196.
Correll, S. J. and Ridgeway, C. L. (2006) 'Expectation States Theory', *Handbooks of Sociology and Social Research*, pp. 29–51. doi: 10.1007/0-387-36921-X_2.
DeLucia-Waack, J. (2011) 'Diversity in Groups', *The Oxford Handbook of Group Counseling*, pp. 1–14. doi: 10.1093/oxfordhb/9780195394450.013.0006.
Dixon, J., Durrheim, K. and Tredoux, C. (2005) 'Beyond the optimal contact strategy: A reality check for the contact hypothesis', *American Psychologist*, 60(7), pp. 697–711. doi: 10.1037/0003-066X.60.7.697.
Ellemers, N. and Rink, F. (2016) 'Diversity in work groups', *Current Opinion in Psychology*, 11, pp. 49–53. doi: 10.1016/j.copsyc.2016.06.001.
Ely, R. J. (2004) 'Ely-2004-JOB-A field study of group diversity, participation in diversity education programs, and performance.PDF', 780(April), pp. 755–780.
European Schoolnet (2014) 'Technology-enhanced collaborative learning', *Briefing Papers*, 8. Available at: http://www.eun.org/documents/411753/817341/Digital+Briefing+Paper+8/2d68991e-36bf-4fed-aa15-18da0ce634da.
Guillaume, Y. R. F. *et al.* (2014) 'Managing diversity in organizations: An integrative model and agenda for future research', *European Journal of Work and Organizational Psychology*, 23(5), pp. 783–802. doi: 10.1080/1359432X.2013.805485.
Harrison, D. A. *et al.* (2002) 'Time, teams, and task performance: a longitudinal study of the changing effects of diversity on group functioning.', *Academy of Management Proceedings*, 2000(1), pp. C1–C6. doi: 10.5465/apbpp.2000.5439172.
Hesse, F. *et al.* (2015) 'Assessment and teaching of 21st century skills', *Assessment and Teaching of 21st Century Skills*, pp. 37–56. doi: 10.1007/978-94-017-9395-7.
Hmelo-Silver, C. E. *et al.* (2013) 'Collaborative learning for diverse learners', in *The International Handbook of Collaborative Learning*. Routledge/Taylor & Francis Group, pp. 297–313.
Inzlicht, M. and Schmader, T. (2012) 'The role of situational cues in signaling and maintaining stereotype threat', in *Stereotype Threat: Theory, Process, and Application*. Oxford University Press, pp. 17–33.
Jehn, K. A., Northcraft, G. B. and Neale, M. A. (1999) 'Why differences make a difference: A field study of diversity, conflict, and performance in workgroups', *Administrative Science Quarterly*, 44(4), pp. 741–763. doi: 10.2307/2667054.
Johnson, D. W. and Johnson, R. T. (1999) 'Making cooperative learning work', *Theory into Practice*, 38(2), pp. 67–73. doi: 10.1080/00405849909543834.
Johnson, D. W. and Johnson, R. T. (2009) 'An educational psychology success story: Social interdependence theory and cooperative learning', *Educational Researcher*, 38(5), pp. 365–379. doi: 10.3102/0013189X09339057.
Kyndt, E. *et al.* (2013) 'A meta-analysis of the effects of face-to-face cooperative learning. Do recent studies falsify or verify earlier findings?', *Educational Research Review*, 10, pp. 133–149. doi: 10.1016/j.edurev.2013.02.002.
Moskowitz, J. M. *et al.* (1985) 'Evaluation of Jigsaw, a cooperative learning technique', *Contemporary Educational Psychology*, 10(2), pp. 104–112. doi: 10.1016/0361-476X(85)90011-6.
Pianta, R. C. *et al.* (2007) 'Opportunities to learn in America's elementary classrooms', *Science*, 315(5820), pp. 1795–1796. doi: https://doi.org/10.1126%2Fscience.1139719.
Rybińska, A. and Zub, M. (2018) *Co-Lab: Final Evaluation and Recommendations Report*. Available at: http://colab.eun.org/c/document_library/get_file?uuid=bb6b6685-7e3c-489f-bc7c-ce24c6a51bb2&groupId=5897016.
Scoular, C. *et al.* (2020) *Collaboration: Skill Development Framework*. Australian Council for Educational Research. Available at: https://research.acer.edu.au/ar_misc/42/.
Stahl, G. K. *et al.* (2010) 'Unraveling the effects of cultural diversity in teams: A meta-analysis of research on multicultural work groups', *Journal of International Business Studies*, 41(4), pp. 690–709. doi: 10.1057/jibs.2009.85.
Stanczak, A. *et al.* (2022) 'Do jigsaw classrooms improve learning outcomes? Five experiments and an internal meta-analysis', *Journal of Educational Psychology*, 114(6), pp. 1461–1476. doi: https://doi.org/10.1037/edu0000730.
Tajfel, H. E. (1978) *Differentiation Between Social Groups: Studies in the social psychology of intergroup relations*. Academic Press.
Todd, A. R. and Galinsky, A. D. (2014) 'Perspective-taking as a strategy for improving intergroup relations: Evidence, mechanisms, and qualifications', *Social and Personality Psychology Compass*, 8(7), pp. 374–387. doi: 10.1111/spc3.12116.

Trilling, B. and Fadel, C. (2009) *21st century Skills: Learning For life in our times*. John Wiley & Sons.
Van Dijk, H. *et al.* (2017) 'Microdynamics in diverse teams: A review and integration of the diversity and stereotyping literatures', *Academy of Management Annals*, 11(1), pp. 517–557. doi: 10.5465/annals.2014.0046.
Van Dijk, H., Van Engen, M. L. and Van Knippenberg, D. (2012) 'Defying conventional wisdom: A meta-analytical examination of the differences between demographic and job-related diversity relationships with performance', *Organizational Behavior and Human Decision Processes*, 119(1), pp. 38–53. doi: 10.1016/j.obhdp.2012.06.003.
van Knippenberg, D. and Mell, J. N. (2016) 'Past, present, and potential future of team diversity research: From compositional diversity to emergent diversity', *Organizational Behavior and Human Decision Processes*, 136, pp. 135–145. doi: 10.1016/j.obhdp.2016.05.007.
Van Knippenberg, D. and Schippers, M. C. (2007) 'Work group diversity', *Annual Review of Psychology*, 58, pp. 515–541. doi: 10.1146/annurev.psych.58.110405.085546.
von Davier, A. A. and Halpin, P. F. (2013) 'Collaborative problem solving and the assessment of cognitive skills: psychometric considerations', *ETS Research Report Series*, 2013(2), pp. i–36. doi: 10.1002/j.2333-8504.2013.tb02348.x.
Williams, K. Y. and O'Reilly III, C. A. (1998) 'Demography and diversity in organizations: A review of 40 years of research', *Research in Organizational Behavior*, 20, pp. 77–140. Available at: https://ils.unc.edu/courses/2013_spring/inls285_001/materials/WIlliams.OReilly.1996.Diversity&demography.pdf.
Worchel, S. and Austin, W. G. (1986) '*The Social Identity Theory of Intergroup Behavior*', in. Nelson-Hall, pp. 7–24.
Zander, L. and Höhne, E. (2021) 'Perceived peer exclusion as predictor of students' help-seeking strategies in higher education', *Zeitschrift für Entwicklungspsychologie und Pädagogische Psychologie*, 51(1–2), pp. 27–41. doi: https://doi.org/10.1026/0049-8637/a000235.
Zander, L. *et al.* (2014) 'How school-based dancing classes change affective and collaborative networks of adolescents', *Psychology of Sport and Exercise*, 15(4), pp. 418–428. doi: 10.1016/j.psychsport.2014.04.004.
Zander, L. *et al.* (no date) 'Entwicklung fachbezogener Hilfenetzwerke unter Peers: Eine soziometrische Analyse unter besonderer Berücksichtigung des Geschlechts [Development of academic help-networks amog peers: A sociometric analysis with a focus on gender]', *PraxisForschungLehrer*innenBildung. Zeitschrift für Schul- und Professionsentwicklung*.
Zander, L., Höhne, E. and Chen, I.-C. (no date) *Too chilly to seek help – Exchange climate as predictor of students' help-seeking in STEM and HEED*.
Zander, L., Chen, I. C. and Hannover, B. (2019) 'Who asks whom for help in mathematics? A sociometric analysis of adolescents' help-seeking within and beyond clique boundaries', *Learning and Individual Differences*, 72(March), pp. 49–58. doi: 10.1016/j.lindif.2019.03.002.
Zander-Mušić, L. (2010) *Perceptions of Ability and Availability of Help Among Classmates – Does Having A Migration Background matter?* Freie Universität Berlin, Germany.

Religion, Education, Science and Technology towards a More Inclusive and Sustainable Future – Rahiem (Ed.)

Radicalization and proximal concerns for society: The case of the United Kingdom

P.M. Henry*
University of Derby, UK

ABSTRACT: Radicalization provides a number of academic and practitioner challenges; these are conceptual, analytical, and practical and are framed methodologically around the desire to better understand the concept of radicalism within political thought. The premise for understanding radicalism is first reflected in the differentiation between nonviolent and violent radicalism and second in the propensity of policymakers in the UK and practitioners working in a multiagency setting to accept a binary approach associated with 'us and them' thinking. In addition, those in practitioner roles, despite a lack of training and skills are expected to successfully intervene where radicalized individuals are subject to state engagement. Drawing on contemporary studies of practitioners on the one hand and radical activists on the other, the study concludes by documenting the ongoing problems with radicalization and identifies that too few studies consider the active promotion of violence as a focal point for researchers' attention. When thinking about the differentiation between violent and nonviolent forms of radicalization, a great deal of academic energy is expended on broad spectra of belief, a few of which have been established as precursors of violence. There is no doubt that radicalization has validity and is a real concern for state agencies and multiagency formations.

Keywords: radicalism, radicalization, counterterrorism, multiagency, prevent strategy, radical activism

1 INTRODUCTION

Radicalization, which was used in the mid-nineteenth century in debates regarding the autonomy of slaves (McLaughlin 2012), in its more recent reintroduction into the English language lexicon, refers to a period in time shortly after the terrorist attacks in New York and Washington, DC, on September 11, 2001, and remains a contested concept associated with negativity that appears to reflect an ill-defined journey for an individual toward extreme views and actions. It is described as a process, and broadly, if crudely framed, as 'what goes on before the bomb goes off' (Neumann 2018). However, what kind of process it is still unclear more than 20 years on. The conceptual understanding of radicalization is challenged largely because of the notion of 'radicalism' (as a general political and philosophical position), which brings with it a range of conceptual debates both in the meaning associated with the term and its use and in the methodology with which academics choose to address this more familiar socially constructed 'ism'-pathway, compared with the newer and emerging noun – radicalization. It is in the context of this debate about how we should better understand radicalization that this paper seeks to briefly examine the case of the UK and its multiagency approach, delimiting the challenges of prevention in a multiagency environment. This is not offered as a template for a globally appropriate method or focus, even though UK prevention methods have been adopted in different forms, in whole or in part,

*Corresponding Author: pm.henry@btopenworld.com

DOI: 10.1201/9781003322054-3

widely across both the global north and the global south. It is an offer open to critical review and comparison by others on the global stage and may or may not resonate with their experiences as academics, policymakers, or practitioners in various fields concerned with radicalism and the prevention of radicalization conceptually and practically toward harms that it might cause. The UK example includes law enforcement, counterterrorism, intelligence networks, and public sector state-sanctioned roles for health, education, social work, and others, which are both encompassed in the laudable position to reduce harm (a statutory requirement), on the one hand, but at odds with the state, that they should even be required to undertake such preventative work on the other. The merits or otherwise of such a challenge are presented here within a threefold analytical framework: (i) in better understanding the language and meaning ascribed in the Anglophone world of the global north (specifically, using the UK as a focal point), (ii) considering the methodological challenges presented by the use of *radicalism* and its relationship to radicalization and extremism, and (iii) in the delineation of theory to praxis and accounting for perspectives from practitioners and radical actors in a multiagency setting in relation to (i) and (ii).

1.1 *Defining the scope of this study*

Radicalism is defined in a number of ways consistent with its etymology, both in historical use and in specific and more general traditional terms. It is also fraught with ahistorical representations that fail to recognize cultural history and its implications for the changes in meaning over time. McLaughlin (2012) defines radicalism 'in terms of a (i) fundamental orientation (toward fundamental objects) (ii) in the political domain (iii) of an argumentative nature.' McLaughlin's (2012) concern with radical political thought does not define it in terms of religious or political ideas, nor does it frame the definition in terms of violence or violent extremism as an outcome of radicalism. Githens-Mazer (2012) similarly developed a definition – but specifically of radicalization – that does not include the words 'religious', 'political', 'ideology', or 'violence,' framing his ideas around an 'individually held, collectively defined moral obligation to participate in direct action, often textually defined.' As the author of this paper, I would like to put forward a more specific definition that could include radicalization within the broader concept of radicalism:

> *It is where actors experience a nonlinear journey providing for a morally negative orientation towards individuals' and groups' concerns, where the influence of those feelings develops emotions from within of which anger, frustration, and injustice (among others) provide fellow feelings that serve to inculcate a moral obligation to act in support of a collective identity* (Henry 2022).

These definitions do not provide for a definitive account of radicalism, or radicalization, as a noun framed within a broader radical tradition of political thought. This remains a contested and often confused space academically and politically. The UK government, for example, with its focus on prevention, has defined radicalization as the process by which a person comes to support terrorism and forms of extremism leading to terrorism' (Home Office 2011). As can be seen from the small sample of definitions examined here, they differ in their attempts to delineate the extent to which conceptually they can reflect a given reality. The significant question raised by states, is one of reducing the opportunities available to any given individual to adopt psychological and social processes in thinking and acting, that the state considers to link the actor to extreme thought and action and potentially to violence as a consequence. As a causal relationship, however, there is no evidence of causation, but rather patterns and trends that present as if they were family resemblances. The obvious challenge being, how does one establish that such a journey could result in acts of terrorism under a range of legal systems globally, when it is obvious from what statistical evidence is available – at least in a UK context (Home Office 2014, 2015, 2016, 2017, 2018, 2019, 2020, 2021); that many who support radical action and actors (individuals or groups) and even violent means as an end to political, religious, or racially motivated outcomes, rarely go on to commit such acts of violence.

The debate, therefore, and the definitional dilemma are thwarted by the lack of an ability to clearly differentiate between nonviolent radicalization and a journey that could result in violent radical action. This is compounded still further by the equally difficult conceptual debate about definitions of terrorism, which, like radicalization, has no universal agreement, but if the link to terrorism is to be understood, a more evidence-based approach is required. The narrower the perspective, the more the definition limits what it stipulates, or within a broad perspective, such stipulations are watered down so much so that the definitions are in danger of becoming meaningless. Epistemologically, what we can know about radicalization and how we frame that knowledge relies on academic and practitioner evidence of working with those individuals who have either been designated by authorities as part of a radicalizing process or, in some rare cases, have self-designated as such as an individual, and they may have undertaken a range of intervention approaches in the UK and elsewhere, which can at least give some credible empirical data to assist in framing the conceptual debates. Albeit that data is limited due to access, accountability, personal data protection, and matters of national security. Finding cohorts of self-designated radical activists without state assistance is a challenge in itself, but one I will refer to later in a rare study undertaken by Derfoufi (2022). In the methods section below the paper examines the challenges alluded to in the introduction conceptually and analytically about our understanding of radicalism and radicalization as one aspect of that broader debate; it also reflects on the ethnographic design of the author's work with practitioners in law enforcement, counterterrorism, communities of interest, and with intervention providers.

2 METHODS

The methods explored here are both conceptual and analytical in terms of the epistemological meaning-making of the language and its use in explaining the phenomenon of radicalism and radicalization. It is empirically framed within a data-gathering process that adopts an ethnographic approach, is socially constructed, and is interpretivist by design. The conceptual and analytical challenges focus on how, where, and when the language and meaning associated with radicalism are used and to what ends. Methodologically, the historical and conceptual approaches help to answer the question: What does it mean to be radical, and what is it that those who might be designated this way are radical about? The second question also poses the significant issue of attribution – who makes the case or justifies who is radical against what criteria? Conceptually, academics and practitioners are contending with a duality, within which there is both a negative and a positive aspect to this phenomenon. Two methodological issues are identified by McLaughlin (2012), the first is the '*elusiveness of meaning*' and the second pertains to the '*contextualization of ideas*,' (McLaughlin 2012) – both ideas have been briefly introduced above. The *elusiveness of meaning* often falls foul of the ahistorical framing, wherein it remains abstract and/or is given a fixed meaning, neither of which bears out the reality within which they are found. For example, if radicalism is both in historical and contemporary use and seen to be progressive, it could be argued to have a positive attribution and equally ascribed with a negative attribution, meaning-making remains elusive, and the stipulated and situational experience of actors becomes more likely to be used to establish meaning in context. It leaves the critic asking, what then are the necessary and sufficient conditions by which either attribution could be understood. For progressive and humanistic radicalism, a position McLaughlin (2012) defends, a positive view of radicalism remains one where the social, political, and/or economic change is endowed with not causing harm, but in seeking a common good. That is not to see *revolution* as the catalyst for change, but a range of activities associated with political thought out of which change may positively affect any given community and has in some cases a bounded legitimacy morally and politically. Conversely, negative views of radicalism, and in this I include *radicalization*, are variously expressed by policymakers and practitioners as psychological and social (behavioral). This position includes a conceptual leap that links radical actors as having the potential to see means–ends outcomes as potentially based on fundamentality and violence as a solution to an in-group problem, be it real or imagined. It is in the *contextualization of ideas* that conformity to appropriate event(s) and/or statement (s) is methodologically part of linguistic conventions where 'the degree of necessity that they capture and theoretical coherence that they yield' (McLaughlin 2012) provides for conditions for evaluative understanding of the often-contestable ideas they seek to explain.

The empirical approach adopts a broadly ethnographic style employing observation and interviewing to investigate the social practices of practitioners and seeks to represent the meaning-making associated with their social interaction (Murphy & Dingwall 2001) in statutory and nonstatutory settings across a number of regions, cities, and towns in England. The researcher spent 14 months in the field, from April 2017 to June 2018, talking to and observing the work of professionals in some of their settings, and has since followed up with professionals in a post-pandemic environment. Observations in professional settings bring with them challenges for the researcher, particularly where radicalization is the topic of interest and people's relationships to managing risk are often associated with it. Even where respondents had no specific responsibility for prevention, there were still apparent sensitivities presented by them (be it real or imagined), and in some settings, a matter of security. During the fieldwork, the focus on how the construction of sensitivity and security plays out both in the professional and community environment was an important consideration. There are significant methods and debates to be addressed in what Eski (2016) talks about in his ethnography of port security, which are also reflected here in the development of a growing 'ethnography of security' in occupational settings. The difference here lies in the variables that shape our understanding of security and ebb and flow in the occupations associated with the Prevent Strategy within the counterterrorism policy, from 'sensitivity to security.' Where there are similarities with Eski (2016) is in the 'socializing, cultivating, and narrating,' where interaction in occupational and community settings is developing 'cultures of commitment' and 'identity formation,' which at times are in opposition to each other (particularly where reluctance, uncertainty, and resilience play a significant part), but have many similar aims.

3 RESULTS

3.1 *Understanding radicalization as a conceptual challenge*

As a contested concept, 'radicalization' is described by many academics (Aly *et al.* 2017; Githens-Mazer 2012; Heath-Kelly 2013; Neumann 2018; Sedgwick 2010; Thomas 2012) and practitioners as a 'process of social and psychological concern.' Yet, such a process seemingly has no defined beginning or end, in a linear sense, and, it can be argued, is in fact 'a nonlinear process in a complex world' (Chandler 2014). If we better understand it in terms of risk at an individual and societal level (Mythen & Walklate 2006), can we reduce the risks from 'radicalization' in individuals and social groups (if indeed we are to understand radicalization conceptually as a risk and/or threat)? 'Manage the risk in order to prevent harm and accept the responsibility on all sides of the debate for such an outcome?' (Shields 2017). This may appear on the face of things to be a laudable response, but in the 21 years since the catalyst of the destruction of the Twin Towers in New York and countless other events described as terrorism under the law, both political and ideological, it leaves us lacking a nuanced understanding.

The concept of radicalization references the negative connotations of a form of knowledge out of which individuals and groups support, promote, and act in opposition to the status quo, yet it has for centuries also had a positive message in terms of radicalism and radical actors seeking change in challenging the social and political norms for the better. Radical views and actions have brought about revolutionary changes, which may be viewed in hindsight as positive. The radical thinkers that changed our world before, during, and after the Enlightenment of the 18th century are not my concern here, nor is the binary thinking associated with classical liberal modernism. From the beginning of human innovation on the planet, change, designed to improve sentient life, has come from myriad sources, not all of which were seen as favorable in their own time. The changing landscape now adopting the concept of 'radicalization' as the *lingua franca* in Anglophone regions of the world forms part of the complexity of life that should be acknowledged as the backdrop to examining and evaluating radicalization and concomitant concerns for global and domestic forms of terrorism (Chandler 2014).

Countering the threat of terrorism under the law has tangible parameters within which, regardless of our views on the type of laws in place, there is a way to measure (in a UK context) how law enforcement, the courts, prison service, and probation (including former Community Rehabilitation Companies) act within their defined roles as actors of the state. The processes of justice function explicitly in the context described under the law. However, what of the function and role of professionals with other statutory and nonstatutory responsibilities, all of whom are connected to the state directly or indirectly, albeit some claiming greater autonomy than others? What of the role of teachers, lecturers (Brown & Saeed 2015), administrators, nursery workers, social workers, doctors, a range of clinicians and nonclinical staff, nurses, ambulance staff, fire service personnel, and charitable organizations (some community-based, others serving communities from outside), many of which, until 2015, stood outside the legal framework affecting terrorism under the law?

In considering the UK counterterrorism provision under the CONTEST Policy (HO 2003, 2009, 2011, & 2018) the counterterrorism policy for the nation (revised in June 2018), the nonenforcement elements of the counterterrorism policy (Prevent, Protect, and Prepare) seem to have a strategic rationale designed around building resilience – regardless of the position you play as either a state or nonstate actor. For this to be possible, legislation like that brought forward under the Counter-Terrorism and Security Act, 2015 (CTSA) requires some scrutiny (partially in the discussion to follow) not least as it presents the following definition of extremism:

> '[as] *sic* vocal or active opposition to fundamental British values, including democracy, the rule of law, individual liberty and mutual respect and tolerance of different faiths and beliefs. We also include in our definition of extremism calls for the death of members of our armed forces, whether in this country or overseas.' (CTSA 2015; Prevent 2011).

This raises the question: Is the state arbitrarily alienating minorities in a community where British values are little known and even less well understood, where freedoms of democracy enshrined in national and international law are seemingly being ignored, and where freedoms of conscience appear to be no longer a contestable position (Kudnanin 2018)? In wanting to avoid binary *us and them* thinking, this definition goes a long way toward inculcating it. If being in opposition to British values makes one an extremist, the fear is where do you draw the line, as most of the population have voiced their disapproval of the rule of law, democracy, liberty, and mutual respect at one time or another. What then would success in countering radicalization, extremism, and terrorism look like if 'resilience-thinking' (to borrow from Chandler) is about 'the complex problem of governance and the policy process suitable to governing this complexity' (Chandler 2014)? In creating arbitrary definitions, like the extremism approach above, is this in any way helpful from a governance perspective? Here, resilience is much more than *bouncing back*; it is to be understood as an analytical and conceptual framework for the discursive management, direction, and development of the learning journeys of professionals and communities within which radicalization is increasingly being seen as a characteristic of complex social and political life. The ill-defined aspect of resilience thinking insofar as radicalization is concerned features in the lack of academic consideration of the very few who go on to become terrorists under the law. Therefore, so-called nonviolent radicalization, surely could turn the tables on the unidirectional debate of escalation from radicalization through extremism to terrorism, by reflecting on those who are exposed to and sympathetic with so-called radical Islamist perspectives, for example, and assessing to what extent they move toward extremism and/or why they do not? (Bartlet & Miller 2012). Rather than seeing a move to political violence, we might take a closer look at resilience to these narratives, as did the recent work of Derfoufi (2022). The empirical data explored briefly in this study below bucks a conceptual and analytical trend in the literature affecting those influenced by radicalism. Participants interviewed provided data that suggests their exposure galvanized their resilience to radicalization and did not increase it. Derfoufi (2022) asks – 'Why participants' radicalism promoted resilience to political violence rather than propel them towards it' (Derfoufi 2022) – a significant shift in the data.

4 DISCUSSION

4.1 *Characterizing radicalization*

The cognitive and behavioral approaches to radicalization (thinking about and acting on a particular course of action) are, it seems, sitting in a morally attributable space within society and have been characterized as outside the *status quo ante* based on societal norms and values. Such pretentions open up debates about what determines the difference between an individual who is in opposition to the *status quo* and one who is seemingly a radicalized individual, *al la* Derfoufi (2022) above, based on what? A state pronouncement? In the UK context that is exactly what happened and as CT policy was developed, the 'Prevent' arm as a strategy document aimed at preventing individuals from engaging in terrorism or supporting terrorists was designed with statutory and nonstatutory actors in mind. Not only as a voluntary model between 2006 and 2015, but in 2015 it became a statutory requirement under controversial legislation, the CTSA (2015). The legislation built on the existing voluntary framework, which had already taken a broadly multiagency approach in professional settings and sought to bring the public along with the sentiments of prevention as a means to reduce the harms of terrorist incidents, events, and group activities, seen as malign by the state.

There are however two fundamental problems in the design of this approach: (1) Professionals in statutory and nonstatutory roles do not have the skillsets to support the ambitions of the strategy, and equally do not, in most cases see themselves as adequately equipped to manage sensitive and potentially life-threatening scenarios in identifying radicalized actors, understanding complex ideologies, and intervening and/or reporting their concerns, to what is a law enforcement arm within the CONTEST policy – the *Prevent Police Teams*. Even though the strategy of Prevent seeks not to criminalize, but to provide support in a safeguarding context for individuals identified to be at risk of radicalization, it also, by dint of the legislation and already agreed safeguarding provisions, presents anyone identified as being in some way vulnerable and/or susceptible to radicalizing and extremist narratives, online and/or offline (McDonald & Mair 2015) where the professional concerned becomes the arbitrator of the decision making about what she/he deems radical material/narratives/images/videos, etc., that warrant a referral. Those being asked to consider their roles in relation to how they might understand and recognize radicalization are, for example, local authority staff and officers, youth workers, social workers, health professionals, teachers, lecturers, police officers, prison or probation workers, and the broader general public within a population; and (2) The process by which individuals are referred to a multiagency panel known as the Channel Panel, is via *Police Prevent Teams*, who act as the lead, even though local authorities are proscribed with chairing and managing such panels. The local authority, however, relies on Police Prevent Teams to undertake the inquiries and therefore provide intelligence about any referred individual(s) so that the panel can make a decision about potential interventions to move a so-called radicalized actor toward a so-called de-radicalization process. Interventions come in many forms, and support similarly could be wide ranging, contingent on a given individual's cognitive, social, economic, and family status, capacity, and capability, and dependent on where the direct link to radicalizing influences might form part of that person's life. If it is within families and/or family members and is seen by the state to be radicalizing others in the family group, it presents greater state actor concerns from a safeguarding perspective. The state may ultimately need to decide if, for example, the family member is a child or young adult, what should be the approach to manage that concern, and would it mean a child being removed from a family setting into *looked after care* provided by the state, via the local authority?

These concerns and others, associated with freedoms of conscience and speech to name but two, concern community narratives in the context of UK Muslim communities, where, until 2020–21 the majority of referrals for radicalization to Channel panels were made up of individuals who would self-designate as religiously Muslim. Since 2020–21 however, referrals of individuals with far and extreme right views and actions adopted as cases in Channel have increased (46%) and overtaken referrals for so-called Islamist radical influenced referrals (22%) (Home Office 2021). A third category of *mixed, unstable, or unclear* ideologies has also figured more readily in the data showing second to far-right referrals in 2020–21 (30%) of adopted cases by Channel. Many referrals during the pandemic period were online postings, which adds to the complexity of disconnected narratives appearing to make a case for violence, often fomented by hate speech or actions, associated with the state, arms of the state, or other communities. The perpetrators are often described by practitioners working in the field of prevention as 'keyboard warrior(s) penning divisive ideological and often racist content online.'

4.2 *Reciprocal radicalization*

There are significant debates in the UK, Europe, and the USA about *reciprocal radicalization* where extremists in militant and radical Islamist groups and/or those in far and extreme right and white supremacist groups are persistently using each other's narratives as *propaganda of the deed* to influence supporters and to galvanize others in an *us and them* environment, which is oppositional at best, divisive, and dangerous with the potential for violence at worst. Subsequently, the securitization of Muslim communities has become a significant issue in academic community and practitioner settings, such that those within radical Islam would fuel the fire of discontent by spreading rumors and conspiracy ideas related to how the state has acted, for example, against Muslims and Muslim youth specifically (statistically, the 15–20 year age range remains the highest proportion of Channel referrals). It is clear, however, from the Home Office figures (2021) that fewer referrals, initially having been triaged by Police Prevent teams, get to a Channel panel and that even fewer, once at Channel, result in direct and sustained action by the panel, and even fewer still become terrorists under UK legislation. This could be because there is a better screening process now in place, but many Muslims would still contend the state is responsible for too many referrals in the first instance due to its perceived lack of tolerance for Muslims rather than a lack of tolerance for radical extremists. This implies that many in Muslim communities have for the last 15 years perceived the state as in opposition to Muslim communities, discriminating against them, and seeing Muslims as both 'risky and at risk' of radicalization, extremism, and terrorism (Githens-Mazer 2012; Heath-Kelly 2013).

4.3 *Empirical evidence*

Empirically, taking a snapshot of professional responses from the author's (Derfoufi 2022) work and a snapshot from Muslims who have been part of the radicalizing process into account, there are a number of converging ideas with which to conclude our discussion. Derfoufi (2022) produces 'research into the political awakening and mobilization of young British Muslims undergoing radicalization who, crucially for the task at hand, were exposed to both nonviolent and violent narratives'. Derfoufi (2022) contends that his study shows that radicalization is not necessarily conducive to terrorism, and his participants demonstrate resilience to it. His claim, as endorsed here, is for 'a stronger focus on empirical research is needed, placing direct action at the heart of radicalization. This is a good starting point because direct action is present in both its nonviolent and violent forms and is, therefore, tangible enough to anchor comparative analyses' (Derfoufi 2022). The research identifies participants' concerns, which are associated with alienation, isolation, police racism, and the impacts of right and left-wing secularists. It talks of relationships where their family members are oppressed, tortured, and displaced, and these create underlying causes associated with radical activism. Defouri (2022) presents a case for the specific concerns of differentiation between radical activism in a nonviolent form and violent radicalism/radicalization. In the former, participants, despite having trigger events associated with racism, family members being killed, or tortured abroad, nonetheless remained opposed to violent action, or if they supported direct action, were not prepared to participate in violence. Quotes from participants' are explained below:

> The characteristics associated with radicalization were not only conceptualized positively but as morally necessary for promoting the "expanded" mindsets necessary to achieve a more "healthy" and just society. Parallels were drawn with formerly marginalized but now iconic figures like Martin Luther King. Given the securitization of young Muslims, this can be interpreted as an important way in which participants internally rationalize and externally articulate the legitimacy of their activities: by positioning themselves as an extension to equality campaigns already familiar to Western societies (Derfoufi 2022).

Extremism, however, was seen as negative by Derfoufi's sample, who states: 'In contrast, extremism was viewed as a negative phenomenon – although not intrinsically violent – for its role in undermining the alternatives envisaged by participants' radicalism' (Derfoufi 2022). The research participants unequivocally condemn both state and nonstate violence as terrorism and are described as developing a *critical consciousness*, resisting dominating structures assumed to hinder social progress.

5 CONCLUSION

Too few studies consider the active promotion of violence as a focal point for researchers' attention when thinking about the differentiation between violent and nonviolent forms of radicalization and spend a great deal of energy on broad spectrums of belief, a few of which have been established as precursors of violence. If we can establish that radicalism is not an obvious precursor to terrorism, at least in its nonviolent form, and better understand the lack of clarity already within policy and practitioner networks, there is a rationale for a review of current practitioner roles that would benefit revisiting our understanding of best practices. This demands that understanding radicalization asks practitioners to consider what type of thinking or even action involves a safeguarding concern and epitomizes the challenges of addressing nonviolent activism. The initial focus on ideological thinking and framing as drivers of radicalization has proved problematic since the inception of the Prevent Strategy in the UK. The emphasis on ideologies to the exclusion of all other forms of promoting violence is difficult to reconcile. There remains some obvious cause for concern based on inciting violent outcomes from radical extremist narratives online, including hatred and racism, which must continue to be the focus of academic attention.

Distinguishing beliefs from actions remains a too-often overlooked academic and practitioner concern. There is, however, justifiable concern among practitioners that radical belief systems have long been the stamping ground for violent actors and should not be ignored. There appears to be a fine line between the debate on national security and what constitutes a breach or potential breach through terrorist activity and where nonviolent extremism sits in terms of a debate about unwanted and inappropriate public integration debates. Should we take Githens-Mazer's (2012) perspective in addressing the many trajectories radicalization can take, and that direct political action is a helpful focus for empirical inquiry? Group dynamics and those with familial connection, *a la* 'band of brothers' (and sisters), is a feature of group mobilization and remains a concern for state actors, but still, there are too few academic studies to help in understanding the dynamics and in promoting single narratives in the relationship between them and how they influence others toward violent direct action. Radicalization as a real concern has empirical validity and affirms the challenges it presents to the existing social and political order. It is also, according to Derfoufi (2022), evidence of rejecting one set of oppressive structures for another in which young Muslims in the West who engage in nonviolent radicalism seek to 'embed themselves in peer-to-peer communities of practice contextualizing their faith within a western context' (Derfoufi 2022).

REFERENCES

Aly, A. *et al.* (2017) 'Introduction to the special issue: Terrorist online propaganda and radicalization', *Studies in Conflict and Terrorism*, 40(1), pp. 1–9. doi: 10.1080/1057610X.2016.1157402.

Bartlet, J. and Miller, C. (2012) 'The edge of violence: Towards telling the difference between violent and non-violent radicalization', *Terrorism and Political Violence*, 24(1), pp. 1–21. doi: 10.1080/09546553.2011.594923.

Brown, K. E. and Saeed, T. (2015) 'Radicalization and counter-radicalization at British universities: Muslim encounters and alternatives', *Ethnic and Racial Studies*, 38(11), pp. 1952–1968. doi: 10.1080/01419870.2014.911343.

Chandler, D. (2014) *Resilience: The Governance of Complexity*. London: HM Stationery Office.

Derfoufi, Z. (2022) 'Radicalization's core', *Terrorism and Political Violence*, 34(6), pp. 1185–1206. doi: 10.1080/09546553.2020.1764942.

Eski, Y. (2016) *Policing, Port Security and Crime Control: An Ethnography of the Port Securityscape*. Abingdon: Routledge.

Githens-Mazer, J. (2012) 'The rhetoric and reality: Radicalization and political discourse', *International Political Science Review*, 33(5), pp. 556–567. doi: 10.1177/0192512112454416.

Heath-Kelly, C. (2013) 'Counter-terrorism and the counterfactual: Producing the "radicalisation" discourse and the UK prevent strategy', *British Journal of Politics and International Relations*, 15(3), pp. 394–415. doi: 10.1111/j.1467-856X.2011.00489.x.

Home Office (2011) *The Prevent Strategy*. London: The Stationery Office.

Home Office (2021) *Official Statistics: Individuals referred to and supported through the Prevent Programme, England and Wales, April 2020 to March 2021*. Available at: https://www.gov.uk/government/statistics/individuals-referred-to-and-supported-through-the-prevent-programme-april-2020-to-march-2021/individuals-referred-to-and-supported-through-the-prevent-programme-england-and-wales-april-2020-to-march-2021 (Accessed: 30 October 2022).

Kudnanin, A. (2018) 'Radicalisation: the journey of a concept', *Race and Class*, 59(4), pp. 34–53. doi: 10.1177/0306396817750778.

McDonald, S. and Mair, D. (2015) 'Terrorism Online: A New Strategic Environment', in Lee Jarvis, Stuart Macdonald and Thomas M. Chen, eds., *Terrorism Online: Politics, Law and Technology*. Abingdon: Routledge, pp. 10–34.

McLaughlin, P. (2012) *Radicalism: A Philosophical Study*. Basingstoke: Palgrave Macmillan.

Murphy, E. and Dingwall, R. (2001) 'The Ethics of ethnography', in Atkinson, P, Coffey, A, Delamont, S, Lofland, J, and Lofland, L, (eds) *Handbook of Ethnography*. London: Sage.

Mythen, G. and Walklate, S. (2006) 'Communicating the terrorist risk: Harnessing a culture of fear?', *Crime, Media, Culture*, 2(2), pp. 123–142. doi: 10.1177/1741659006065399.

Neumann, P. R. (2018) 'Introduction.', in *Perspectives on Radicalisation and Political Violence – Papers from the First International Conference on Radicalisation and Political Violence*. London: International Centre for the Study of Radicalisation and Political Violence. Available at: http://icsr.info/wp-content/uploads/2012/10/1234516938ICSRPerspectivesonRadicalisation.pdf.

Sedgwick, M. (2010) 'The concept of radicalization as a source of confusion', *Terrorism and Political Violence*, 22(4), pp. 479–494. doi: 10.1080/09546553.2010.491009.

Shields, J. (2017) *Countering Online Radicalisation and Extremism: Baroness Shields' Speech, Delivered to the George Washington University Centre for Cyber and Homeland Security's Extremism Programme*. Available at: https://www.gov.uk/government/speeches/countering-online-radicalisation-and-extremism-baroness-shields-speech (Accessed: 7 October 2022).

Thomas, P. (2012) *Responding to the Threat of Violent Extremism – Failing to Prevent*. London: Bloomsbury Academic.

Religion, Education, Science and Technology towards a More Inclusive and Sustainable Future – Rahiem (Ed.)

Pakikipagkapwa (fellowship) as mode for an inclusive and sustainable future

F. Del Castillo*
De La Salle University Manila, Philippines

ABSTRACT: This study utilizes the value of Pakikipagkapwa (Fellowship) and the four models of dialogue. Dialogue takes the form of a *shared life* and *common action*. The Filipino ideal of *pakikipagkapwa* (having positive and sincere relations with one's brethren) promotes positive inclusive relations and social cohesion. The author argues that although religion compelled many believers to help their neighbors, it is the value of *pakikipagkapwa* that undergirds the care and concern of many people who are suffering. *Pakikipagkapwa* enabled cordial cooperation between faith-based institutions and individuals, which promoted the physical well-being of members of the community. This study contributes to a deeper understanding of inclusive and sustainable future in the contemporary Philippines.[1]

Keywords: inclusivity, culture, higher educational institutions, inter-religious

1 INTRODUCTION

The prevalence of religious diversity indicates that a homogeneous understanding of faith or even God is untenable (Tran 2018). Through interfaith dialogue and inclusivity, each faith group can make its unique contribution to peaceful co-existence that fosters social solidarity (United States Institute of Peace 2004). A broad range of interreligious dialogue practices and initiatives exists (Halafoff 2013).

Interreligious dialogue is one of the pillars of mission in the Asian region, along with a dialogue with the poor and dialogue with Asia's cultures (Tran 2018). The interreligious encounter requires "sincere openness of mind and heart, generous hospitality toward the other, inclusivity and a genuine willingness to adapt and change. Moreover, the dialogue must speak truth to power and maintain the integrity of creation" (Phan 2018). The foremost goal of interreligious dialogue is to build communities and inclusivity (Francis 2015).

In the Philippines, the existence of different religions, the prominence of faith for many Filipinos, and the freedom to practice their faith indicate a religious and faith-diverse society. Around 81% of 110 million Filipinos identify as Roman Catholic (Highest Catholic Population 2022). Approximately 6% of Filipinos are Muslim. And the remaining are from other religious denominations.

Many Filipinos regard the concept of relating with one's *kapwa* (brethren) as the essence of human interaction. *Kapwa* is a relationship-based Filipino value that refers to seeing and considering the other person the same way one does oneself. In this way, a person cultivates a humane relationship with another. It can be challenging to describe the nuances of the Filipino word *kapwa* using an English term since it has a socio-psychological component and a moral connotation.

*Corresponding Author: fides.delcastillo@dlsu.edu.ph
[1]A part of this study was submitted and being processed for publication in the HTS Theological Studies (Open access journal). The author retain copyright and non-exclusive right to use of the intellectual property.

DOI: 10.1201/9781003322054-4

Nevertheless, the term "brethren" can be used. Being a *kapwa* is to situate oneself within the other's situation and extend oneself to your brethren's existential condition, especially when the situation requires it. An individual's concern for *kapwa* can serve as a motivation for having an altruistic relationship with others. The Filipino prefix *pakiki-* signifies a commitment to the other person according to the principles of *kapwa*, hence *pakikipagkapwa*. As such, *pakikipagkapwa* is the act of treating others as significant members of one's existential circle.

2 METHODS AND OBJECTIVES

This article is a qualitative study on interreligious dialogue and inclusivity. The essay addresses how different religions can help build a more inclusive and sustainable future in faith-based organizations in the context of the Philippines. It utilizes the Filipino value of *pakikipagkapwa* and the Catholic models of dialogue. It also hopes to contribute to a deeper understanding of an inclusive and sustainable future in the contemporary Philippines and in Asia.

3 THE VALUE OF PAKIKIPAGKAPWA (FELLOWSHIP)

An essential part of Filipino culture is the notion of *kapwa*. The Filipino term *kapwa* is translated into English as "neighbor," "fellow man/woman," or "others" (Tagalog English Dictionary n.d.). However, in Filipino indigenous psychology, *kapwa* is conceptualized as the "recognition of shared identity, an inner self shared with others" (Enriquez 2004). It is *kapwa* that "bridges the deepest recess of a person with anyone outside him or herself—even total strangers" (De Guia 2013 p. 180). As such, *kapwa* "reflects a perspective that links (includes) people rather than separates (excludes) them from each other" (Macaraan 2019).

Throughout the Philippines, *kapwa* permeates every ethnolinguistic and religious group (Desai 2016). A community that is so connected and committed to one another that they function as one strongly manifests the value of *kapwa* (Desai 2016). "*Kapwa*—the basis of Filipino sense of interpersonalism—is expressed in *pakikipagkapwa-tao*" (Aguas 2016). Pakikipagkapwa builds communities, supports social cohesion, and helps people find meaning in difficult situations (Canete & del Castillo 2022). It has been characterized by a deep acknowledgment of the existence of other people and a way to provide peaceful dialogue with others (del Castillo & Eder 2022).

4 RELIGIOUS DIVERSITY, THE SALIENCE OF RELIGION, AND THE ROLE OF RELIGIOUS INSTITUTIONS

In the Philippines, many religious institutions and organizations show compassion to the needy. For instance, Caritas Manila is the leading social service and community development organization of the Catholic Church in this predominantly Christian country. One of its programs, Caritas *Damayan* (compassion for others), provides "preventive healthcare services, crisis intervention, emergency response, disaster recovery, and rehabilitation to needy Filipinos" (Manila 2022). An important ministry of Caritas *Damayan* is the integrated nutrition program *Hapag-Asa* (Table of Hope), which feeds numerous hungry and malnourished children throughout the country (Patinio 2019).

Other religious groups also perform benevolent acts for the poor and hungry. These include, but are not limited to, the indigenous Christian church *Iglesia ni Cristo* (Church of Christ), whose outreach program called *Lingap sa Mamamayan* (Aid to Humanity) contributes to the welfare of those in need, irrespective of their religious affiliations (Cornelio 2017).

Also, Islamic Relief Worldwide works closely in Mindanao to empower the poor and provide food packs to the region's most vulnerable communities (Islamic Relief World 2016). Moreover, the Buddhist Tzu Chi Foundation has provided food relief for people in impoverished areas in the country (Buddhist Tzu Chi Foundation 2015). Furthermore, the Khalsa Diwan Sikh Temple's *langar* (community kitchen) provides free meals to people regardless of gender, religion, or race (Divakaran 2014; Hutter 2012). Also, in partnership with the American Jewish Joint Distribution Committee (JDC), the Jewish Association of the Philippines engages in emergency relief and ongoing sustainability efforts in disaster-stricken provinces in the country (Sandler 2019).

5 PAKIKIPAGKAPWA AS A MODE OF INTERRELIGIOUS DIALOGUE THAT FOSTERS A "NEIGHBORLY SPIRIT" AND ADVANCES PEOPLE'S INTEGRAL DEVELOPMENT

Undoubtedly, religion has driven many Filipinos to address the issue of hunger during the ECQ. However, the generosity of many secular individuals and organizations indicates that charity surpasses religious boundaries. Contemporary culture—"the practices and resources mediated and used in everyday life to make sense of being and living in this world" (Ross & Bevans 2015)—significantly influence how people interact with others and society at large.

Various interfaith dialogue activities have been noted as a religious response to care for the people, especially during the pandemic. Different religious groups discussed the role of faith-inspired organizations in a forum to respond to the global health crisis (Thompson 2020). The participants shared their thoughts on the crucial role of interfaith partners in providing relief for suffering communities. Likewise, the World Council of Churches (2020) published a document offering a Christian basis for interreligious solidarity that can inspire and confirm the impulse to serve a world wounded by the virulent disease and many other wounds.

The Filipino proverb "*Ang sakit ng kalingkingan, damdam ng buong katawan*" ("The entire body feels the pain of the little finger") also point to *pakikipagkapw*. The concern for the welfare of the Other encompasses *pagtutunguhan* (interaction) (Aquino 2004). *Pakikipagkapwa* and *pagtutunguhan* are evinced by Filipinos who sensed the hunger of the poor and generously supported the community pantries (del Castillo 2021). Indeed, if the plight of the poor is not alleviated, the social fabric will fray and weaken over time. From the perspective of *pakiki-pagkapwa* and *pagtutunguhan,* the community pantries reveal that every Filipino is:

1) expected to regard other members of the community with respect;
2) involved in the problems faced by the community and is part of the solution; and
3) expected to show concern for others.

Consequently, *pakikipagkapwa* requires that persons respect and value the dignity of the Other and build meaningful relationships with them (Reyes 2015).

Pakikipagkapwa can also be conceptualized as "love of neighbor"—central to many religious traditions. Among Christians, a good neighbor shows compassion and helps those in need (Luke 10:25–37). Giving food to the hungry is a corporal work of mercy (Catholic Church 1994). *Pakikipagkapwa* is the embodying spirit that compelled grassroots Filipinos, different religious institutions, and non-religious actors to provide for the needy during the ECQ.

6 CONCLUSION

Many Filipinos consider their brethren to be a part of themselves, as evinced by the emergency food programs during the COVID-19 pandemic. In Philippine society, it is *pakikipagkapwa* that undergirds positive interpersonal relations that are critical to social cohesion. Moreover, *pakikipagkapwa* can be a mode of interreligious dialogue, particularly in the discourse of life and action. *Pakikipagkapwa* enabled cordial cooperation between religious institutions and individuals (whether believers or not), which promoted the physical well-being of the poor.

More importantly, by participating in the dialogue of life and action, the different religious traditions have demonstrated that all strive toward integral human development.

Hence, *pakikipagkapwa* can facilitate openness, inclusivity, and sustainability toward flourishing human life.

REFERENCES

Aguas, J. J. S. (2016) 'The Filipino value of pakikipagkapwa-tao vis-à-vis gabriel marcel's notion of creative fidelity and disponibilitè', *Scientia: The Research Journal of the College of Arts and Sciences*. Available at: http://scientia-sanbeda.org/wp-content/uploads/2017/05/vol-5.2-jj-aguas.pdf.

Buddhist Tzu Chi Foundation (2015) *Tzu Chi Provides Food Relief To The Victims In Philippines*. Available at: https://reliefweb.int/report/philippines/tzu-chi-provides-food-relief-victims-philippines (Accessed: 9 October 2022).

Canete, J. J. and del Castillo, F. (2022) 'Pakikipagkapwa (Fellowship): Towards an Interfaith Dialogue with the religious Others', *Religions*, 13(5), p. 459. doi: https://doi.org/10.3390/rel13050459.

del Castillo, F. (2021) 'Community pantries: Responding to COVID-19 food insecurity', *Disaster Medicine and Public Health Preparedness*, Cambridge University Press, pp. 1–1.

del Castillo, F. and Eder, C. J. (2022) 'Pakikipagkapwa (Fellowship) toward positive collective health behavior', *Journal of Public Health*. doi: https://doi.org/10.1093/pubmed/fdac137.

Catholic Church (1994) *Catechism of the Catholic Church, Liguori: Liguori Publications. CNN Philippines, 2021*. Available at: https://www.youtube.com/watch?v=xQRFNVGt28Y (Accessed: 17 June 2022).

Cornelio, J. S. (2017) 'Religion and civic engagement: The case of Iglesia ni Cristo', *The Philippines, Religion, State and Society*, 45(1), pp. 23–38. doi: 10.1080/09637494.2016.1272894.

Desai, M. (2016) 'Critical kapwa: Possibilities of collective healing from colonial trauma', *Educational Perspectives: Journal of the College of Education/University of Hawai'i at Manoa*, 48, pp. 34–40. Available at: https://coe.hawaii.edu/wp-content/uploads/2020/04/2017_5_16_Tinalak-book.pdf.

Divakaran, M. (2014) *Philippines: Khalsa Diwan Sikh Temple; leaving Manila for Hong Kong*. Available at: http://mithunonthe.net/2014/01/28/philippines-2013-khalsa-diwan-sikh-temple-leaving-manila-to-hong-kong/ (Accessed: 16 July 2022).

Enriquez, V. (2004) *From Colonial to Liberation Psychology: The Philippine Experience*. Manila: De La Salle University Press, Inc.

Francis (2015) *Interreligious General Audience*. Available at: https://www.vatican.va/content/francesco/en/audiences/2015/documents/papa-francesco_20151028_udienza-generale.html (Accessed: 26 June 2022).

Halafoff, A. (2013) *The Multifaith Movement: Global Risks and Cosmopolitan Solutions*. Dordrecht: Springer.

Highest Catholic Population (2022) No title. Available at: https://worldpopulationreview.com/country-rankings/highest-catholic-population (Accessed: 10 August 2022).

Hutter, M. (2012) '"Half Mandir and half Gurdwara": Three local Hindu communities in Manila, Jakarta, and Cologne', *Numen*, 59, pp. 344–65.

Islamic Relief World (2016) *IR Philippines' First Ramadan*. Available at: https://islamic-relief.org/news/a-time-of-faith-and-forgiveness/ (Accessed: 9 October 2022).

Macaraan, W. (2019) 'A kapwa-infused paradigm in teaching Catholic theology/catechesis in a multireligious classroom in the Philippines', *Teaching Theology and Religion*, 22, pp. 102–13.

Manila, C. (2022) *Caritas Damayan*. Available at: https://www.caritasmanila.org.ph/our-work-1/ (Accessed: 16 July 2022).

Patinio, F. (2019) *Faithful Urged to Help Those in Need*. Available at: https://www.pna.gov.ph/articles/1063778 (Accessed: 9 October 2022).

Phan, P. (2018) Introduction. In *Christian Mission, Contextual Theology, Prophetic Dialogue*. Edited by D. T. Irvin and Peter Phan. Maryknoll, NY: Orbis Books.

Reyes, J. (2015) '"Loób and Kapwa: An Introduction to a Filipino Virtue Ethics."', *Asian philosophy*, 25(2), pp. 148–171.

Ross, C. and Bevans, S. (2015) *Mission on the Road to Emmaus: Constants, Context, and Prophetic Dialogue*. Markynoll, New York: Orbis.

Sandler, A. (2019) *Recent JDC Study Mission to the Philippines*. Available at: http://www.jewishnewsva.org/recent-jdc-study-mission-to-the-philippines/ (Accessed: 16 June 2022).

Thompson, S. (2020) An interfaith dialogue on religious responses to COVID-19 in Bangladesh. *Berkeley Center for Religion, Peace and World Affairs*. Available at: https://berkleycenter.georgetown.edu/posts/an-interfaith-dialogue-on-religious-responses-to-covid-19-in-bangladesh (Accessed: 9 October 2022).

Tran, A. Q. (2018) 'Experience Seeking Faith', in Irvin, D. T. and Phan, P. (eds) *Christian Mission, Contextual Theology, Prophetic Dialogue*. Maryknoll, NY: Orbis Books.

United States Institute of Peace (2004) *What Works? Evaluating Interfaith Dialogue Programs*. Available at: https://www.usip.org/sites/default/files/sr123.pdf (Accessed: 9 October 2022).

Religion, Education, Science and Technology towards a More Inclusive and Sustainable Future – Rahiem (Ed.)

Muslim Nobel Laureates in science: The legacy of redefining the nexus of Islam and modernity

S. Naim*
IPMI International Business School, Jakarta, Indonesia

F.R. Widjayanto
Universitas Airlangga Surabaya, Indonesia

ABSTRACT: Modernity is closely tied to the rationalism movement that led to the culmination of the rampant spread of materialism and hedonism. Other prominent properties of modernity include a tenet that accentuates the importance of believing in rationality and science rather than tradition or myth, alongside a belief that stresses humans' adroitness in controlling nature as well as the advancement of market capitalism. The origin of the modernity concept itself is intrinsically at odds with humans' imagined future of a sustainable world. All relevant features prescribed in modernity predominantly disparage the relationship between humans and nature, which can be seen from how the primacy of humans over nature is perpetually inculcated. The mainstream discourse of modernity integrates almost no perspectives from Islamic scholars. Scientists par excellence from the Muslim tradition represented by Salam, Zewail, and Sancar, whose evident credentials are shown by their achievement as Nobel laureates, offer alternative views of modernity with a more nuanced approach aiming at striking the balance between Islamic values and modernity through more dialectics. Copious extracts of valuable thoughts gained from each Nobel laureate present us with a sort of selected high-quality rare corpus containing extensive enlightenment on the redefined thinking of humans' authentic affinity with nature.

Keywords: modernity, science, Muslim Nobel Laureates

1 INTRODUCTION

1.1 *Modernity in its essence*

The term modernity is usually understood to refer to a significant collection of cultural, political, economic, and spatial relationships that have profoundly impacted the way society functions, the state of the economy, and how time and space are used and experienced. A focus on humanism, individuality, and self-awareness; confidence in human mastery over nature; an emphasis on rationality and science over tradition and myth; a belief in progress and improvement; a close association with the emergence and development of market capitalism; and a significant reliance on the state and its legal and governmental institutions are just a few of the general characteristics of these relationships. Modernity's development has been gradual throughout history (Linehan 2009).

Central to the proliferation of modernity, whose blossoming is characterized by the discovery of scientific knowledge and technological invention, is rationalism, which

*Corresponding Author: sidrotun.naim@ipmi.ac.id

DOI: 10.1201/9781003322054-5

fundamentally underpins all of this. Built upon the perspicacity of René Descartes's rationalism, the mind and body duality theory also lays a foundation for the dualistic relationship between the two, thus sustaining the dichotomy of humans and non-humans. Descartes' (1981) depiction of scientific knowledge stressed the unique and essential role of the human being. He suggested that a strong mind and body dualism means that the human person is the sole type of entity with mind or soul, the only being who can think, known as the *res cogitans*, or the 'thinking thing.' Beever (2018) outlined that this paradigm distinguished the human being from the rest of the natural world both descriptively (in terms of an *a priori* argument about the nature of the universe) and normatively (as a matter of the moral fallout of this nature). Human beings have a special and privileged position in the natural world. Yet, this view tends to place the human being in a solitary and separated sphere from other nonhuman beings.

Descartes believed that humans learned about the universe and other beings in it through *a priori* reasoning. We do not learn that fellow human beings are also 'the thinking thing' through our direct contact with them, but we gain this knowledge from the process of meticulous reasoning that we go through internally. By the same token, human beings understand their surroundings by identifying other living organisms as something empty, meaningless, and merely a mechanism or corporeal. In its own notoriety, this flawed Cartesian view is linked to heinous anthropocentrism, as it reinforced the belief, for instance, that the noises of an animal (and other living organisms, for example, plants) in agony were likened to the sound of a clock rather than to the sounds of distress felt by a sentient person.

Following the rising prominence of anthropocentrism, which is inextricably related to the paramountcy of human being's pursuit of materialistic happiness, human being's material happiness peaked in hedonism. It became the foremost imperative for all moral precepts underlying modernity. Consequently, human beings make the pursuit possible by exploiting nature, which is already bolstered with a gendered perspective on nature – whereby its existence stems from the deep-rooted dualism of nature and culture. Human beings' needs are everything, so other things are deemed subordinate and unimportant. The act of how humans belittle the natural world runs deep through centuries-old rationalism. Since the world outside one's reasoning is devoid of meaning and value, human beings are considered the most flawless, pristine, and noble to do everything toward nature, which is an empty entity with no soul, not a sentient being, and passively motionless.

Hence, it implies that rationalism is somewhat fixated with the 'pure, objective, and universal' knowledge that came from an 'unmarked rational gaze, which is already de-corporealized as well as de-contextualized' (Sundberg and Dempsey 2009), which later forged the difference between 'self' and 'other,' leading to the creation of normalcy critical to the advent of modernity, as mentioned formerly, known as the dualism between nature and culture. According to Latour (1991), most modern science is based on an ongoing 'purifying process' that aims to construct nature and culture as two different ontological zones. The main issue that lies in the purification is that it only blurs the reality we live in, which, in fact, is evidently made up of hybrid beings (humans and nonhumans continuously making interactions), and that it is too far-fetched to divide it into discrete ontological categories. He also asserted that epistemologies of purification led to the escalation of environmental issues ascribed to the superiority of human beings over the natural world and precluded the possibility of democracy for all worldly things – humans and others.

Plumwood (1993) even elaborated on the consequences of the nature and culture binary on the environment and human welfare. She contended that nature and culture were made ontologically distinct and independent spheres because of a flaw in Western philosophy. Nature and culture, in the conception of other feminist theorists such as Merchant (2006) and Haraway (2003), are not equal but rather historically determined by dynamic (and contextual) relations between privilege and oppression, particularly those that identify gender differences. Nature is construed as an entity associated with femininity, thus rendering it close to the trait defined as soft, welcoming, and passive. On the contrary, culture is perceived in a masculine way, whose innate quality is being active (Sundberg and Dempsey 2009), even excessively aggressive. Mother Earth has been conceived as a care provider, while culture is understood as the living entity that enjoys and takes benefit from the care provider. The contrasting feature is highly influenced by the heteronormativity concept and perpetually justified to drive sprouting practices of industrially extractive business and massive commercialization of natural resources, leading to the degradation of the natural environment. As a result, the dualistic view of nature and culture is created by and contributes to the perpetuation of phallocentrism and patriarchy in modern society.

Casassas & Wagner (2016) noted that the outset of modernity took place at least since the late 18th century revolutions ensued because this event is viewed as the beginning of the roll-out of the foundations for the institutional structures that backed modern society, specifically evidenced by the exponential growth of a capitalistic market economy and modern democratic state. Bayly (2004) even pointed out that the birth of modern societies is followed by the arrival of a new outlook toward human life and the moment where it served as a springboard for a new re-interpretation of the world. In this regard, modernity ushered in a new set of social and political vision, primarily centered on the concept of autonomy.

Deeply entrenched in enlightenment thought, the concept of autonomy vastly informed the postulation of human rights, civil liberty, freedom of expression, sovereignty, and democracy, as well as, most importantly, the liberty to engage in unbridled trade. Autonomy is arguably the backbone of the strong ties between the capitalist economy and modern society, even though there is no clear visionary picture pertinent to the conceptions of a sustainable world under this term. This denotes that the idea of sustainability itself was not inimically embedded in modernity until a trove of debates addressing the importance of a more sustainable relationship between the human and the natural world increasingly resurfaced, drawing on the massive impact of the market economy on the sustainability of the earth. Arnason (2015), for example, emphasizes the modern and democratic state's commitment to autonomy, which infers the devotion to prowess arising from 'pseudo-rational, pseudo-mastery,' and thus tantamounting to capitalism.

For decades, the Western world's thought of modernity has not adequately incorporated the view on the relations between modern society and the natural environment from the perspectives of Muslim scholars or Nobel laureates, making it untenable to harsh contemporary criticisms of modernity per se. This paper intends to expand the understanding on how the long-neglected narrative of Islam and modernity could contribute more to the theoretical analyses of how human beings can be able to form a better, authentic, and thoughtfully redefined sustainable relationship with nature, as it is essential for creating a robust foundation and Islamic thought relevance for a stronger modern civilization.

1.2 *Modernity and Islamic intellectualism*

An interpretive approach is used in this paper. The data collection technique uses a literature review from two primary sources, namely official biographies or autobiographies on the Nobel website or books, as well as speeches when receiving the Nobel Prize, which is the primary data with high originality and authenticity. The Nobel Prize is widely recognized as the world's highest award (although all may not agree) in recognition of scientific and humanitarian contributions. The analysis is done by looking for commonalities and differences between the three. Abdus Salam, Ahmed Zewail, and Aziz Sancar in general have the same views on science, religion, and modernity.

All three were raised in the Western scientific tradition with strong roots in the Islamic tradition, as they were born and spent before young adulthood in a once highly developed civilization. They inherited that civilization even when the nation was in a relatively low state. They combine wisdom with intellect to explore the scientific roots of Islamic or non-Western traditions so that they are unique. This is where the scientific character emerges from wisdom, namely spirituality. All three show a love and seriousness for science and make it an integral pillar of spiritual practice. Salam, Zewail, and Sancar are proof that Islamic intellectualism is in line with the idea of modernity, that Islam and the West are mutually enriching, and should be able to synergize with one another.

2 MUSLIM NOBEL LAUREATES WORK ON SUSTAINABILITY

2.1 *Abdus Salam*

On the Nobel pages, Abdus Salam did not write an autobiography, but Miriam Lewis, one of his staff at the ICTP (International Center for Theoretical Physics) in Trieste, Italy. Lewis (1979) wrote that Abdus Salam is a devout Muslim, whose religion did not occupy a separate space in his life but was integrated and inseparable from his work and family life. Salam views that the Qur'an encourages people to think about the truth of the natural laws created by Allah. Here it is seen that Salam views nature as an integral part of human life, and therefore attention to the laws of nature also means that nature becomes an important reference for human thinking in living their lives.

Salam sees that the main foundation for achieving truth is to live a life that is in harmony with the way nature works. Interpreting the teachings of Islam also means reading the instructions given by nature. Consequently, modernity is perceived as a way of life that follows the development of an advanced era; however, that progress should not be contrary to the laws of nature. The most basic manifestation, for example, is how human faith in Islam must be maintained without making oneself consumed by excessive desires leading to greed. Almost all human wants and needs are provided by nature, and nature – in the view of Western modernity – is seen as a passive entity, even though the laws of nature themselves have indicated that nature is not passive. Salam had attempted to promote a kind of spiritually enlightened view that sees beyond nature and cultural dualism. Rather, it is clear that his argument favors the premise that human and nonhuman entities are equally important, mutually dependent, and dynamically interactive. A scientific and transcendental mechanism exists that allows nature to work according to human behavior. If humans are full of humility and do not base their actions on greed, natural disasters can be avoided. On the other hand, if one takes too much from nature without considering the balance of the ecosystem, the damage and destruction of nature will undermine human life. The conception developed by Salam implicitly rejects anthropocentrism because it is not in accordance with the essence of faith, which is bound by peaceful Islamic values and brings mercy to the entire universe.

Salam (1994) places great emphasis on his Islamic identity and spirituality, as well as acknowledges that East and West, North and South, all have important and equal contributions to science, especially physics. This is in line with the spirit and narrative outlined by Nasr (2006) about Eastern Nations, including the Islamic world, exploring their intellectual traditions so that they have appreciation and pride in what has been neglected.

At the peak of his career on receiving the Nobel Prize in the field of physics he quoted from the Qur'an (surah al-Mulk verses 3–4), which included relevant interpretations of at least two things: (1) that the Qur'an has relevance to science, and the Qur'an encourages Muslims to develop intellectualism; (2) that the essence of physics based on what he has experienced and interpreted through his work is about the balance of nature, harmony, order, and how humans are able to observe it. Salam continued the Islamic intellectual tradition in the golden age. It is also very relevant for all audiences to reflect on social and economic systems that are increasingly moving away from the principle of natural balance. Being religious also means getting closer to and understanding the harmonious relationship between humans and the universe (Fraser 2008). This is such an overlooked idea of Salam, even though he conveys an important message so that humans can learn from their own mistakes and history and can continue to be in harmony with nature.

Salam sees that the main foundation for achieving truth is to live a life that is in harmony with the way nature works. Interpreting the teachings of Islam also means reading the instructions given by nature. Consequently, modernity is perceived as a way of life that follows the development of an advanced era; however, that progress should not be contrary to the laws of nature. The most basic manifestation, for example, is how human faith in Islam must be maintained without making oneself consumed by excessive desires, leading to greed. Almost all human wants and needs are provided by nature, and nature – in the view of Western modernity – is seen as a passive entity, even though the laws of nature themselves have indicated that nature is not passive. A scientific and transcendental mechanism exists that allows nature to work according to human behavior. If humans are full of love and do not base their actions on greed, natural disasters can be avoided. On the other hand, if you take too much from nature without considering the balance of the ecosystem, the damage and destruction of nature will obstruct human welfare. The conception developed by Salam implicitly rejects anthropocentrism because it is not in accordance with the essence of faith, which is bound by peaceful Islamic values and brings grace and continuity of life throughout the universe.

2.2 *Ahmed Zewail*

Zewail also conveys the important message that science is for humanity. Science for humanity means an entity rooted in social justice and human well-being. Meanwhile, human welfare cannot be achieved without paying attention to the sustainability of nature. At this point, Zewail argues that science must be able to provide hope for answering the problems of life and the environment. Zewail understood that the growth of modernity goes hand in hand with the emergence of environmental problems due to the complex relationship between humans and industrial businesses based on natural resources that has been going on for so long. He shows how modern life, driven by the economic system of capitalism, needs to find its antithesis. In this context, Zewail (2003) puts Swedish civilization as an example of the embodiment of good modern life. This, of course, refers to an economic system that is based on the Scandinavian welfare state tradition adopted by Sweden.

In his speech in 1999, Zewail consciously reviewed the heyday of Egypt and Islamic civilization. If the Nobel Prize was awarded when Egypt was victorious or when Islam was in its prime, then surely the children of that civilization would be the winners (Zewail 2002). By bringing this message, Zewail (2000) emphasized the universality of science, which is related to the civilization of a nation. The universality of science is indeed one of the characteristics of modernity. However, for Zewail, universality is not the homogenization of a blind standard of the traditions and culture of a society. Zewail emphasizes the care and caution that scientists must hold when they carry out their roles, including how to find a middle point between universality and local cultural values that pay attention to how nature is treated as a living entity that is integral to human life itself. That is, science is developed for the welfare of a community, which in the process will always involve nature, so that this affects the perspective of a development plan. How Zewail shows Sweden as an example – can be interpreted that modernization in the development of a country must not sacrifice the environment because it involves very high social costs regarding the safety and welfare of humans and future generations.

2.3 *Aziz Sancar*

Sancar himself was heavily influenced by the idea of modernity, especially regarding the attainment of knowledge and enlightenment through education. After completing his medical education at Istanbul University, Sancar took a doctoral program at the University of Texas, Dallas. His research that won the Nobel Prize was regarding DNA repair. This can be done because of the DNA mapping beforehand. When describing his work (a short map of the human genome), Sancar makes an analogy with the Piri Reis Map, which is a well-known monumental work from Turkey (Zaimeche 2010). Piri Reis was a Turkish admiral and cartographer who drew a map of the world in 1513 with a level of accuracy unmatched by any other cartographer of his time.

Here, we can see how Sancar was moved to bring about change and breakthroughs in the field of science that were driven by the intention to solve human problems. From his education and career as a scientist, he understands that as a scholar he is bound by bioethics. Moreover, for him who is a chemist and biologist, and very much involved with experiments, science becomes worthy of being developed when faced with greater interests. Sancar also composes his highly acclaimed masterpieces, and this shows how his bright vision of science also puts forward ethical considerations, including the issue of human relations with nature. Sancar traces human DNA to understand the position and limitations of humans, whose growth and development cannot be separated from natural conditions. This message is implied in his great work and is widely read by chemists and biologists all over the world.

3 CONCLUSION

Modernity is driven predominantly by rationalism, which has developed over the centuries. The way of thinking contained in the rationalism paradigm is already rooted in Western civilization and is institutionalized in various fields. The problem is the dichotomy between nature and culture, resulting from rationalism, which has in fact become the main driver of the capitalistic economic system and the growth of modern society living under a democratic state system. The over-emphasis on the superiority of humans – in the concept of anthropocentrism – as the most important creature so that their needs must be met (only because of the consideration that humans are the most intelligent creatures) is what gives rise to various chronic environmental problems. In Western scientific discourse, modernity and the Islamic world are also often seen as contradictory, the consequence of which is the lack of seriousness and deep interest in efforts to bridge modernity and Islamic values. In fact, the repertoire of Islam that is not well and completely explored makes Islamic thought increasingly marginalized from the dialectic of mainstream modernity. From our analysis, which shows the relationship between Islam and modernity through a review of the thoughts of the three Nobel Prize winners, it turns out that the chronicles of their thoughts are extremely relevant in addressing the relationship between humans and nature, which has indeed been so destructive. The three Nobel laureates are proof that they accept the ideas of modernity and even become a part of it, although to some degree they are also critics of modernity. However, all three clearly criticize anthropocentrism because it is contrary to noble and Islamic values. These three figures have shown how voices against the exploitation of nature provide strategic insight into modernity because it is concerned with the safety and survival of humans themselves. Selective modernity must also be critically considered and contemplated. Similar searches related to constructively bridging Islamic thought and modernity become very important, especially how the Islamic world can continue to be critically involved in scientific conversations and remind us about the harmony of life between humans and nature as their living space.

ACKNOWLEDGMENT

The authors thank Sabil Mokodenseho for editing the paper. Part of the paper was originally a final assignment for ‘Contemporary Islamic World’ on modernism and intellectualism, a doctoral program mandatory course at UIN Syarif Hidayatullah Jakarta for Sidrotun Naim. The course was taught by the late Professor Azyumardi Azra, PhD, CBE, a historian, a prominent Muslim scholar, and a Beacon of Knowledge. After reading the first draft, he provided reading materials from his works as feedback and encouraged S. Naim to publish it in a journal. S. Naim will forever be indebted to him. For his dedication to knowledge and humanity. May Allah forgive him and grant him the highest level of Paradise.

REFERENCES

Arnason, J. P. (2015). Theorizing capitalism: Classical foundations and contemporary innovations. *European Journal of Social Theory*, *18*(4), 351–367. doi: 10.1177/1368431015589153

Bayly, C. A. (2004). *The Birth of the Modern World, 1780 – 1914: Global Connections and Comparisons* (1st ed.). Malden, MA: Oxford Blackwell.

Beever, J. (2018). Obsolete: Anthropocentrism. In *Reference Module in Earth Systems and Environmental Sciences*. Elsevier. doi: 10.1016/B978-0-12-409548-9.10454-3

Casassas, D., & Wagner, P. (2016). Modernity and capitalism. *European Journal of Social Theory*, *19*(2), 159–171. doi: 10.1177/1368431015600016

Descartes, R. (1981). *The Philosophical Writings of Descartes* (2nd ed.; J. Cottingham, R. Stoothoff, & D. Murdoch, Eds.). Cambridge: Cambridge University Press.

Fraser, G. (2008). *Cosmic Anger: Abdus Salam – The FIRST Muslim Nobel Scientist*. New York: Oxford University Press.

Haraway, D. J. (2003). *The Companion Species Manifesto: Dogs, People, and Significant Otherness*. Chicago: Prickly Paradigm Press.

Latour, B. (1991). The Impact of science studies on political philosophy. *Science, Technology, & Human Values*, *16*(1), 3–19. doi: 10.1177/016224399101600101

Lewis, M. (1979). Abdus Salam, The Nobel Prize in Physics in 1979. Stockholm.

Linehan, D. (2009). Modernity. *International Encyclopedia of Human Geography*, 155–161. doi: 10.1016/B978-0-08-102295-5.10202-1

Merchant, C. (2006). The scientific revolution and the death of nature. *Isis*, *97*(3), 513–533. doi: 10.1086/508090

Nasr, S. H. (2006). Spirituality and science: Convergence or divergence. *The Essential Sophia, 214*.

Plumwood, V. (1993). *Feminism and the Mastery of Nature*. London & New York: Routledge Taylor & Francis Group.

Salam, A. (1994). Weak and electromagnetic interactions. *World Scientific Series in 20th Century Physics-Selected Papers of Abdus Salam*, 244–254. doi: 10.1142/9789812795915_0034

Sancar, A. (2015). *Aziz Sancar – Biographical*. Retrieved October 12, 2022, from The nobel prize in chemistry website: https://www.nobelprize.org/prizes/chemistry/2015/sancar/biographical/

Sundberg, J., & Dempsey, J. (2009). Culture/Natures. *International Encyclopedia of Human Geography*, 458–463. doi: 10.1016/B978-008044910-4.00936-6

Zaimeche, S. (2010, February 13). *Piri Reis: A Genius 16th-Century Ottoman Cartographer and Navigator*. Retrieved October 12, 2022, from Muslim Heritage website: https://muslimheritage.com/piri-reis-16th-c-cartographer-navigator/

Zewail, A. (2000). *Femtochemistry: Atomic-Scale DYNAMICS of the Chemical Bond Using Ultrafast Lasers (Nobel Lecture). Angewandte Chemie International Edition*, *39*(15), 2586–2631. doi: 10.1002/1521-3773(20000804)39:15<2586::AID-ANIE2586>3.0.CO;2-O

Zewail, A. (2002). *Voyage Through Time: Walks of Life to the Nobel Prize*. Cairo: American University in Cairo Press.

Zewail, A. (2003). Light and life. *Current Science*, 84(1), 29–33.

Religion, Education, Science and Technology towards a More Inclusive and Sustainable Future – Rahiem (Ed.)

Revitalization of uṣūl al-fiqh through iḥtiyāṭī principles

M. Rosyid*, A.N. Kholiq, F. Bintarawati & M.K. Rofiq
Universitas Islam Negeri Walisongo, Semarang, Indonesia

L. Lutfi
University of Canberra, Canberra, Autralia

M.N. Irfan
UIN Syarif Hidayatullah Jakarta, Ciputat, Indonesia

ABSTRACT: The imbalance of understanding between the text (*naṣṣ*) and the context causes Islamic legal products to become rigid, extreme, and radical. *Uṣūl al-fiqh* is one way to present Islamic law that is flexible, humanist, and in accordance with the needs of the times. This paper aims to revitalize *uṣūl al-fiqh* through the *iḥtiyāṭī* principle as the main method for finding moderate Islamic law. This article uses descriptive-qualitative method with library data. This paper finds two main factors in revitalizing *uṣūl al-fiqh*. First, the existence of the *iḥtiyāṭī* principle in the three *uṣūl al-fiqh* frameworks, namely 1) a balance of thinking patterns, 2) process flexibility, and 3) a holistic approach. Second, the existence of the *iḥtiyāṭī* principle in the source of the law. These two factors are strong indicators that *uṣūl al-fiqh* is the right method to realize benefit. The authors suggest to further researchers to examine *uṣūl al-fiqh*, especially regarding the *iḥtiyāṭī* principle in other sources of law.

Keywords: uṣūl al-fiqh, iḥtiyāṭī, moderate of Islamic law

1 INTRODUCTION

Moderation applies to all fields, religion, politics, society, culture, economy, and law, from ideas and attitudes to actions (Leelakulthanit 2017). Etymologically, the word moderation means *tawassuṭ* (middle), *i'tidāl* (fair), and *tawazzun* (balanced). Meanwhile, in terminology, moderation is interpreted as a perspective, attitude, and behavior that is always in the middle and fair and not extreme (Kementerian Agama Republik Indonesia 2019). Moderate law is close to the realization of human benefit. Because without it, the law has no soul and spirit (Rosyid & Irfan 2019). It will only become a rigid rule and tend to force. Moderate directs the perpetrator (*mujtahid*) not to *ifrāṭ* (exaggeration) and *tafrīṭ* (reckless). Both are not justified in the formation and application of the law. Excessive law impacts forcing its implementation so that it ignores common sense, humanity, and goodness. While the reckless impact on the loss of authority and obedience to it.

Moderate requires balance in all its aspects, including the law. The discourse between revelation and reason, as well as the *naqlī* proposition and the *'aqlī* argument, is balanced. Religious texts require reasoning so that it becomes applicable. This harmony requires a set of work tools called *uṣūl al-fiqh*. Its position is needed so that goodness and benefits are right on target.

*Corresponding Author: masykurxrejo@walisongo.ac.id

 DOI: 10.1201/9781003322054-6

Uṣūl al-fiqh, citing the definition of al-Bayḍawī, as quoted by Aldepo (2011), is a set of knowledge about the arguments of fiqh in general, how to draw law from these arguments, and the mujtahid. Based on this meaning, *uṣūl al-fiqh* examines the process of seeking legal sources (*istidlāl*), the process of drawing law from it (*istinbāṭ*), and the role of *mujtahids*. Fair in seeking legal sources, proportionate in drawing legal conclusions, and professionalism of the *mujtahids*, bring the law to a moderate level.

The orientation of *ḥalāl* vs. *ḥaram* or *sunnah* vs. heresy in law only leads to the blurring of its primary projection, namely human benefit. Law is flexible (Shehada 2009). Conditions and situations very much determine it. When both change, the law can change too. However, caution is required at every stage of its formation. Carelessness in undergoing the process hurts the effectiveness of the law itself. Therefore, it is essential to reiterate the urgency of *uṣūl al-fiqh*. More than that, prudence as a principle of legal discovery is necessary so that the law does not lead to extremes and liberalities.

2 LITERATURE REVIEW

Studies of moderation always go hand in hand with religion and politics. In the study of religious moderation, it is directed to present the principles of freedom of religion and non-violence. Islam, in many studies, is referred to as a moderate teaching, upholds freedom of choice of religion and belief and prohibits any acts of violence. This is as stated by Dauda (2020) and Faqihuddin (2021). While the study of moderation juxtaposed with politics requires that moderate political parties are those who are willing to accept the concept of democracy. This is as stated by Adnan and Amaliyah (2021), Afriansyah et al. (2020) and Islam and Khatun (2015).

As for *iḥtiyāṭī*, as far as we are concerned, it is studied in three fields of knowledge. First, fiqh al-Shafi'ī. This is stated by Arsyad *et al.* (2020) Second, in the study of Islamic astronomy (*falak*). In this field, the precautionary principle is applied to the calculation of the initial prayer times, as stated by Jayusman (2012). Third, *iḥtiyāṭī* in the field of economics and shariah banking and is known as the prudential principle which is applied through the five c's of credit analysis, namely character, capacity, capital, condition, and collateral. This is as mentioned by Putera (2020), and Almaududi (2021).

While this paper is different from these studies. This paper can be called the first writing to present the principle of *iḥtiyāṭī* in the field of *uṣūl al-fiqh*. As for the objectives to be achieved, this paper supports the writings of Dahlan *et al.* (2019) and Said and Rauf (2015) which state that Islamic law is projected for the benefit of mankind. Moderate Islamic law means inclusive and humanist.

3 RESEARCH METHOD

This paper is qualitative research with the descriptive-analytic method. This research data comes from library data, both in the form of books and journals. The data that has been collected is sorted according to the theme of this paper. The *iḥtiyāṭī* principle in building *uṣūl al-fiqh* as a moderation step in Islamic law is the main idea.

4 DISCUSSION

4.1 *Iḥtiyāṭī concept discourse*

Iḥtiyāṭī is a principle that -is almost always-is addressed to al-Shāfi'ī, especially in the field of methods of finding law and *fiqh* of worship. Although not explicitly mentioned, this principle can be seen through his decisions. His acceptance of *ḥadīth aḥad* as a source of law with strict conditions and his rejection of *istiḥsān* and *istiṣlāḥ* as a legal basis are two real examples of this principle being applied (Al-Shāfi'ī 2009). This rejection is based on the argument that the law should be returned to the text. While the texts never command to use *istiḥsān* and *istiṣlāḥ* reasoning. If the Prophet PBUH allows both, of course, the Prophet provides himself an example.

The *iḥtiyāṭī* principle has also been applied by al-Shāfi'ī in his *ijtihād* product. He stated that when a man and a woman who are not *maḥram* touch the skin, the *wuḍū'* (ablution) is invalidated. Even though the touch of the skin does not have the effect of stimulating lust, according to him, it still invalidates *wuḍū'*. The law of impurity of dogs, their hair, skin, and saliva is another example of applying this principle.

Recently, at least through internet searches, the principle of prudence refers more to the field of Islamic astronomy and Islamic banking. In the study of astronomy, *iḥtiyāṭī* is often used in the case of determining the time of prayer. Whereas in Islamic banking, *iḥtiyāṭī* is called the prudential principle, for example, through feasibility studies before financial institutions provide financing to their customers.

Iḥtiyāṭī is a principle in all disciplines. The study of *uṣūl al-fiqh* is interpreted as turning away from something doubtful, taking the most substantial arguments and *wasīlah* (medium), which aims to prevent someone from violating the Shari'a. *Iḥtiyāṭī* has an impact on minimizing errors in the *ijtihād* process. Caution leads the *mujtahid* to a balanced scientific process to reject rigidity and frivolity.

4.2 *Iḥtiyāṭī principles in the uṣūl al-fiqh framework*

Apart from being a method of legal discovery, *uṣūl al-fiqh* also serves to provide open-mindedness. It is here to introduce various arguments and rules as guidelines. The slogan back to the Qur'an and Ḥadīth, which only has a literal meaning, can cause legal chaos. The Qur'an and Ḥadīth indeed occupy the leading position in the strata of the legal foundation building, but many new problems are not found in detail in these two sources. In this condition, both are placed as the central moral idea.

Uṣūl al-fiqh plays a vital role in moderating the law. This moderation can be seen from the principle of prudence (*iḥtiyāṭī*) in the legal discovery process. This principle can be seen through at least three frameworks.

First, the balance in understanding the texts with the context that surrounds them (Asroor 2019). Most scholars agree upon religious texts (*naṣṣ*) as the primary source of law (Saeed 2008). The linguistic rule approach is thus necessary. A verse, through this linguistic approach, is seen from the side of *'am* and *khaṣṣ*, *muṭlaq*, and *muqayyad*, *muḥkām* and *mutashābih*, etc. The first approach is essential because both use Arabic with rules and structures whose contents can only be understood when one knows these rules. Through this reasoning, someone who can only know the text is limited to translation; it is not appropriate to carry out legal excavations to refer directly to these two sources.

The Qur'an and Ḥadīth, the main sources of Islamic law, are limited in quantity. Many new problems have no answers in both. A substance approach is thus needed to understand the spirit brought by these two sources. *Uṣūl* scholars then offer various approaches to understanding it, such as *istiḥsān*, *istiṣlāḥ*, *istiṣḥāb*, *sadd al-dharī'ah*, and so on. This substantive approach is often referred to as the *maqāṣid al-sharī'ah* approach which is the main purpose of promulgating Islamic law. (Rifai 2021) While the core of *maqāṣid al-sharī'ah* is the human benefit (Dusuki & Abdullah 2007). In this second approach, the text is the main moral idea, and the reason is the reasoning spirit of the text.

Second, flexibility in the process of finding Islamic law. A problem that the texts have conclusively determined, then the law is determined according to these provisions (*ma'lūm min al-dīn bi al-ḍarūrah*). While in a case where the text only provides global guidance, reasoning finds its urgency. More than that, a rule states that the law can change due to changes in space, time, situation, and conditions. Laws can also change when the causes have changed. It means the application of Islamic law is very flexible.

The law of cutting hands, for example, is determined by referring to the textual sound of the QS. 5:38. meanwhile, the purpose of punishment is to anticipate the occurrence of the same crime, whether committed by the same person or by different people. Cutting hands is a means to achieve that goal (Afzal & Khubaib 2021). Referring to al-Ṭūfī's opinion that goals should take precedence over means (Rosyid & Hafidzi 2020). So, the actual means can change dynamically following the development of situations and conditions. Thus, cutting hands as a means can turn into prisons and other types of punishment.

The third is a holistic approach to the process of finding Islamic law. It requires innovation through a comprehensive study and a process of scientific collaboration (Buladi 2012). *Mujtahid*s initially carried out the *ijtihād* process. However, along with the dynamics of the development of science, *ijtihād* must be carried out jointly by involving many fields of science (Karčić 2001). Being stuck in reasoning in one field of science is the same as denying flexibility in Islam.

Ijtihād jamā'ī (collective) (Kausar & Nazar 2020) is needed to answer the challenges of modernity. For example, a collaboration between religious and health and medical authorities is a must in health and medical matters. With a comprehensive study involving many elements of science, legal problems in this field can be answered, and their validity is accounted for.

The three indicators prove that *uṣūl al-fiqh* is the proper method for moderating the law. Its balance, flexibility, and novelty become a counter to the realization of *fiqh*, which is haphazard, rigid, and not up to date. Law should be responsive to changes in space and time. When both change, the law adjusts the change to become practical and applicable.

4.3 *The iḥtiyāṭī principle in the uṣūl al-fiqh source; istiṣḥāb and sadd al-dharī'ah*

As mentioned earlier, *iḥtiyāṭī* is a principle that must be adhered to by *mujtahid*s. It is applied to every scientific work process in the *istidlāl* and *istinbāṭ* processes. In the process of *istidlāl* work, the compatibility between the proposition and the case becomes an important part. The involvement of many experts in various disciplines is needed to know this relevance. Mistakes in taking arguments have an impact on errors in the legal withdrawal process (*istinbāṭ al-aḥkām*). Carefulness and foresight in undergoing this process are thus very much needed.

Although the use is disputed, one of the arguments is *sadd al-dharī'ah*.(Abdulaziz 2010) The argument is interpreted by closing everything that has the potential or is strongly suspected of having a negative impact. Simply put, *sadd al-dharī'ah* is a preventive measure. The precautionary principle is evident in this approach, which the Indonesian Ulema Council conveyed.

Sadd al-dharī'ah necessitates prohibitions on actions or preventing specific actions. It is done because if an act is done, it is strongly suspected of causing harm. The classic example is that Islam forbids its adherents to curse the God of other religions. This prohibition is enforced because it can negatively impact; followers of other religions and will retaliate against the God of Muslims. Negative impacts are the primary consideration in everything.

Sadd al-dharī'ah is the most appropriate method to be used as a point of view. Every word and action should consider the positive and negative impact. When words and actions have a negative impact, they must be stopped. The *mujtahid*s (*mufti*s, *qādi* [judge], government, and parliament) need to pay attention to this aspect. If not, then apart from the legal product being ineffective, the most dangerous is being a source of conflict. Including when the government and parliament make policies, the impact should be a concern. If not, then it will not only impact the waste of the budget, it will even cause harm to the people. Punishing criminals, and blasphemy against symbols of sacred majesty, for example, must be enforced.

It is a preventive effort against the occurrence of similar actions, both by the first perpetrators and other people.

In addition to *sadd al-dharī'ah*, another argument that reflects the *iḥtiyāṭī* principle is *istiṣḥāb*. (Rosyid 2018) It is interpreted as applying the law based on its initial circumstances. This initial law is enforced as long as no conditions change it. (Muslimin & Kharis 2020) The classic example is the persistence of a holy state until there is clear evidence that the situation has changed. In such circumstances, a person may legally perform worship, for example, prayer. However, when conclusively the initial state has changed, because of exhaling, for example, the sacred state also changes.

Istīṣḥāb as a reasoning process is necessary and very relevant at this time. Under this approach, the initial state remains in effect indefinitely until there is evidence to change it. The good that exists in a person will last forever until there is strong evidence that changes it. Therefore, a person should not be accused of having committed a crime, a drunkard, a thief, a molester, or other negative things. That is because he is a good person. However, this situation may change when there is evidence to the contrary. As for the proof that someone has committed these actions, it must be carried out by competent authorities, for example, investigators from the police. This *istiṣḥāb* method is in line with the presumption of innocence in criminal law.

The two approaches, *istīṣḥāb* and *sadd al-dharī'ah*, prove that the legal discovery process is carried out in-depth and not rashly. The completeness of the search for arguments, evidence, and the involvement of many experts, is evidence that legal discovery is carried out with the principle of prudence. Carelessness, rigidity, and errors in deciding the law of a case so that it results in harm can thus be minimized.

5 CONCLUSION

Based on the description above, this paper has found two main factors in revitalizing *uṣūl al-fiqh*. First, the existence of the *iḥtiyāṭī* principle in the three *uṣūl al-fiqh* frameworks, namely 1) a balance of thinking patterns, 2) process flexibility, and 3) a holistic approach. Second, the existence of the *iḥtiyāṭī* principle in the source of the law. These two factors are strong indicators that *uṣūl al-fiqh* is the right method to realize benefit. The author suggests to further researchers to examine *uṣūl al-fiqh*, especially regarding the *iḥtiyāṭī* principle in other legal sources.

REFERENCES

Abdulaziz, A. binti, 2010. Al-Dharā'i' and Maqāṣid al-Sharī'ah: A case study of Islamic insurance. *Intellectual Discourse*, 18 (2), 261–281.

Adnan, M. and Amaliyah, A., 2021. Radicalism vs extremism: The dilemma of Islam and politics in Indonesia. *Jurnal Ilmu Sosial*, 1 (1), 24–48.

Afzal, M. and Khubaib, M., 2021. Flexibility in the implementation of Islamic criminal law in modern Islamic society in the light of Qur'ān and Sunnah. *Journal of Islamic Thought and Civilization*, 11 (1), 396–410.

Al-Shāfi'ī, M.I.I., 2009. *Al-Umm*. 1st ed. Beirut: Dār al-Kutub al-'Ilmiyyah.

Aldepo, I.A., 2011. Mawqif al-Shāfi'ī min 'Ilm al-Kalām wa Manāhij al-Mutakallimīn. *Al-Tajdīd*, 15 (29), 51–79.

Almaududi, A., 2021. Formulasi prudential principle dalam kolaborasi antara bank dan fintech lending. *Menara Ilmu*, 15 (2).

Arifinsyah, A., Andy, S., and Damanik, A., 2020. The urgency of religious moderation in preventing radicalism in Indonesia. *Esensia: Jurnal Ilmu-Ilmu Ushuluddin*, 21 (1), 91–108.

Arsyad, A., Ibtisam, I., and Asti, M.J., 2020. Konsep ihtiyāṭ imam syafi'i terhadap anjuran menutup aurat bagi anak-anak; analisis tindakan preventif pelecehan anak. *Mazahibuna*, 2 (2), 255–269.

Asroor, Z., 2019. Tekstualitas Vis-À-Vis Kontokestualitas (Studi Kritis Penafsiran Ayat-Ayat Politik Muhammad Asad [1900–1992]). *Ilmu Ushuluddin*, 18 (2), 152–172.

Buladi, K., 2012. The Importance of Holistic Approach and the Relation of Meaning and Context for Understanding Qur'an. *İstanbul Üniversitesi İlahiyat Fakültesi Dergisi*, 21, 27–58.
Dahlan, M., Baidlawy, Z., and Sugiono, S., 2019. Gus Dur's Ijtihād paradigm of contemporary fiqh in Indonesia. *Al-Ahkam*, 29 (2), 167.
Dauda, K.O., 2020. Islamophobia and religious intolerance: Threats to global peace and harmonious co-existence. *QIJIS (Qudus International Journal of Islamic Studies)*, 8 (2), 257.
Dusuki, A.W. and Abdullah, N.I., 2007. Maqasid al-Shari`ah, Maslahah, and corporate social responsibility. *American Journal of Islam and Society*, 24 (1), 25–45.
Faqihuddin, A., 2021. Islamic moderate in Indonesia. *Ar-Risalah*, 12 (1), 107–118.
Islam, T. and Khatun, A., 2015. "Islamic Moderation" in perspectives: A comparison between oriental and occidental scholarships. *International Journal of Nusantara Islam*, 3 (2), 69–78.
Jayusman, J., 2012. Urgensi Ihtiyath dalam perhitungan awal waktu salat. *Al-'Adalah*, 10 (1), 279–290.
Karčić, F., 2001. Applying the Shari'ah in modern societies: Main developments and issues. *Islamic Studies*, 40 (2), 207–226.
Kausar, S. and Nazar, S., 2020. Legal status of collective Ijtihād (Ijtihād al-Jamā'ī) in contemporary era. *Al-Azhār*, 6 (1), 106–120.
Kementerian Agama Republik Indonesia, 2019. *Moderasi beragama*. Jakarta: Badan Litbang dan Diklat Kementerian Agama RI.
Leelakulthanit, O., 2017. The factors affecting life in moderation. *Asian Social Science*, 13 (1), 106–113.
Muslimin, J.M. and Kharis, M.A., 2020. Istisḥsān and istiṣḥāb in islamic legal reasoning: towards the extension of legal finding in the context of Indonesia. *Al-Risalah*, 20 (2), 163–179.
Putera, A.P., 2020. The prudential principle as the basis in implementing banking transaction. *Hang Tuah Law Journal*, 4, 52–60.
Rifai, S.L., 2021. Maqāsid al-Sharī'ah. Origins and definitions of the general philosophy of Islamic law. *SSRN Electronic Journal*, 1–18.
Rosyid, M., 2018. Istiṣḥāb sebagai solusi pemecahan masalah kekinian. *Syariah Jurnal Hukum dan Pemikiran*, 18 (1), 45–64.
Rosyid, M. and Hafidzi, A., 2020. Paradigma dan alienasi konsep maslahat al-tufi sebagai legalitas sumber syariah. *Al-Banjari: Jurnal Ilmiah Ilmu-Ilmu Keislaman*, 19 (2).
Rosyid, M. and Irfan, M.N., 2019. Reading fatwas of MUI a perspective of maslahah concept. *Syariah: Jurnal Hukum dan Pemikiran*, 19 (1), 91–117.
Saeed, A., 2008. Some reflections on the contextualist approach to ethico-legal texts of the Quran. *Bulletin of the School of Oriental and African Studies*, 71 (2), 221–237.
Said, H.A. and Rauf, F., 2015. Radikalisme Agama dalam Perspektif Hukum Islam. *Al-'Adalah*, 12 (1), 593–610.
Shehada, N., 2009. Flexibility versus rigidity in the practice of Islamic family law. *Political and Legal Anthropology Review*, 32 (1), 28–46.

Religion, Education, Science and Technology towards a More Inclusive and Sustainable Future – Rahiem (Ed.)
© 2024 the Author(s), ISBN: 978-1-032-56461-6
Open Access: www.taylorfrancis.com, CC BY-NC-ND 4.0 license

Seyyed Hossein Nashr's criticism of problems of modernity

B. Badarussyamsi* & E. Ermawati
UIN Sulthan Thaha Saifuddin, Jambi

M. Ridwan
Islamic University of Indragiri

ABSTRACT: The crisis of modernity, which impacts human life, is still a topic of serious discussion among scientists. It is not enough to criticize the objectives of modernity with anti-global warming demonstrations, go-green slogans, and the like. Some scientists prefer to conduct in-depth studies and philosophical contemplation to produce sketches of scientific paradigms that reveal the causes of the crisis of modernity and solutions that can be implemented to overcome them. This article examines Seyyed Hossein Nasr's ideas about the worldview of modernity, with its potential problems, and the construction of his critical paradigm of the problems of modernity. The method in this article is qualitative analytical, which examines Nasr's thought map, determines his domain, and analyzes it. According to Nasr, sticking to positivist premises has been the reason for the loss of human intuitive sensitivity to the point of spiritual emptiness. Nasr's serious criticism of the worldview of modernity revolves around the carelessness of modernity in justifying all means so that modernity itself becomes detached from the 'Transcendent' as the most important foundation of human life.

Keywords: crisis of modernity, positivism paradigm, spiritual emptiness, worldview of modernity

1 INTRODUCTION

As a process of history, modernity is an inevitable development. However, occasionally, the modernism that was developing in the West brought with it negative effects in the form of confusion and deviation of values (Azra 1998, p. 8). The lives of modern humans continue to make them feel increasingly anxious and hollow. They have lost the vision of divinity or transcendental dimension and are susceptible to a spiritual void. As a result, modern humans experience alienation from their own selves, their social environment, as well as their God.

Nashr saw modern cultures in the West. The representation of the Renaissance era was a failed experiment because it reduced quality to quantity, all the essential definitions in metaphysics to material and substantial physical definitions (Nashr 1993, p. 38).

The study of the problems of modernity from the perspective of Seyyed Hossein Nashr has sparked interest. Thus, it is common to find many studies on the topic. The author has found similar studies on the topic by Irfan Safrudin (dissertation paper titled: *Criticism on Modernism: A Comparative Study of Jurgen Habermas and Seyyed Hossein Nashr Perspectives 2003*). Safrudin aimed to point out the similarities and differences between both perspectives: he compiled the emancipatory paradigm of Jurgen Habermas and the transcendental paradigm of Nashr.

*Corresponding Author: badarussyamsi@uinjambi.ac.id

 DOI: 10.1201/9781003322054-7

The other study is by Elya Munfarida (titled: *The Concept of Human According to Seyyed Hossein Nashr 2004*). Elya explained Seyyed Hossein Nashr's criticisms of the modern world as well as the solutions proposed by Nashr. Moreover, she described the nature of humans from various aspects, i.e., creation, potential, and spirituality. Similar to Safrudin, Elya also discussed the philosopher Seyyed Hossein Nashr from a traditional perspective, his epistemology, and his criticism of modern humans. However, the discussion was not in-depth.

There is also a similar study, but more focused on Sufism (titled: *Neo-Sufism and Modernity Problems: A Study of Seyyed Hossein Nashr Perspectives*), by Ujang Safruddin. He discussed Nashr's postmodernism as a solution to the aridity of modern humans. Besides that, a more detailed thesis was written by Ali Maksum (titled: *Sufism as a Liberation of Modern Human: A Study of the Significance of Islamic Traditionalism Concept of Seyyed Hossein Nashr 2003*). This study discussed Nashr's Islamic traditionalism concept as an anti-Western modernism movement with the revival of Sufism as a spiritual way to the real nature of life. Again, although Ali Maksum's study has several similarities, it did not provide an in-depth discussion about the tradition of Islamic thinking and Islamic philosophical study, which is the root of modern science. As a result, Islamic spiritual tradition cannot be defined as abandoning actions that are separate from modern world problems.

The topic that will be discussed in this article is Seyyed Hossein Nashr's criticisms of the problems of modernity. The author has not found studies about this topic, whether in the library or in internet references. Since the discussion on the problems of modernity and the solutions for modern humans is very broad and constitutes a significant portion of Seyyed Hossein Nashr's thinking, the author has decided to focus on specific areas of the study: the paradigm construction of modernity that leads to Nashr's criticism, the paradigm of Nashr's criticism on the problems of modernity, and the construction of scientific solutions proposed by Nashr.

2 RESEARCH METHODOLOGY

To discuss the main topics of the study, the author used the qualitative method, which is a descriptive analysis method. The author described the main problems according to data collected from literature references and then analyzed them into a concept. The author not only discusses Seyyed Hossein Nashr's background but also analyzes his criticism of the problems of modernity. The author has used the literature review and the internet to gather the data needed. However, the author was aware of insufficient data on the topic apart from Nashr's discussion.

3 DISCUSSION

3.1 *Biography, background, and life activities of Seyyed Hossein Nashr*

Sayyed Hossein Nashr was born in Teheran in 1933 in an educated Ahlul Bayt family. His father was a doctor and an educator who became a minister-level official in the Reza Pahlevi reign. In Iran, he received Eastern education, which was full of esoteric and strong Islamic traditions. On the other hand, he also received Western education. Therefore, he experienced two different civilizations. In Iran, he studied philosophy, kalam, and Sufism from Thabathaba'I for around 20 years. In Iran, he founded Husayniyah Irsyad together with Murtadha Muththahari and Ali Syariati (Nashr 1983, p. 24).

After the Islamic revolution in Iran, Nashr chose to live in the United States. While living in the United States, he studied at MIT (Massachusetts Institute of Technology) and Harvard University, where he learned the philosophy of Fritjof Schuon, Rene Guenon, and others. He received B.S. in physics and mathematics, M.A. in geology and continued to study philosophy, science, and history. He received education from H.A.R. Gibb, George Sarton, and Bertrand Russell (Yanuri 2022 June 17).

Nashr's academic records show that he acquired vast knowledge, not only from the Eastern world but also from the Western world.

3.2 *Worldview of modernity by Seyyed Hossein Nashr*

According to Nashr, modern society is a group of humans who are organized in their intellectual structure through positivistic premises without trying to find a connection between nature and human beings. The implication is that both nature and human beings have built their own world. Eventually, today's humans are living in an urbanized environment, leading to suffocation. They have lost intuitive sensitivity to natural phenomena. Humans seem to have lost their freedom of movement and expression because of their exploitation of nature. Therefore, modern humans are living outside of their own existence (Chittick 1981, p. 90).

In Nashr's opinion, modern humans do not have a positivistic character, and they always exploit nature arbitrarily, for example, exploiting oil and gas continuously for energy sources, deforestation logging, and illegal logging for construction. This condition worsens with humans' tendency to make life easier with the help of technology, which increases pollution. Hence, it has also thinned the ozone layer, causing global warming on the Western side of the world specifically and the Eastern side generally.

However, the modernist framework of the Eastern world does not pose a serious threat to natural phenomenon. Nashr assumed that there is a spiritual and social life synergy between Islam and modernity. Islam is a religion of revelation; its main source is the Quran, and it was revealed in the context of the local Arabians at that time. When Islam started spreading around the world, it met various perspectives that were different from the local Arabian tradition. For this reason, according to Nashr, the revelation from the Quran became the root that strengthened the Islamic dimension (Nashr 1990, pp. 75–76).

Nashr believed that the ecological crisis that happened in the Western world was an impact of a spiritual drought. The balance of the ecosystem is not maintained; the forests that were originally a place for animals were deforested for humans' living space, threatening the lives of animals. Universally, Nashr thought that humans should have the awareness to create harmony among living organisms and establish a balanced ecosystem. Nashr hinted at a high spiritual awareness to get to the level of the preservation of nature.

Modernity is a pioneer of materialistic and hedonistic lifestyle in the doctrine of humanism. The logical consequence of this is that humans are the main controllers of the reality of life (Hollingdel 1968, p. 535). According to Nashr, this attitude has distorted the true nature of human beings. In his opinion, the dimension of humanity is located in the relationship between humans and the faraway transcendent world. Hence, to establish a vital spirit of life, humans should start from this sacred assumption (Nashr 1989, p. 31).

3.3 *Paradigm construction of Seyyed Hossein Nashr's criticism on problems of modernity*

Seyyed Hossein Nashr is an Islamic intellectual and professor at several universities in the Western world. The term Islamic traditionalism in the modern era was born from his thoughts. In short, it can be defined as bringing out Islamic traditional values in modern era. Let us examine the three important statements explained by Nashr:

(1) Modern science is not the only science that is absolute, representing knowledge about the universe. Modern science is only a simplification of natural science. It is only valid for assumptions and premises of thoughts.

(2) Islamic civilization should not only imitate the Western knowledge and technology that could destroy it. There are some specific facts on which modern science contradicts Islamic perspectives.
(3) Human science is not free of values.

Nashr has criticized modern Western civilization many times, and the interesting thing is that his criticisms arose from his relationship with the works of Classic Muslims such as Ibnu Sina and his Peripatetic philosophy, Suhrawardi and his Illuminationism philosophy, and Ibn Arabi with his Irfaniya thinking.

Therefore, Islamic traditionalism in the modern era does not have the meaning of bringing traditional or classic cultures to the modern era, but it is more about bringing genuine human values that position humans as honorable creatures, and not as destroyers.

This is because Nashr believed that the concepts of humanity in the thoughts of Classic Muslims can be used as a map for modern Western societies when they face a dead-end and aridity because of the confinement of modernity. According to Nashr, Islamic science is not a part of Western science but an independent way to see how nature works. Western science is mostly related to Western civilization, while Islamic science is mostly related to Islamic civilization. The principle is that science is not a value-free activity. Besides, it is proper and possible for all civilizations to be able to learn science from other civilizations (Nashr 2013, April 12). Therefore, there are differences between Western science and Islamic science, epistemologically and axiologically. Epistemologically, the sources of study of each science originated from different civilizations, while axiologically Islamic science does not acknowledge a "value-free science".

Nashr's perspectives have a close relationship with Muslim scientists who consider Western knowledge to have originated from a specific historical background; this is related to other institutions in Western civilization. Regardless of its claim to be universal, it still is a creation of Western civilization. Thus, it is deeply rooted in different perspectives of Islam. In fact, not only science but all modern knowledge requires an epistemological correction (Iqbal 2007, p. 167). Muslim scientists took the initiative to establish an Islamic-related science – a kind of initiative and thinking that created the Islamization of science. Pioneered by Ismail al-Faruqi (1921–1986), this movement was based on the assumption that what caused the deterioration of Islam was the "education system" that was divided into two subsystems – modern and Islamic. To resolve this concern, al-Faruqi tried to combine both education systems and Islamize science (Zaidi 2011, p. 54).

For Nashr, the term "modern" did not present the success of the domination of nature, but it means something that is cut off from The Transcendent, from eternal principles managing every single thing, which were known through revelation (Maksum 2003, p. 21). Thus, in Nashr's perspective, the term modern refers to Western societies' point of views and lifestyles. The Western tradition is defined as a quality of life that is rationalistic, capitalistic, secularist, and tends to neglect religious perspectives. Besides, although Nashr did not mean to differentiate the two regions geographically, he defined Western as a different region (geographically) than Eastern, for example, Asia, China, Japan, and India. The Western constituted the European countries undergoing the Renaissance in the 17th century (Nashr 1994, p. 2).

As stated by Osman Bakar, Nashr wanted Muslims to go against modern science and technology with broader knowledge and moral responsibilities, as well as with integrity under the Islamic Scientific Tradition. Nashr wanted Muslims to be teachers of modern science and not avoid it. Nashr also stated that Muslims should develop positive Islamic criticism of modern science. The criticism should be based on Islamic scientific tradition and should be clearly understood. This is a sacred duty of Muslim clerics, intellectuals, and scientists to establish an authentic contemporary Islamic science (Bakar 2022, June 17).

For that reason, it is important to consider Nashr's criticisms on modernity, which is glorified by Western society. On the one hand, it should be acknowledged that modernity has contributed to the improvement of human civilization; on the other hand, it should be recognized that modernity also has its dark side. According to Ach. Maimun, the ecological crisis that has reached its lowest point became one of the matters that made Nashr restless. Deforestation and water and air pollution are the most prominent examples (Cusdiawan 2022, June 17).

In his opinion, Nashr thought that those consequences are inevitable because humans use technology and science to exploit nature for short-term material benefits. Nashr stated that it happened because physical reality dominated awareness (modernity); thus, every metaphysical matter was unimportant, even nonsense.

In Nashr's point of view, the root of the problem was the fundamental error of modern epistemology in particular. A life based only on the knowledge of physical reality would be dry and limited. Ecological damage is one of the impacts seen and felt by people around the world.

In *Religion and the Order of Nature* (1996), Nashr stated that the crisis has reached a concerning level, citing examples such as global warming, the damage of the ozone layer, etc. According to him, those ecological crises would be a serious threat to all structures of life in the world's societies.

In *A Young Muslim's Guide to the Modern World* (1994), Nashr criticized the development of modern science that turned into scientism, considered a new absolute belief, refusing truths from other sources. This overview of Nashr's perspectives strengthens his position as one of the most reputable intellectuals of his time, criticizing science and modern technology.

According to Nashr, some of the characteristics of the modern world are: First, it is anthropocentric, where the locus of the universe is located to humans. Everything is measured with human standards. Second, because measurement is human, the modern world is a world without eternal principles. Third, modern humans can simply be defined as a type of human who has lost their sense of sacredness, and fourth, the aspect of metaphysics has disappeared in the modern world. The constructions of Western civilization have created consequences that have caused various crises in its dimensions (Anas 2012, p. 22).

In Nashr's perspective, the root of all crises in the modern world is due to the mistake in conceptualizing humans. A modern world civilization founded on the fundamental concept of humans does not include the most essential things for humans; this is what causes the failure of the modernity project. In fact, the most essential thing for humans is the spiritual dimension. As a result of that mistake, the world is facing tragedy and crises, i.e., spiritual crisis, environmental crisis, anxiety of war, etc.

3.4 *Construction of scientific solutions proposed by Seyyed Hossein Nashr*

Nashr used Sufism to resolve the spiritual crisis in the Eastern world. In this case, Nashr is an adherent of perennial philosophy. In short, perennial philosophy is a view that acknowledges that there is a basic divinity knowledge, which in its principle is achieved by everyone and has the characteristics of cross-religion and cross-history.

He believed that all religions have an esoteric aspect, where it can be a meeting point of all religions. Sufism was believed to be able to give alternative solutions in facing the aridity of the modern world, where modern humans should return to the center of their existence. They have to find the meaning of life that has been eroded by modernity. Nashr even criticized some Muslim scholars who were not close enough with Sufism such as Jamaluddin Al-Afghani, Muhammad Abduh, etc. Establishing the tradition of Sufism as a face of Islamic esotericism is an important effort to reach the objectives of Islam.

Nashr also proposed perennial philosophy as an alternative. However, in his personal opinion, the author does not quite agree with the alternative paradigm developed by Nashr. The author puts Nashr's criticism of modernity discussed in this article as awareness that science also has limitations. The development of modernity created a big gap that can be a serious threat to our lives. To resolve the problem, we need common wisdom and a mindset that we should not be captivated by modernity promises. According to the author, to build a civilization we cannot rely on one discipline of knowledge, but on various disciplines that complement each other to improve the quality of our civilization.

4 CONCLUSION

Nashr's criticism of the problems of modernity was constructed by exploring and tracing the history of Western science because Western science is the main component that supports the development of modernity. In fact, Western science has lost transcendental reference. The loss of reference to the Absolute was caused by the separation of science and theology. The separation of these two was indicative of the scientific revolution that encouraged the appearance of Descartes, Galileo, Newton, and others. In this period, there was a change in the meaning of science that was very different from what happened in the Islamic world. Contrary to the Western world, science in Islam is continuously and constantly based on monotheism as the basic foundation of all existence. However, initially, these sciences came from outside of the Islamic tradition, such as India, Babylon, Greece, and others. When these sciences reached the Muslims, they gained their 'sacredness.' This means that Islam did not recognize the term 'dichotomy,' or the separation of religion and science. Nashr also supposed the creation of a progressive world presented with self-awareness about the Absolute existence. Nashr believed that modern humans could live with the idea of a perennial philosophy that connects everything in this world to the sacred world. According to him, the virtues of modernity will be tarnished if humans abandon all sacred matters.

REFERENCES

Anas, Moh. (2012). "Kritik hossein nashr atas problem sains dan modernitas," *Jurnal Kalam* 6, No 1 (2012), 21–37. https://doi.org/10.24042/klm.v6i1.391.

Azra, Azyumardi. (1994). "Pasca modernisme, Islam dan Politik", in *Ulumul Qur'an*, No. 1, Vol. V.

Bakar, Osman. (2022). *Seyyed Hossein Nashr and Muzaffar Iqbal: Islam, Science, Muslims, and Technology: Seyyed Hossein Nashr in Conversation with Muzaffar Iqbal*, sumber: http://i-epistemology.net/osman-bakar/31-seyyed-hossein-Nashr-and-muzaffar-iqbal-islam-science-muslims-and-technology-

Chittick, William C. (1981). *The Philosophy of Seyyed Hossein Nashr*. New York: Unwinn Press.

Cusdiawan. (2022). *Menimbang Kritik Seyyed Hossein Nashr terhadap Modernitas*, Sumber: Https://Iqra.Id/Menimbang-Kritik-Seyyed-Hossein-Nashr-Terhadap-Modernitas-231556/.

Hollingdel, R. J. (1968). *Twilight of Idol and The Anti-Christ*. New York: Pinguin Books.

Iqbal, Muzaffar. (2007). *Science and Islam*. New York: Greenwood Press.

Maksum, Ali. (2003). *Tasawwuf sebagai Pembebasan Manusia Modern: Telaah Signifikansi Konsep Tradisionalisme Seyyed Kossein Nashr*. Surabaya: PSAPM dan Pustaka Pelajar.

Nashr, Hossein, (1983). *Islam dan Nestapa Manusia Modern, terj. Anas Mahyuddin*. Bandung: Pustaka.

Nashr, Hossein, (1989). *In Search of the Sacred*. New York: Sunny Press, 1989.

Nashr, Hossein, (1990). *Traditional Islam in The Modern World*. New York: Columbia University Press.

Nashr, Hossein, (1993). *Sprituallitas dan Seni Islam, terj.* Sutejo, Bandung: Mizan.

Nashr, Hossein. (1994). *Menjelajah Dunia Modern: Bimbingan untuk Kaum Muda Muslim, terj. Hasti Tarekat*. Bandung: Mizan.

Pakistan Study Group. (2013). *Nashr*. accessed, 12 April 2013. http://www.muslimphilosophy.com/ip/ Nashr1.

Yanuri, Yusuf. (2022). *Sayyed Hossein Nashr & Nestapa Manusia Modern*. Accessed, 17 June 2022. https://Kalimahsawa.Id/Sayyed-Hossein-Nashr-Nestapa-Manusia-Modern/.

Zaidi, Ali. (2011). *Islam, Modernity, and The Human Sciences*. New York: Palgrave Macmillan.

Religion, Education, Science and Technology towards a More Inclusive and Sustainable Future – Rahiem (Ed.)
© 2024 the Author(s), ISBN: 978-1-032-56461-6

"Scenting the heart" and "two wings" of sufism

A.M. Usep*
UIN Syarif Hidayatullah, Jakarta, Indonesia

ABSTRACT: This article investigates how four Sufi masters, who were ajami or non-Arab, responded to certain political issues. This article found that they had no similar collective sentiments about a true ingredient (meaning) of "scenting the heart" in Sufism. The present study of Sufism is comprehensive, viewing Sufism as a religious institution which includes not only inner mystical experience, but also its social and political aspects. This study argues that these features contributed to making Sufi masters agents of change.

Keywords: Scenting the heart, two wings of messengership and sainthood, Sufism

1 INTRODUCTION

The term "ajam" refers to those who were neither articulate nor eloquent in Arabic as opposed to those who were native Arabic speakers. Ahmad Bouhsane called such persons "Ariba Arab." He designated as "Mustariba (Arabized) Arab" those non-Arabs (ajam) who learned Arabic as a second language, and who were brought up alongside Arabs (Bouhsane, no date; Shaikh, no date; Khaldun, 1958)This definition has led this study to categorize the four Sufi masters discussed as Arabized non-Arabs. Even when they claimed to be descendants of the Prophet Muhammad, they learned Arabic as a second language (Ernst and Lawrence 2002). Their original works were composed in languages other than Arabic, such as Urdu, Pashto, Persian, and Turkish. Moreover, their birth place was outside the Arabian peninsula or the core Arab regions of the Nile-to-Oxus (Bouhsane no date; Ngom & Kurfi 2017).

2 METHOD

This article employed polythetic method to study the terms "scenting the heart" and "two wings" of Sufism of four ajami (non-Arab) Sufi masters. This method positions people (Sufi masters) as the agent of change. This approach led this study to go beyond theological perspective and to grasp any peculiarity in Sufism (Ernst 2005). It is different from orientalist essentialist approach that stresses Sufism itself as the agent of change (Ernst 2005). Next, the first two Sufi masters in this article were of the pre-modern time: Mawlana Jalaluddin Rumi (1207–1273), and Shaykh Nizam ad-din Awliya (1243/1244-1325). The other two were of the modern time: Bediüzzaman Said Nursi (1873–1960), and Sheikh Muzaffer Ozak Al-Jerrahi (1916–1985). This research is qualitative by referring to Sufi master's texts as evidences that support the argument of this paper. Next, this article enriched the polythetic method by interweaving its argument with interdisciplinary (social and religious) theories as a way of inquiry.

*Corresponding Author: usep.abdulmatin@uinjkt.ac.id

 DOI: 10.1201/9781003322054-8

3 FINDINGS: RESULTS AND DISCUSSION

3.1 *"Scenting the heart"*

Newman said that our innermost thoughts and feelings are inevitably related to the character of the culture in which we live in (Newman 2002). Rumi and Awliya were experienced the same historical trauma as a result of the Mongols' devastation. Chingiz Khan attacked ruthlessly both Balkh in Afghanistan in 1220, and Bukhara in 1219. He massacred their inhabitants (Awliya 1992; Newman 2002). A year or two prior to this devastation, Rumi, with his family, migrated from Balkh to Laranda (Karaman, Turkey), where he became a Sufi master (Kennedy 2002; Thackston 1992;). Here, he founded the *Mevlewi* Sufi Order (Thackston 1992). Khwaja 'Ali and Khwaja 'Arab escaped from Bukhara to Badaon, India. There, Khwaja Ahmad (Khwaja 'Ali's son) married Bibi Zulaikha (Khwaja 'Arab's daughter), and she bore a future Shaykh (1243/1244), who became a Sufi master of the Chisti Order, and ran his mystic activity at Ghiyathpur, India (Nizami 1992).

Both Rumi and Awliya had the same reactions to the Mongols' domination. In his *Fîhi mâ fîhi*, Rumi responded critically with claims that those whose properties were robbed by Mongols had the right to take them back because they were taken without any permission (Rumi 1994, p. 66–67). Rumi suggested further combatting Mongols by "scenting the heart" or "dwelling in musk" not in its "scent" (Thackston 1992, p. 61, 133). The term "musk," for Rumi, was a metaphor for the heart, and the term "scent" was a metaphor for worldliness. For Rumi, musk was a substance of perfume, from which the scent emanated. The heart was substance, and this world was to become its scent. So, those who lived by the musk would not die because their hearts were not attached to the scents during their lifetimes (Rumi 1994). This was a latent and transcendental service of Sufis, added Rumi, and so their tombs deserved glorification (Rumi 1994).

In *Fawa'id al-Fu'ad*, Awliya brought Rumi's notion of "scenting the heart" to escape from the Mongol's domination by referring to the example of the story of his predecessor, Shaykh Farid ad-din 'Attar (b. 1142) (Awliya 1992, p. 142; Ernst & Lawrence 2002, p. 59). 'Attar was killed during the Mongols' invasion in Nishapur in 1221 (Awliya 1992, p. 142). When the Mongols reached Nishapur, its ruler asked 'Attar to petition God about the Mongols' invasion. In advance, 'Attar responded that the time for petitionary prayer had passed. He then advised the ruler not to lose his courage due to this invasion. 'Attar regarded this invasion as a misfortune that God had ordained. What he meant by courage was contentment; that was, the ruler should say that this misfortune was descending from God, and regard it as if it had never happened to the ruler personally. 'Attar added that the ruler should still offer a petitionary prayer when the misfortune descended. This action, said he, could lessen the hardship, even though it could not prevent a misfortune, such as the Mongols' invasion (Awliya 1992, p. 142, 215–216).

Awliya brought the notion of "scenting the heart" to the saints' tombs by typifying the saints' death as individuals having a dream: if they were suddenly startled awake from their dreams, they could see themselves lying beside the Beloved, God. Throughout his life, the Awliya went to his mother's tomb whenever he had problems (Awliya 1992, p. 134–135; Nizami 1992, p. 20).

Rumi's *Fîhi mâ fîhi* and Awliya's *Fawa'id al-Fu'ad* were pre-modern Sufi treatises in Persian transcribed, but intended for select groups. They were essentially conversations with questions from audiences, and answers from the master. These were both written down. Both masters used poetry to awaken within their audiences' realities beyond the intellect. The difference is that Rumi's *Fîhi mâ fîhi* is more philosophical as its master was influenced by the theosophical philosophizing of Ibn-Arabi (1165–1240). While it may have been written down during his lifetime, the collection as a whole was not made until after his death. So, Rumi did not check all his written sermons. Awliya's *Fawa'id al-Fu'ad* had been recorded and completed (from 1308 to 1322) by Amir Hasan Sijzi from Awliya's utterances at his assembly in Ghiyathpur. The master revised some of its pages, and filled lacunae before it was published (Awliya 1992; Nizami 1992; Thackston 1992).

3.1.1 *"Two wings"*

Nursi and Jerrahi were among the numerous Sufi masters who deeply reflected on sainthood, and implicitly quarrelled with each other about what true sainthood was. Nursi concerned himself with politics by building a political movement called *Nurculuk*. The key concept of this movement was the term "sainthood," referring to his idiom of "two wings of messengership and sainthood" (Nursi 1993a, p. 30). Derrida's and Foucault's social theories offer further clarification of this idiom. Both Derrida and Foucault state that to get a true meaning of a word is to grasp the meanings lying within it ((Derrida 1967, p. 31); (Lemert 1993, p. 451–455)). Therefore, there were supposed to be a discourse or a specific sphere of social relations behind the term "sainthood" for Nursi, and the phrase "a pair of wings" for Jerrahi.

For Nursi, the term "sainthood" referred to an emanation of God's knowledge through the Prophet to the purified scholars, like al-Ghazali, al-Sirhindi, Ibn al-'Arabi, and Jilani (Nursi 1993a, p. 30), from whom God's knowledge emanated to others. The basic duty to attain this knowledge was "to act and embark on all things in God's name [*bismillah*], like a soldier; to take and to give on God's behalf" (Nursi 1993b, p. 4). Nursi regarded *bismillah* as an inexhaustible strength, and those who recited it as the blessed people (Nursi 1993c, p. 1). Nursi claimed that those who neglected putting their trust in God were heedless people (Nursi 1993c, p. 2). The knowledge of these people was not that of God, but 'objectivism:' they worshipped merely the externals of this world, said Nursi (Nursi 1993d, p. 9). According to Nursi, this objective knowledge emanated from European ideas, Western materialism and atheism, from which the Muslims should deviate (Nursi 1993d, p. 2).

Seeing this idea of Nursi vis-à-vis objectivism, this study proposes that he legitimized his concept of sainthood to go against Western materialism and atheism, both in the Turkish Republic and outside of it. In fact, since 1906, Nursi aimed at putting forth *shari'ah* (Islamic law) into the constitution. This is discernible because Nursi often mentioned the term "soldier," which has a political meaning, in his words. This term may refer to Nursi's group of disciples, and Nursi himself might represent their government or state. The root of Nursi's anger originated in the rebuff of Sultan Abdül Hamid II in 1896 to Nursi's proposal reform within his own country, Kurd (Mardin 1989, p. 19). In return, in that same year, the Sultan offered both position and wealth as a reward for Nursi's services to the Ottoman Empire in the Balkans. However, Nursi refused the Sultan's offer (Nursi 1993e, p. 5).

Calculating from the year of his birth, this study can say that Nursi made this proposal for reform when he was 23 years old. This hints that, even at this early age, he was demonstrating his skills as a great debater. The people called him "Bediüzzaman" (Wonder of the Age) (Nursi 1993f, p. 1). Additionally, what Şerif Merdin said was verified: Nursi "became fluent in Turkish only after the age of twenty" (Mardin 1989, p. 177). His mother tongue was Kurdish, and his second tongue was Arabic (Mardin 1989, p. 177). Nursi also mentioned that he read and wrote in Persian (Nursi 1993d, p. 4). This must have become his fourth language after Turkish.

The Sultan's rejection of Nursi's proposal made him even more determined to be more desirable to the *shari'ah* (Islamic law) campaign. Together, with army officers, who had little experience in politics and political administration, Nursi founded a revolutionary secret society. This organization changed its name to the Committee of Union and Progress (CUP) in 1908. At that time, Nursi was a guest in the house of Manyasijade Refik Bey, who was to be a Minister of Justice in the first cabinet following the proclamation of the constitution. Refik Bey was, at that time, Chairman of the Central Committee of the CUP (Nursi 1993g, p. 1).

The history of CUP has allowed this study thus far to make the connection between Nursi and Jerrahi. It was evidenced by the fact that CUP accepted the positive function of Islam in society. This organization freed itself from Sultan Abdulhamid on March 31 [1909]. They justified this action to save the crumbling empire by reinforcing *shari'ah* (Islamic law). They used rebellion and murder to attack the opposition: whoever supported the Sultan. CUP killed oppositional reporters on the Galata Bridge in Istanbul, and exiled unsympathetic intellectuals to Sinop (a town in present Northern Turkey) ((Nursi 1993g, p. 1); compare to (Mater 2002)).

That tragedy was supported by the fact that in his introduction to Jerrahi's *Irshad*, Holland quotes what Jerrahi said that there were seven-hundred sheikhs and theologians who were exiled by the revolutionaries of the CUP to the port of Sinop on the Black Sea. Jerrahi's father (Hajji Mehmed Efendi) of Konya, and his Qur'an teacher (Nevshehirli Haji Hayrullah) were among the banished. His father was an Islamic scholar and teacher at the court of Sultan Abdul Hamid. He was also the first scholar in a long line of warriors. The banishment of these religious public figures persisted until 1914 and the First World War (Holland 1988). So, Jerrahi's father was deported by Nursi's CUP in Sinop for about six years (1909–1914).

Until this point, this study suggests that in return, Jerrahi justified implicitly his own notion of "a pair of wings" to debase Nursi's focus on one wing (*shari'ah* reinforcement). Jerrahi explained that the first wing was external, and symbolized the sacred law (*shari'ah*); while the second wing was internal, and represented the mystic path (Sufism). Jerrahi added that Muslims could not fly with only one wing: "to be outwardly dirty, though inwardly clean, is to resemble a water container in the toilet; to be outwardly clean, though inwardly dirty, is to be like a toilet bowl" (Jerrahi 1988, p. 507–508). Here, Jerrahi wanted indirectly to say that Nursi's notion of [two wings of messengership and] sainthood tended to emphasize only *shari'ah* reinforcement, and this tendency led one to harm other sheikhs, including Muzaffer's father. For Jerrahi, only a pure heart and good deeds are the true ingredients of sainthood because they were companions worth having in the world of the tomb. Whereas, power, sovereignty, rank, and status of property could not be taken with people into the graves (Jerrahi 1988, p. 478).

Nursi wanted to engage his students, mainly common people, in his state or *Nurculuk* to fight his first enemy, objectivism of science, and to reinforce *shari'ah* (Islamic law) (Mardin 1989). For this reason, Maryam Jameelah (Nursi 1993g) and Mustafa Tuna (2017) said that Islamic faith still remains in Turkey to this day because of Nursi's efforts. What Jameelah and Tuna said reinforces the finding of this study; that is, Nursi agreed with the implementation and enforcement of *shari'ah*. In contrast to Nursi, Jerrahi did not agree with the implementation and enforcement of *shari'ah*, but Jerrahi encouraged instead the silent form of Islam: the maintaining of a pure heart, and the practice of good deeds. This behavior fitted the term of the Sufi order he led, "*halveti*" (*khalwatiyya*, literally, seclusion or contemplation, solitude) (Holland 1988, p. 19–20).

4 CONCLUSION

This polythetic study on Sufism found that the collective sentiments of those four non-Arab or ajami Sufi masters were not the same. For Rumi, Awliya, and Jerrahi, the true ingredient of sainthood was the "scent the heart" meaning: keep a pure heart, and do good deeds without having to reinforce the external wing, the sacred law (*shari'ah*). Both components are the only companions needed in the world of the tomb. Here, the tomb has a social function: spiritual endurance, a virtue through which the Sufis released their bitter feelings. This is what Berger called a masochistic attitude (Chernus 1990). From Nursi, this study conceived that the tarcing the external wing; that is, the Islamic law (*shari'ah*).

If the external wing (*shari'ah*) is to subsist in Sufism in this twenty-first century, then this study recommends that this type of Sufism become Islamicate cosmopolitan spirit (ICS) that Bruce B. Lawrence defines. That is, the creative impulse/ethos/activity that embodies the egalitarian urge for justice (*adab*) with a global awareness of difference and hierarchy (*tas-khir*) "or reciprocity that works across social, cultural, and linguistic boundaries" (Lawrence 2021). By the ICS, this Sufism, which contains of its external wing, might still become a true solution for the chasm of civilization between the East and the West (Ernst 2003, p. 15).

REFERENCES

Awliya, N. (1992) *Nizam ad-din Awliya: Morals for the Heart*. New York, Mahwah: Paulist Press.

Awliya, N. ad-din (1992) *Nizam ad-din Awliya: Morals for the Heart*. New York, Mahwah: Paulist Press.

Bouhsane, A. (no date) No Title. Available at: http://www.sfb541.uni-freiburg.de/Termine/Bouhsane.html.

Chernus, I. s. (1990) *Summary of Peter Berger, The Sacred Canopy*. Available at: http://www.colorado.edu/ReligiousStudies/chernus/2400/Readings/SummariesOfTheSacredCanopy.htm.

Derrida, J. (1967) *De La Grammatologie*. Paris: Les Éditions de Minuit.

Ernst, C. W. (2003) *Following Muhammad: Rethinking Islam in the Contemporary World*. Chapel Hill & London: The University of North Carolina Press.

Ernst, C. W. (2005) *'Between Orientalism and Fundamentalism: Problematizing the Teaching of Sufism'*, Brannon M. Wheeler, Teaching Islam. New York: Oxford: University Press.

Ernst, C. W. and Lawrence, B. B. (2002) *Sufi Martyrs of Love: The Chisti Order in South Asia and Beyond*. New York: Palgrave Macmillan.

Holland, P. W. (1988) *'Causal INFERENCE, Path Analysis and Recursive Structural Equations Models'*. doi: https://doi.org/10.1002/j.2330-8516.1988.tb00270.x.

Jerrahi, S. M. O. A.- (1988) *Irshad: Wisdom of a Sufi Master*. Westport, Istanbul: An Ashki Book of Pir Press.

Kennedy, H. (ed.) (2002) *An Historical Atlas of Islam*. Leiden, Boston, Köln: Brill.

Khaldun, I. (1958) *The Muqaddimah: AN Introduction to History*. New York: Pantheon Books.

Lawrence, B. B. (2021) *Islamicate Cosmopolitan Spirit*. New Jersey: Wiley Blackwell.

Lemert, C. (1993) *Social Theory: The Multicultural and Classic Readings*. Boulder: Westview Press.

Mardin, Ş. (1989) *Religion and Social Change in Modern Turkey: The Case of Bediüzzaman Said Nursi*. New York: State University of New York Press.

Mater, T. (2002) *Armenian Deportation*. Available at: http://www.bianet.org/diger_eng/documents3381.htm.

Newman, D. M. (2002) *Sociology: Exploring the Architecture of Everyday Life*. Thousand Oaks, California, London, New Delhi: Pine Forge Press.

Ngom, F. and Kurfi, M. H. (2017) *'Ajamization of Islam in Africa'*, *Islamic Africa*, 8(1–2), pp. 1–12.

Nizami, K. A. (1992) *'Nizam Awliya: Morals for the Heart'*, in. New York, Mahwah: Paulist Press, pp. 3–59.

Nursi, B. (1993a) *'Chapter Seven: War and Captivity'*. Available at: http://www.risaleinur.com.tr/rnk/eng/tarihce/1chap7.html.

Nursi, B. (1993b) *'Chapter Six: Service in the Balkans, and in the 'Special Organization'*. Available at: http://www.risale-nur.com.tr/rnk/eng/tarihce/1chap6.html.

Nursi, B. (1993c) *'Chapter Three: Freedom and Constitutionalism'*. Available at: http://www.risaleinur.com.tr/rnk/rng/tarihce/1chap3.htmll.

Nursi, B. (1993d) *'The First Word'*. Available at: http://www.nur.org/newfiles/risale-inurcollection/words/word01.htm.

Nursi, B. (1993e) *'The Seventeenth Word'*. Available at: http://www.nur.org/newfiles/risale-inurcollection/words/word17a.htm.

Nursi, B. (1993f) *'The Sixth Word'*. Available at: http://www.nur.org/newfiles/risaleinurcollection/words/word06.htm.

Nursi, B. (1993g) *'The Thirty Third Word'*. Available at: http://www.nur.org/newfiles/risale-inurcollection/words/word33.htm.

Rumi, J. (1994) *Signs of the Unseen: The Discourses of Jalaluddin Rumi*. Vermont: Threshold Books.

Shaikh, H. al- (no date) *Man Hum al-'Arab al-Musta'ribah*. Available at: https://mawdoo3.com (Accessed: 27 January 2022).

Thackston, W.W., J. (1992) *'Signs of the Unseen: The Discourses of Jalaluddin Rumi'*, in. Vermont: Threshold Books. xiii–xxxiv.

Tuna, M. (2017) 'At the vanguard of contemporary muslim thought: Reading said nursi into the Islamic Tradition', *Journal of Islamic Studies*, 28(3), pp. 311–340

Religion, Education, Science and Technology towards a More Inclusive and Sustainable Future – Rahiem (Ed.)

The term ‘adl in the Qur’an perspective of tafsir Maqāṣidī Ibn ‘Āsyūr

J. Azizy*, Sihabussalam, Dasrizal, Asmawi, S. Jannah & G. Wasath
UIN Syarif Hidayatullah, Jakarta, Indonesia

ABSTRACT: One of the proofs of the Qur’an’s miracles and beauty is the variety of pronunciation and uniqueness of meaning. In fact, each of these words has a distinct meaning and function that is not apparent in other words. This fact necessitates the disclosure of the *maqāṣid* of each lafaz used in the Qur’an, one of which is the term *‘adl*. The purpose of using the term *‘adl* in the Qur’an from the perspective of Ibn ‘Āsyūr is explored in this paper. What is the meaning of the *maqāṣid* term *‘adl* in the Qur’an? This article employs a qualitative method with content analysis and interpretation of Ibn ‘Āsyūr maqāṣidī. According to the data, the meaning of *‘adl* in general includes *al-kayl* (measurement), *al-jaza’* (reward), *al-farīḍah* (obligation), *al-nāfilah* (sunnah), *al-fida’* (ransom), *al-sawiyyah* (equal), and *al-istiqāmah* (straight). The term *‘adl* and its derivation appear 28 times, while *ism maṣdar* appears 14 times. According to the author, the *maqāṣid* term *‘adl* includes the following: (1) the existence of similarities between objects is confirmed. The use of *ism maṣdar* on the term *‘adl* in the Qur’an aims to restore the original function of the *ism maṣdar* term *‘adl*, which can be applied to the lafaz *mu’annats* and *mużakkar*; (2) to build peace; and (3) to serve as a universal concept in managing affairs.

Keywords: ‘adl, Ibn ‘Āsyūr, tafsīr maqāṣidī

1 INTRODUCTION

The Qur’an’s status as a holy book serves not only as a way of life but also as proof of the Prophet Muhammad’s truth. This viewpoint emphasizes the Qur’an’s miraculous nature. In general, the miracles of the Qur’an include three aspects: linguistic aspects, scientific signs, and supernatural preaching (Al-Qurṭubi 1964). The revelation of these three aspects of the miracle is not easy, especially given the language of the Qur’an. The variety of lafaz in the Qur’an is not only for decoration and beauty but also has a special significance and purpose. This is what ensures that debate over the text of the Qur’an will not die out. Furthermore, it is related to the commentator’s sociocultural background. The evolution of contemporary problems will have consequences for the evolution of interpretation itself. *Ṣāliḥ likulli zāmān wa makān* is a popular adage among modern-contemporary commentators. That is, the Qur’an must be positioned as a book that is timely and relevant (Azizy et al. 2022).

The Qur’an contains 77,437 words (Al-Suyūṭī 2017), one of which is the term *‘adl*. From the Mecca to Medina periods, the term’s meaning changed. When presented in *fi‘il muḍāri‘* derivation, the characteristics of the term *‘adl* that are oriented toward social community relations must be reviewed. This can be seen in the interpretation of Qs. Al-An‘ām/6: 1. Even in this verse, it is interpreted as the act of associating with Allah Subhanahu wa ta’ala (Swt)’s partners.

This research focuses on the term *‘adl* as *ism maṣdar*. The author chose the derivation of *ism maṣdar* because it is the origin of another derivation; the Qur’an mentions it more than other derivations, so the meaning will be broad. The purpose of this article is to explain the meaning

*Corresponding Author: jauhar.azizy@uinjkt.ac.id

DOI: 10.1201/9781003322054-9

of the term *'adl* in the Qur'an through Ibn 'Āsyūr's concept of interpretation. What is the meaning of the *maqāṣid* term *'adl* in the Qur'an? After the *maqāṣid* is revealed, society can be built in such a way that It is fair, moderate, and wise in deciding problems. On a theoretical level, this article can be used to support the revelation of *maqāṣid* lafaz in the Qur'an.

2 METHODS

According to the method of analysis, this article falls under qualitative research. Based on data sources, this paper falls under the category of library research, namely inventorying and collecting data on the term *al-'adl*, whether found in the Qu''an or interpreted from it. Furthermore, to see the distinction and identify maqid *al-'adl*, the author will convey the term that is considered the same as *al-'adl*, namely the terms *al-qisṭ* and *al-mīzān*. The author employs this research method to investigate the term *al-'adl* while also elucidating its meaning. Two types of data sources have been used: primary and secondary. The primary source for this paper is Ibn 'Āsyūr's commentary book, *al-Taḥrīr wa al-Tanwīr*. Other references, whether in the form of books, journal articles, or research results related to the topic being studied, become secondary sources in the meantime.

The subject of this study is the term *'adl* in the Qur'an as *ism maṣdar*. In the first stage, the author collected data using the library research method. In this paper, several phases of data collection are carried out: (1) determining the problem to be studied, in this case, the word *'adl*; (2) searching for verses containing the word *'adl* using the book *al-Mu'jam al-Mufahras li al-Fāẓ al-Qur'ān al-Karīm*; (3) verifying these verses in the Qur'an. This verification is required to ensure that the number of verses contained in *al-Mu'jam al-Mufahras li al-Fāẓ al-Qur'ān al-Karīm* and the Qur'an is correct; (4) examining the verses that have been collected into the makkiyah madaniyah division of categories and classifying them according to the theme of the verse; (5) creating tables based on the topic, context, and general interpretation of the verse under consideration to make data access and processing easier. Furthermore, it makes it easier for writers to find transitions, the development of the meaning of the term *'adl*, and the disclosure of its *maqāṣid*; (6) analyzing data with content analysis and linguistic approaches; and (7) drawing conclusions from data analysis results so that the *maqāṣid* term *'adl* is revealed;

This research employs content analysis. Content analysis, according to Krippendorff, is "a research technique for making replicable and valid inferences from texts (or other meaningful matter) to the contexts in which they are used" (Krippendorff 2004). This method is used to analyze existing data by revealing details about the problem and research goals. One of them is to investigate and analyze the verse contained in the word *'adl*. The author employed Ibn 'Āsyūr's language approach, which is a component of the *maqāṣid al-qur'ān* approach. This language is used to reveal the meaning of a word. Ibn 'Āsyūr considers linguistic analysis to be the first step in determining the *maqāṣid* of verses and words. Furthermore, an inductive method is used to uncover issues concerning the *maqāṣid* surah and to support the disclosure of the *maqāṣid* term *'adl*.

3 RESULT

3.1 *Meaning of the term al-'Adl in the Qur'an*

The Qur'an mentions the term *al-'adl* and its derivation 28 times (Al-Bāqī 1944). Some examples are *fi'il māḍī*, *fi'il muḍāri'*, *fi'il 'amr*, and *ism maṣdar*. According to M. Quraish Shihab, the term *'adl* and its derivation in the Qur'an have four meanings: (1) the word *al-'adl* means to equate, as in Qs. Al-Nisā'/4: 58, which talks about equal rights; (2) it means balanced, as in Qs. Al-Infiṭār/82: 7, which talks about a balanced position and physical human being; (3) the fulfillment of rights to each individual; and (4) fairness, which is Allah Swt's nature (Shihab et al. 2007).

3.2 *Maqāṣid term al-'Adl in the Qur'an*

The Qur'anic use of the term *'adl* in various contexts serves the following purposes:

(1) As confirmation that the object of *'adl* has similarities. This can be deduced from God's purpose by applying the *ism maṣdar* term *'adl*. Ibn 'Āsyūr restores the original function of the *ism masdar* term *'adl*, which can be used with *mu'annats* and *mużakkar* words. This is the basis on which his interpretation of ransom or substitutes on the Day of Resurrection will be rejected ('Āsyūr 1984). It will not be accepted if neither the redeemer nor the redeemed have anything in common. As a result, Allah Swt employs the term *al-'adl* to indicate that the object has the same position and influence.
(2) As a means of fostering peace. Ibn 'Āsyūr stated that the "term "*al-'adl*" contains benefits, a pleasant attitude, and honesty ('Āsyūr 1984). This second *maqāṣid* is relevant to the verse that calls for social responsibility, particularly in legal matters. The Qur'an mentions "a just person or witness," which means that the person must be honest, content, and concerned with the common good.
(3) As a general concept in business management. The term *al-'adl* includes in the Qur'an everything to govern the affairs of the creatures in this world and the next ('Āsyūr 1984). This universality can be seen in the context of Qs. Al-Mā'idah/5: 115, which explains the Qur'an's position and nature. In addition to being a proof of truth, the Qur'an is also a representation of Allah Swt's justice.

Table 1. The themes of the verses studied (Al-Zuhailī 1418).

Verse	Theme
Qs. Al-Baqarah/2: 48	An example of the bad morals of the Jews
Qs. Al-Baqarah/2: 123	Remembering the blessings and being afraid of the afterlife
Qs. Al-Baqarah/2: 282*	Concerning debt and pawn
Qs. Al-Nisā'/4: 58	Islamic law method
Qs. Al-Maidah/5: 95*	The reward for hunting during ihram
Qs. Al-Maidah/5: 106	Witnesses when death approaches
Qs. Al-Maidah/5: 115	Al-Qur'an as proof of the truth of the Prophet's messages.
Qs. Al-An'ām/6: 70	Avoiding gatherings of people who make fun of the Qur'an; a type of torment
Qs. Al-Naḥl/16: 76	Two parables about idols and statues
Qs. Al-Naḥl/16: 90	A collection of Qur'anic verses on goodness, ugliness, trust (sincerity), promises, guidance, and goals
Qs. Al-Ḥujurāt/49: 9	Morality in general (how to solve problems that fall under the law of persecution)
Qs. Al-Ṭalāq/65: 2	The law of *talaq*, *'iddah*, the impact of *taqwa*, and *tawakkal*

The term "'adl*" appears several times in these verses

Table 2. The maqāṣid surah containing the term *'adl* ('Āsyūr 1984).

Surah	Maqāṣid 'Aṣliyyah
Qs. Al-Baqarah/2	The big theme of this letter of Al-Baqarah is religion and shari'a. There are two points in the maqāṣid of this letter: (1) establishing the greatness and the greatness of Islam than in the previous religions; establishing the majesty of Islamic guidance and the foundations for purifying the soul; (2) explaining the shari'a (laws) of religion to follow and to aim to improve people's lives.
Qs. Al-Nisā'/4	Surah al-Nisā'/4 has a great theme of sharia and morals. The maqāṣid of the surah is to establish a moderate attitude in the practice of good while constructing benefit.

(continued)

Table 2. Continued

Surah	Maqāṣid 'Aṣliyyah
Qs. Al-Maidah/5	The dominating theme of this surah is shari'a. The purpose of presenting this surah is as a perfection of Shari'a or Islamic laws.
Qs. Al-An'ām/6	In general, this surah speaks of creed and faith. The form of the maqāṣid surah is to establish the godhead (Allah Swt.), revelation, apostolate, the day of resurrection, and vengeance.
Qs. Al-Naḥl/16	This letter focuses on the theme of tawhid. The purpose of this letter is to prove the oneness of God as well as to show the postulates of corruption, the deeds of enslaving God and revealing his ugliness.
Qs. Al-Ḥujurāt/49	The great theme of this letter is about laws and ethics. The purpose of this letter is to shape one's morals, either with others or with Allah Swt.
Qs. Al-Ṭalāq/65	The existence of this letter aims to explain in detail how the law of talaq and matters related to the family; e.g., nursing, 'iddah, and residence.

4 DISCUSSION

4.1 *Initial meaning of the term Al-'Adl*

The word *'adl* is not the only one in the Qur'an that represents "justice." At least two words in Arabic mean just (justice), namely *al-'adl* and *al-qisṭ* (Ngaisah 2015). The word *al-'adl* can be interpreted as *al-fidyah* (ransom) as well as fair (Manẓūr 1414 H). Another term from the meaning of *al-fidyah* is *al-'awḍ* (replace). This is consistent with Ibn 'Āsyūr's interpretation of a signal Qs. Al-Baqarah/2:48, *"wa al-'adl bi fatḥ al-'ayn al-'awḍ wa al-fidā"* (the term *al-'adl* with-line fat-the letter *'ayn* means replacement and ransom) ('Āsyūr 1984).

The antonym of the term *'adl* is the term *al-jawr* (injustice/oppression) (Al-Zabīdī 2001). Fairness is explicitly defined as an impartial inner attitude intermediate between *ifrāṭ* and *tafrīṭ* (Al-Zabīdī 2001; Manẓūr 1414). That is, people who do not take sides in terms of attitude or problem/law resolution. Furthermore, the meaning of *al-'adl* can be expressed with *al-musāwāmah fī al-mukāfa'ah* (equality in reward) (Al-Zabīdī 2001). This meaning can be understood by the expression "if someone does a good deed, he gets it back in kind;" the same is true for evil. According to al-Fayruz Ābādī (d. 817), *'adl* means both *al-mitsl* and *al-naẓīr* (which is similar or the same) (al-Fayruz Ābadi 2005). The word *qisṭ* also defines the term *al-'adl*; in fact, the two words are synonymous. The meaning is derived from one of the terms of *asmā' al-ḥusnā* (good names of Allah Swt) *al-muqsiṭ*. Ibn Manẓūr (w. 711) claims that *qis* is included in Allah Swt's names, which means *al-'ādil* (Manẓūr 1414). This meaning is associated with Allah Swt's will when considering a servant's deeds. As a result, there is a link with the term *al-mīzān* (to weigh).

The plural form of the term *al-'adl*, namely *a'dāl* or *'udalā'*, explains this richness of meaning. *al-kayl* (measure), *al-jazā'* (reward), *al-farīḍah* (obligation), *al-nāfilah* (sunnah), *al-fida'* (ransom), *al-sawiyyah* (equal), and *al-istiqāmah* (straight) are among these meanings (al-Fayruz Ābadi, 2005). This variety of initial meanings can be used to construct the meaning of the term *al-'adl*. Furthermore, if it is related to the context of various verses, it will undoubtedly give rise to new meanings.

4.2 *Ibn 'Āsyūr and his concept of interpretation*

Ibn 'Āsyūr's full name is Muḥammad al-Ṭahir ibn Muḥammad ibn Muḥammad al-Ṭahir ibn Muḥammad ibn Muḥammad al-Shażily ibn 'Abd al-Qādir ibn Muḥammad Ibn 'Āsyūr. Ibn 'Āsyūr was born at Al-Marsa, Tunisia, in Jumād al-Awwal 1296 H/1897 AD (Iyāzī 1996; Ṣāliḥ, 2003). Ibn 'Āsyūr began to study the Qur'an (*tajwid, qira'at*, and memorize it) at the age of six. At the age of 14 (1310 H/1893 AD), he continued his studies at al-Zaytoonah University (Iyāzī 1996). Ibn 'Āsyūr gave an introduction to the science of interpretation before entering into his interpretation. He explained the introduction to the science of interpretation in 10 discussions: (1) the explanation of interpretation and takwil. He understood the interpretation to refer to the meaning of the *żāhir* verses of the Qur'ân, while ta'wīl refers to the various possible meanings that the verses of the Qur'an contain; (2) *istimdād* (the device of science/knowledge on which it relies) was used as a tool for interpreting texts before the science of interpretation existed; (3) the validity of interpretation without historical excerpts (*ma'thūr*) and interpretations that rely on reason (*ra'yi*); (4) the purpose of a mufassir in interpreting the Qur'an; (5) *asbāb al-nuzūl.* Ibn 'Āsyūr places great emphasis on referring to a clear and valid history related to the verses from the Qur'an that have sabab al-nuzul; (6) the variety of Qira'at; (7) the explanation of the stories of the Qur'an; (8) the name, the number of verses, the number of surahs, the arrangement, and other names of the Qur'an; (9) meanings of the Qur'an; and (10) the miracles of the Qur'an ('Āsyūr, 1984).

In addition to being a concept in his interpretation, the above ten points are also the phases carried out by Ibn 'Āsyūr in revealing the three *maqāṣid*, namely *al-a'lā* (the general/highest presence of the Qur'an), *al-'aṣliyyah* (a specific purpose that includes the conformity of the theme with the surah), and *al-tafṣīliyah* (the *maqāṣid* contained in each verse and term). Ibn 'Āsyūr traced the three *maqāṣid* using two theories: (1) linguistics (position, derivation, or beauty of words) and the identification of the meaning of origin; and (2) drawing conclusions.

5 CONCLUSION

The term *'adl* in the Qur'an has various meanings according to its context and *maqaṣid*: (1) the affirmation that there is an equation in the object. The use of *ism maṣdar* on the term *'adl* in the Qur'an aims to restore the original function of the *ism maṣdar* term *'adl*, which can be applied to the terms *mu'annats* and *mużakkar*. Allah Swt uses the term *al-'adl* as that the object has a common position and influence; (2) as a way to build peace; the term *al-'adl* contains benefit, rida attitude, and honesty. This second *maqāṣid* has relevance to the verse that commands us to have an attitude of social responsibility, even more so in legal matters; and (3) as a universal concept in regulating affairs; the term *al-'adl* covers all things in the Qur'an to regulate the affairs of beings in the world and the hereafter. Besides the Qur'an as proof of truth, it is also a representation of the justice of Allah Swt. This paper is only limited to the term *al-'adl*. The author suggests further studies to explore the *maqāṣid* terms that are considered to be synonymous with the term *al-'adl.* The study aims to show that the variety of terms in the Qur'an has different *maqāṣids*.

REFERENCES

'Āsyūr, I., 1984. *Al-Taḥrīr wa al-Tanwīr*. Tūnis: Al-Dār al-Tūnisiyyah.

Al-Bāqī, M. 'Abd, 1944. *al-Mu'jam al-Mufahras li Alfāẓ Al-Qur'an al-Karīm*. Kairo: Dar al-Hadis.

al-Fayruz Ābadi, M.I.Y., 2005. *al-Qāmūs al-Muḥīṭ*. Beirūt: Muassisah al-Risālah.

Al-Qurṭubi, M. bin 'Ahmad al-'Anṣārī, 1964. *al-Jāmi' li al-Ahkām al-Qur'an*. Kairo: Dār Kutub al-Mishriyyah.

Al-Suyūṭī, J. al-D., 2017. *Al-Itqān fī 'Ulūm al-Qur'ān*. Kairo: Dār al-'Alamiyyah.

Al-Zabīdī, M.M., 2001. *Tāj al-'Arūs min Jawāhir al-Qāmūs*. Dār al-Hidāyah.

Al-Zuhailī, W., 1418. *al-Tafsīr al-Munīr Fī al-'Aqīdah wa al-Syarī'ah wa al-Manhaj*. Damaskus: Dār al-Fikr al-Mu'aṣirah.

Azizy, J., Syarifuddin, M.A., and Ubaidah, H.H., 2022. Thematic presentations in Indonesian Qur'anic commentaries. *Religions*, 13 (2), 140.

Iyāzī, M. 'Ālī, 1996. *Al-Mufassirūn: Ḥayātuhum wa Manhajuhum*. Teheran: Mu'assasah al-Ṭibā'ah wa al-Nashr Wazārat al-Thaqāfah al-Irshād al-Islamī.

Krippendorff, K., 2004. *Content Analysis: An Introduction to Its Methodology*. Thousand Oaks: SAGE Publications.

Manẓūr, I., 1414. *Lisān al-'Arab*. Beirūt: Dār Ṣādir.

Ngaisah, Z.F.N., 2015. *Keadilan dalam Al-Qur'an (Kajian Semantik atas Kata Al-'Adl dan Al-Qist)*. Universitas Islam Negeri Sunan Kalijaga Yogyakarta.

Ṣāliḥ, A. al-Q.M., 2003. *Al-Tafsīr wa al-Mufassirūn fī al-'Aṣr al-Ḥadith*. Beirūt: Dār al-Ma'rifah.

Shihab, M.Q. *et al.*, 2007. *Ensiklopedia Al-Qur'an: Kajian Kosa kata*. Jakarta: Lentera Hati.

Religion, Education, Science and Technology towards a More Inclusive and Sustainable Future – Rahiem (Ed.)

Contextualizing the transformations of Swahili Qur'anic translations within the contemporary religio-cultural-political dynamics of the world

H.I. Mohamed*
Umma University, Kenya

ABSTRACT: The contentions about the translatability of the Qur'an to the Swahili language ended with the publication of the first Swahili Qur'anic translation in 1926 by a German missionary, Godfrey Dale (1861–1941). Since then, many translations have been produced, either in response to the previous one or as necessitated by the emerging needs of the Swahili Muslim community, resulting in the transformation of these works. These transformations come at different levels; those that are a result of theological discourses among scholars of different schools of Islamic thoughts, i.e., Salafi, Shia, Ibadi, etc., and other discourses that were about the tools used to translate the Quran to Swahili, with traditional Swahili Muslim clerics adhering to the vernacular Swahili or Swahili of the Arabic version while others adopting standard Swahili, which was largely considered by mainstream Muslims at the time as a way of advancing the colonial agenda in East Africa. There were shifting discourses thereafter, each time becoming narrower and more internalized. This paper aims to analyze these transformations and relate them to contemporary world dynamics. The researcher used a qualitative method to analyze the various contexts within which these Quranic works were produced. The researcher primarily based his deductions on the translation itself, as well as employing the arguments of many writers associated with the studies of Swahili Quranic translations. The historical sequence was considered for a better analysis. The results showed that the variety of translations was influenced by various religious, cultural, and political dynamics of the world.

Keywords: translations, transformations, Qur'an, Swahili, contextualization

1 INTRODUCTION

1.1 *General background*

Ever since Godfrey Dale published the first Swahili Qur'anic translation in 1926, many other translations have followed. These translations have undergone many transformations dictated by various religious, cultural, and political dynamics that the Muslim world underwent, given the fact that the Swahili community is strongly connected, culturally and religiously, to the Arab Muslim world. The use of the Swahili standard by Dale in his translation sparked reactions among Swahili Muslim clerics who accused him of promoting a colonial agenda for the colonialists who were championing the introduction of Standard Swahili against Swahili of the Arabic version or vernacular Swahili. The reaction to Dale's work led by the Islamic reformist movement resulted in the resurgence of several other translations on the one hand and the disappearance of Dale's work from local libraries on

*Corresponding Author: haseyow14@gmail.com

 DOI: 10.1201/9781003322054-10

the other. The retraction of the reformist movement during the second half of the 20th century created space for the influx of Salafists and Shi'a, supported by the Saudi and Iranian governments, respectively, into the east coast of Africa, and both met with Sufis, which was already a dominant culture of the Swahili Muslim community. Consequently, the discourse within Swahili Qur'anic translations shifted from the type of Swahili language used or the identity of the translator to an intellectual genealogy. On the other hand, the rising need among the young Muslim generation for independent interpretation due to the growing level of literacy, coupled with the growing calls from among the Salafists for pure Islamic teachings and the insisting of scholars with reformist ideas to have translations that have more contextual relevance necessitated the availability of more advanced works both in terms of quality and quantity. While the theological discourse became more internalized as more works were produced, the use of standard Swahili also became more prevalent because of its dominating nature, with many Swahili clerics showing flexibility in using it.

Although many studies have been done on Swahili Quranic translations, there remains a gap regarding a specific area of contextualizing the transformations of these works. Therefore, this research is important as it lays the foundation for further study of contemporary Islamic thought in East Africa and the dynamics involving it. The research is also important as it answers questions about the relationship between culture, politics, and Islam and shows that Islamic teachings have always been influenced by the prevailing cultural and political dynamics of a given time.

1.2 *Methods and techniques used*

In this paper, a qualitative method is used to describe and analyze the various contexts within which these Swahili Quranic works were produced. This involves a literature study to review certain written material on the topic, observations by the researcher to get insight into certain settings and patterns, and semi-structured interviews that the researcher conducted with some Swahili clerics who are well conversant with Swahili Quranic translations. This method, according to Shenton (2004), will ensure trustworthiness in qualitative research.

2 DISCUSSIONS AND FINDINGS

2.1 *Colonial missionaries, indigenous Swahili clerics, and Qur'anic translation*

The contention about the translatability of the Qur'an came to an end immediately after Dale wrote his Swahili Qur'anic translation in the context of the European colonization of East Africa. Dale's preference for the Roman script was seen by Swahili clerics as an attempt to advance the colonial agenda (Omar and Frankl 1997), since the same thing was done by German and British colonialists in Tanzania, where they suppressed the use of Swahili in Arabic script (Ibid. 1997). The efforts by Mubarak (1910–2001), a leading Ahmadiyya preacher in Africa of Indian origin, led to 'refuting Dale's deviations, delusions, and falsehood'(Mazrui 1980). His 1953 translation of the Holy Quran with Swahili translation and interpretation (*Qur'an Tukufu Pamoja Na Tafsiri Na Maelezo Kwa Kiswahili*) met with suspicion among Swahili Muslim clerics for using the Swahili standard and diffusing his extreme religious views. The use of standard Swahili, at the time, was seen as showing disloyalty to the Swahili Arabized culture and strengthening the colonial Christian mission in East Africa, calling into question Mubarak's "identity" (Mohamed 2017). For some critics, Mubarak's Indo-Pakistani origin was the reason why he betrayed Swahili–Arab culture. Thus, the first two Swahili translations of the Qur'an failed because of political and religious factors associated with the translators.

Having accepted Qur'anic translation, the Swahili Muslim clerics confined either to the Swahili of Arabic script or to old Swahili literary language (dialects), with Al-Buhriy

(1889–1957) opting for the first one (Chande 2021), while Mazrui (1891–1947) went for the second (Mvita/Mombasa Swahili dialects) (Mazrui 1980) in their efforts to respond to Dale's translation. The choice of the two scholars clearly indicates their attempt to defend and reclaim the already threatened religious and cultural peculiarities of the Christian proselytization mission. The incompleteness of the two translations and the publication of the Ahmadiyya one forced Alfarsy (1912–1982) to produce a complete translation in Kiunguja Swahili dialect titled *Qur'ani Tukufu* (or *The Holy Qur'an*). It was published in 1969 in 7,000 copies and later printed eight times in 123,000 copies (Chesworth 2013). Alfarsy's work became so popular that it seemed to have won against previous works written in Swahili Standard. It also became a point of reference for all other translations that were produced later and is seen as a major turning point in the linguistic and theological shifting within the followed translations.

2.2 *Encountering of rivals: Salafists, Shi'a, and Sufists*

By the time Alfarsy published his translation, there were some religious and political changes across the Muslim world. The turning of the powerful Islamic reform activities in Egypt during the first half of the 20th century to a movement style encompassing among its agendas political change caused a collision between it and the political authority of the time, leading to a series of crackdowns on the leaders of the movement, leading to the shrinkage and retreat of its influence in Egypt and across the Muslim world, including East Africa (Loimeier 2016). At the same time, the influence of Ahmadiyya has been dwindling in East Africa and the world at large due to the strong opposition that the movement met with from the Jamaat-e Islami organization in Pakistan (Chande 2021). and more effectively, the decision by the Pakistani parliament in 1974 to hereticize the movement (Qadir 2014).

This comes at a time when the rich petrodollar country's funded Salafists are almost all over the Muslim world and also not far from the 1979 Iranian revolutions that led to the spread of Shi'a Islam all over the world (Ostebo 2015). In addition, the emergence of the Salafists also awakened the Sufis, which is a dominant Swahili Muslim culture (Nimtz 1980). Historically, Salafists, Sufis, and Shi'as had a long standoff, and they all met on the east coast of Africa. Salafism spread very quickly, and because of the amount of money remitted from Gulf countries, e.g., Saudi Arabia, several institutions, including schools, madrasas, mosques, etc., were established. Hundreds of youths were taught at universities in Saudi Arabia and returned home completely indoctrinated with Salafi ideologies. Within the shortest period of time, this movement brought about changes that shifted the theological discourse from one level to another. The Ansaru Sunnah movement in Tanzania is an example of the most purist and fast-growing Salafi organization in East Africa (Gilsaa 2015).

At the time, Alfarsy's translation became very popular among the Swahili Muslim community, and despite having adopted an Ash'arite theological interpretation in his translation (Calis 2022), which Salafists consider a deviation, Alfarsy's inclinations toward Salafism at the later age of his life (interview, Sheikh Abdalla Nassir 2017) made the Salafists see him as one of their own. They therefore started building on his achievements. However, the deviations contained in Alfarsy's work must be corrected. At one point, the *King Fahd Center* for printing the Holy Qur'an interfered by requesting some Swahili clerics to work on Alfarsy's translation to conform to the Salafi interpretation (interview, Al-khatib 2017). In this regard, three translations appeared to review Alfarsy's work, especially the area of Salafist concern; interpretations of the names of Allah and his attributes. These are: (1) *Tarjama Ya "Al Muntakhab" Katika Tafsiri Ya Qur'ani Tukufu* (*Translation of Al-Muntakhab in Interpreting the Holy Qur'an*) by Sheikh Ali Bin Muhsin Albarwani (1919–2006). It is a translation of a 'Tafsir' exegesis originally written in Arabic. (2) *Qur'ani Tukufu'* (*The Holy Qur'an*) by Sheikh Abdulrahman Mohamed Abubakar (1939–1992). (3) *Qur'ani Tukufu Na Tafsiri Ya Maana Yake Kwenye Lugha Ya Kiswahili* (*The Holy Quran and the Explanation of its Meaning in Swahili*) by two prominent Swahili scholars, Dr. Abdalla al-khatib (b.1952) and Sheikh Nasir Khamis.

The Sufists, on the other hand, were irked by Alfarsy's Salafi-style criticisms of some of their religious orders, e.g., *Dhikr* (a form of Islamic mediation in which phrases or prayers are repeatedly chanted to remember God), and in response, three books carrying the same title, *Fimbo Ya Musa* (*The Rod of Moses*), were published between 1970 and 2010, each of which critically investigates Qur'anic translations and vernacular religious texts in Swahili. The example of these three polemics over the last four decades shows the shifting concerns in reaction to the translated Qur'an in Swahili. The act of translation from Arabic to the vernacular is no longer attacked, but rather the theological implications of a deficient translation are at the heart of the more recent discussions (van de Bruinhorst 2013).

The Shi'a presence grew along the east coast of Africa with the growing political influence of Iran across the world. The efforts to produce the Shi'a readings of the Quran in African Swahili contexts to counter the Salafi translations intensified (Mazrui 2016). Ali Jumaa Mayunga (b. 1947) best represents what the standoff between the Sunnis and Shi'as looks like. In his *Qurani inayobainisha*, Mayunga explores three layers of history: personal, Islamic, and contemporary (van de Bruinhorst 2019). His extreme language against his critics and frequent provocations, e.g., changing his name, are seen as attempts in revolutionizing his work and finding ground for his sect in a predominantly Sunni culture. He always related his persecution to the history of the Prophet's descendants, who underwent the same situation to comfort himself (Ibid 2019). It is interesting to note that all Shi'a translations opted for standard Swahili, denoting another divergence from the mainstream witnessed with missionary and Ahmadiyya translations. They are all culturally and politically connected to Euro–Indo–Persia and they never bothered about what dialect of Swahili is used.

2.3 *Changing needs and more comprehensive translations*

Religious learning is no longer primarily transmitted through the well-established links of personal authorities but can increasingly be derived from private study and reading. As a direct result of this opening up of a wide field of knowledge for a non-Arabic reading audience, the potential number of discussants increases: each new Swahili Qur'an translation reveals more of the enigmatic character of the Qur'an and fuels new debates (van de Bruinhorst 2013).

In the introductory pages of his 18-volume translations titled: *Asili Ya Uongofu Katika Uhakika wa Mteremsho Na Ubainisho Wa Tafsiri* (*The Sources of Guidance in the True Revelations and the Interpretations of the Qur'an*), published between 1992 and 2014, Alkindy (b.1943) expressed the need for a more comprehensive work that could serve a growing number of literate people who cannot read Arabic (Alkindy 2010). Although this was Alkindy's main goal, it cannot be ruled out that he, as an Ibadi scholar (Gaiser 2014), with good connections and relationships with the Oman government, which has been the custodian of the Ibadi school of thought for a long time, was equally attempting to serve his divergent theological ideas that can be seen vividly across the pages of his work, e.g., his belief that Muslims who commit major sins will remain in hellfire for eternity, which is against what the mainstream Sunni scholars believe (Alkindy 2010).

In the meantime, a call by Salafists for the spread of Islamic knowledge, not only in terms of quantity but also quality, reached every corner of the world. They adopted a selective approach in dealing with classical Islamic literature, and therefore emphasized on some specific references that were seen, from their perspective, as most purist. In this regard, a group of Tanzanian Muslim clerics (Doga et al. 2006) endeavored to translate the *Mukhtasar Tafsir Bin Kathir* (abbreviated Qur'anic exegesis by Ibn Kathir), which is the most favored tafsir by Salafists, for adopting the methodology of *Tafsir Almathurat* (explaining the Qur'an by itself or by the narration of prophet Muhammad). A 14-volume translation published in

2014 emphasized including the aspects of explaining the verses of the Qur'an by the prophetic tradition as a way of protecting the texts of the Qur'an from human interference, which is always a Salafist concern. The Salafists also emphasized the issue of methodology as a way of defending Islam against religious deviations, which have manifested itself, among many things, in the form of religious extremism from within the Salafists themselves. It is therefore the belief of the mainstream Salafi denomination that to erase such deviations a pure Islamic knowledge must be taught through the Qur'an as explained by the prophetic traditions. The same concern was also emphasized by Alkindy (Gerard 2011). In *Tafsiri Ya Maana Ya Qur'ani Tukufu Kwa Lugha Ya Kiswahili* (*Explanation of the Meaning of the Qur'an in Swahili Language*) published in 2009 by an unknown author, the same concern was reiterated (unknown author 2009).

On the other hand, while the Salafists were classically handling Muslim matters, some Swahili clerics endeavored to continue the reformist agendas to deal with contemporary matters. They maintained connections with Muslim reformists around the Muslim world the same way scholars like Mazrui did in the past. Sheikh Mwalupa's 30 volumes of *Tarjuma ya al-Kashif* (*Translation of Tafsir Al-kashif*) from the Arabic *Tafsir* of a Lebanese scholar is such a good example of what the continuing shift in Swahili Qur'anic translation discourse means. Published between 2003 and 2010, Mwalupa cited the unique nature of this translation and why he chose it; the translation was made in light of modern sciences, both human and physical sciences, and attempts were made to balance between the texts and the contexts and characteristics of the reformists works (Mwalupa 2003). Although Sheikh Mwalupa did this translation because of the prevailing needs among Swahili readers, it is not something to rule out that he did so to popularize one of the great intellectual achievements of a Shi'a scholar, the Islamic faith that Sheikh Mwalupa confesses. In general, the changing needs of the Swahili Muslim community have a great deal of influence on the shifting discourse within Swahili Qur'anic translation and, as such, is expected to continue in the endless future.

3 CONCLUSION AND RECOMMENDATIONS

The Swahili Qur'anic translations went through various stages of transformation. These transformations were dictated by various religio-cultural-political factors in the world. The Swahili Muslim community, being a religious community with a deeply inherited Arab culture, made various political and religious actors believe in the importance of a religious tool in bringing about changes within such communities. Hence, in the competition to translate religious texts into the local language, each work of translation carries the signs of one's religious, cultural, and political background.

The use of the Swahili standard by Dale was seen as acting on behalf of Western colonialists who were then occupying East Africa. Mubarak's translation faced problems because it was seemingly disloyal to the Arabic language for using the same standard of Swahili that colonialists were using to advance their agendas. The encountering of rivals, Salafists, Shi'as, and Sufis along the east coast of Africa shifted the discourse within Swahili Qur'anic translations to another level, with the linguistic issue being maintained at the same level, i.e., the use of Swahili standard versus Swahili vernacular or Swahili of Arabic version continued to be used equally. While the debate about intellectual genealogy became more internalized as it moved forward. Also, these transformations were fuelled by other factors, such as the growing literacy among the Swahili Muslim community that made them independent readers. Hence the need for more comprehensive works in the local language. Not forgetting other dynamics such as a call by Salafists to put more emphasis, not only on the quantity the readers need but also on the quality of what they read, as well as the efforts by some Swahili clerics to produce translations that are less classical but more contextual so as to make sure the youth of this century get a better understanding of Islam through the light of the emerging branches of knowledge today.

The discourse regarding Swahili Qur'anic transformations is ever-changing, and these transformations will likely shift given the dynamics that are currently taking place within the Muslim world, i.e., some countries that have been associated with financing Salafists, e.g., the UAE, are opting for Sufism because of the involvement of some members of the former in acts of extremism and terror. Whatever the case, these transformations will continue as long as the dynamics within the Muslim world continue to change. With this level of Swahili Qur'anic achievement, a study that will deeply analyze all aspects of it is required to make sure that all it entails is put forward. This therefore lays the foundation that will consequently pave the way for more elaborative studies in the future.

Some of the limitations of this study include: untraceable references, lack of proper preservation, which makes it difficult for referencing, a lack of enough studies previously done on the topic, etc.

REFERENCES

Alkindy, S.M., 2010. *Asili ya Uongofu*. Distributor in East Africa: uongofu shop, Tanzania.

Calis, H., 2022. The theoretical foundations of contextual interpretation of the Qur'an in Islamic theological schools and philosophical sufism. *Religions*, 13 (2), 188.

Chande, A., 2021. Shaykh Ali Hemed al-Buhriy's Mrima Swahili translation of the Qur'ān and its place in Islamic scholarship in East Africa. *Australian Journal of Islamic Studies*, 6 (4), 111–132.

Chesworth, J., 2013. Holy Scriptures and their Use by Christians and Muslims in East Africa. *Transformation*, 30 (2), 82–95.

Doga, M.J. Girangiza, K.M, C., A.J, M., M.H.A, Kilemile, S.I, and Alma, 2006. *Tafsir bin kathir, mukhtasar (utangulizi): 1st edition*. Tanzania: Brighthouse Ltd.

Gaiser, A.R., 2014. The Essentials of Ibadi Islam.

Gilsaa, S., 2015. Salafism (s) in Tanzania: Theological roots and political subtext of the Ansār Sunna. *Islamic Africa*, 6 (1–2), 30–59.

Loimeier, R., 2016. *Islamic Reform in Twentieth-Century Africa*. Edinburgh University Press.

Mazrui, A.M., 2016. *Cultural politics of translation: East Africa in a global context*. Routledge.

Mazrui, M.K., 1980. *Tafsiri Ya Qurani Tukufu; Sheik AL-Amin bin Ali Al Mazrui*. shungwaya publications limited.

Mohamed, H.I., 2017. *Swahili translations of the Qur'an in East Africa*. Thesis. International University of Africa.

Mwalupa, H.A., 2003. *Al-kashif, 1st edition*. Tanzania: Al-itra foundation.

Nimtz, A.H., 1980. The history of Islam in East Africa" Islam and politics in East Africa: the Sufi order in Tanzania.

Omar, Y.A. and P. J. L. Frankl, 1997. An historical review of the Arabic rendering of Swahili together with proposals for the development of a Swahili writing system in Arabic script (based on the Swahili of Mombasa). *Journal of the Royal Asiatic Society*, 7 (1), 55–71.

Ostebo, T., 2015. African Salafism: religious purity and the politicization of purity. *Islamic Africa*, 6 (1–2), 1–29.

Qadir, A., 2014. Parliamentary hereticization of the Ahmadiyya in Pakistan: The modern world implicated in Islamic crises. *Religion in Times of Crisis*, 135–152.

Shenton, A.K., 2004. Strategies for ensuring trustworthiness in qualitative research projects. *Education for information*, 22 (2), 63–75.

unknown author, 2009. *Tafsiri ya maana ya Qurani tukufu. Taasisi ya daaru-salaam*. 1st edition. Riyadh.

van de Bruinhorst, G.C., 2013. Changing Criticism of Swahili Qur'an Translations: The Three 'Rods of Moses'. *Journal of Qur'anic Studies*, 15 (3), 206–231.

van de Bruinhorst, G.C., 2019. 'A confirmation of what went before it': Historicising the Shi'i Swahili Qur'an Translation by Ali Jumaa Mayunga. *Approaches to the Qur'an in sub-Saharan Africa*, 189–230.

Religion, Education, Science and Technology towards a More Inclusive and Sustainable Future – Rahiem (Ed.)

The ecology of the Qur'an: Religion-based environmental preservation efforts

J. Azizy*, B. Tamam, N.A. Febriani, Sihabussalam, H. Hasan & H.H. Ubaidah
UIN Syarif Hidayatullah, Jakarta, Indonesia

ABSTRACT: Environmental damage is a common problem that must be addressed with real solutions and real evidence. One of the causes of environmental damage is the anthropocentric paradigm. This paradigm states that humans are the center of the universe; hence, humans have the right to exploit natural resources because nature was created for them. This article will reveal the preservation of the environment based on religion in the scriptures, including whether humans are the center of the universe or partner with nature. This article will also explore the concept of Qur'an-based nature conservation so that it can be used in real-world conservation efforts. This research uses qualitative research methods with a content analysis approach. The conclusion of this article suggests that man and nature are partners with different roles and goals with their fellow creatures of God's creation. Humans can use nature and the environment to meet their needs, but they must also conserve and prosper nature. Humans are given various potentials/abilities to prosper and conserve the earth.

Keywords: environmental damage, conservation, preservation

1 INTRODUCTION

Environmental damage is becoming increasingly of concern today. The arrogance of man in exploiting nature without regard to the environment is to be blamed for this environmental damage. According to the Intergovernmental Panel on Climate Change IPCC), the earth's temperature rises by 1.5 degrees Celsius (Hasan 2003), leading to global warming and melted icebergs in the Arctic, causing flooding because the sea cannot hold the water discharge. High air pollution, excessive use of forest land, and forest fires, among others, contribute to global warming (www.ipcc.ch 2019).

Humans are indeed creatures designed to harness the power of nature for their own purposes. However, one of the factors causing such damage is the irresponsible attitude toward the use of natural resources, which is said to be rooted in the anthropocentric paradigm. This paradigm states that man is the most important component of the universe. Humans are also considered the center of the universe, which allows them to freely exploit natural resources because the world was created for them. This exploitation of nature affects nature and the environment negatively.

Scientific debates about ecology and environmental damage have sprung up, with various solutions offered. Admittedly, however, classical and modern commentators have not yet explored the meaning of the greatness of the *Kauniyah* verses. This paper tries to make a scientific contribution to exploring the ecological concept of the Qur'an, which is read from

*Corresponding Author: jauhar.azizy@uinjkt.ac.id

 DOI: 10.1201/9781003322054-11

the perspective of interpretation. This paper tries to provide an applicative solution on how to make a connection between humans and the environment/nature (*ḥabl min al-alam*).

2 METHODS

Based on the perspective of its analysis, this article is included in qualitative research, which is a type of research that is descriptive and uses analysis based on the data that has been collected. However, based on data sources, this research is included in the category of library research, namely inventorying and collecting data related to the word *khalīfah* and the role of humans in its creation, both contained in the Qur'an and its interpretation. The data sources used include two types, namely primary and secondary. The main source of the paper is a book of classical and modern interpretations. The classical tafsir is represented by *Tafsīr al-Qurṭubī* and *Tafsīr Ibn Kathīr*, while the modern tafsir book is represented by Muḥammad 'Ālī al-Ṣābūnī, *Ṣafwah al-Tafāsīr*. Secondary sources of this study are information related to the topic being studied, both in the form of books and journal articles.

The object of this study is the word *khalīfah* and the role of humans in his creation in the Qur'an. In this paper, several phases have been carried out in the collection of data: (1) determining the problem to be studied, in this case, the word *khalīfah* and the role of humans in his creation; (2) look for verses in which the word *khalīfah* is there and the role of humans in his creation using the kitab *al-Mu'jam al-Mufahras li al-Fāẓ al-Qur'ān al-Karīm*; (3) reveal the interpretation of the verses of *khalīfah* and the role of humans in his creation, both from the works of classical and modern interpretation; (4) analyze the data with content analysis and linguistic approaches; and (5) infer from the results of data analysis the answer to the formulation of predetermined problems.

3 RESULTS

3.1 *Ecological discourse*

The word "ecology" comes from the Greek words "oicos" (household) and "logos" (science), which was first introduced in biology by a German biologist named Ernts Hackel (1869). It means the study of the relationship between one organism and another and between these organisms and their environment. William H. Matthews states that ecology focuses on the inter-relationships between living organisms and their environment (Hardjasoemantri 1994).

In Arabic, ecology is known as *'Ilm al-Bī'ah*. Etymologically, the word *bī'ah* is taken from the word *bawwa'a*; it means: stay, stop, and stay. *Al-bī'ah* means house/dwelling. Whereas the terminology *'Ilm al-Bī'ah* is the study of the environment. Mamduḥ Ḥamid 'Atiyyah states that *'Ilm al-Bī'ah* is a deep knowledge of the interaction of living beings with the surrounding environment. But in a more specific and comprehensive sense, 'Atiyyah states that ecology (*'Ilm al-Bī'ah*) is the study of the constant interconnection between humans and the entire ecosystem contained in the world ('Atiyyah 1998; al-Jirah 2000; Duwaidiri 2004).

Ecology is also understood as the entire ecosystem in which humans live with other beings; these ecosystems are interrelated with each other in carrying out their respective activities (al-Suḥaibanī 2008; Rice 2009). One of the core concepts in ecology is ecosystems, which are ecological systems formed by the mutual relationship between living beings and their environment (Hardjasoemantri 1994). Ecosystems are formed by living components and do not live in interacting places to form an organized unity.

Humans are part of the ecosystem; humans are also managers of systems. Environmental damage is the side effect of human actions to achieve goals that have consequences for the environment. This expression is in line with the concept of the Caliphate as implied in the Qur'an in Surah al-An'ām [6]: 165. The purpose of the creation of humans on earth is, in addition to being a submissive servant (*'abid*), obedient, and devoted to Allah Almighty, to become a Caliph on earth (*khalīfah fī al-Arḍ*). So with the sign that there is an interconnection between nature and humans who need and influence each other, as implied in QS. Al-Rūm [30]: 9 and Hūd [11]: 61.

Not only does it use the function of humans as a caliph (substitute) in the world, but the ecological discourse based on the Qur'an is sustained by the realization that humans are the *'āmir* (builder, manager, or regulator) of this earth. It is also a follow-up to humans' abilities as God's "substitute/representative" on earth. It is reinforced by the gesture that God has taught Adam (man) all the things that are on this earth. At its peak, these knowledge and abilities entrust the existence of processes and practices in environmental management and conservation.

4 DISCUSSION

4.1 *Human as Khalīfah*

Allah has given humans the authority to manage and maintain the earth in proportion (Surah Al-Baqarah [2]: 30 and Hūd [11]: 61). Humans are granted the right to use natural resources, explore natural materials to benefit and save human life and the environment, enjoy the beauty of God's creation, manage the earth, obtain a beautiful environment, and eat animal and vegetable nutrients.

Surah al-Baqarah [2]: 30

"And [mention, O Muhammad], when your Lord said to the angels, "Indeed, I will make upon the earth a successive authority." They said, "Will You place upon it one who causes corruption therein and sheds blood, while we declare Your praise and sanctify You?" Allah said, "Indeed, I know that which you do not know."

Surah Shad [38]: 26

"[We said], "O David, indeed We have made you a successor upon the earth, so judge between the people in truth and do not follow [your own] desire, as it will lead you astray from the way of Allah." Indeed, those who go astray from the way of Allah will have a severe punishment for having forgotten the Day of Account."

The word *khalīfah* is mentioned twice in the Qur'an: in QS. Al-Baqarah [2]: 30 and QS. Ṣād [38]: 26 (al-Bāqī 1364 H). The word *khalīfah* comes from the word *khalfun*, which means behind. From the word *khalfun* are formed various other words, such as the words *khalīfah* (substitute), *khilaf* (forget or mistake), and *khalafa* (replace) (al-Bāqī 1364 H). The word *khalīfah* linguistically means "substitute." This meaning refers to the original meaning of "behind." It is called *khalīfah* because the person who replaces is always behind or comes from behind, after which he replaces (al-Aṣfahānil 1997).

The word *khalīfah* is mentioned in the Qur'an in two contexts: (1) In the context of Allah's talk to angels regarding the creation of Adam (QS. Al-Baqarah [2]: 30). This dialogue informs us that God will create man as God's caliph (representative) on earth who is in charge of prospering or building civilization on earth with the concept that God has ordained to assign him. (2) In the context of Allah's kalam to the Prophet Daud (QS. Ṣād [38]: 26). Allah designated Prophet Daud as the caliph who was given the task of being a just leader and managing his fiefdoms in proportion. Both contexts of the term caliphate point to God's representatives to administer the earth.

The use of the word *khalīfah* in QS. Al-Baqarah [2]: 30 and QS. Ṣād [38]:26 can be understood to have four interrelated aspects, namely (1) the assignor, which is God; (2) the assignee, that is, the human being; (3) the place or environment in which humans live; and (4) assignments to be carried out (Raya 2007). The duties of the assigned caliphate will be of no value if the assignment material is not carried out. Allah gives supplies to humans before he performs the duties of the caliphate, such as science (Al-Baqarah [2]: 31 and 32) and the tasks to be performed, as mentioned in QS. Ṣād [38]: 26.

Humans have obligations to nature as well, including the obligation to protect and serve the earth. Some of these obligations include humans' obligation to research nature. The leaders have the obligation to make policies with environmental insights and persistence in carrying them out. It is the obligation of humans to empathize with nature and maintain natural productivity. Whereas nature is completely submissive to Allah (*sunnatullāh*). Nature has natural rights that must be protected by humans in order to be a source of knowledge for human life. Natural rights include nature's right to worship, the right to live in a community, and the right to regenerate. Man should regard this right as a form of respect for the existence of Allah's creatures.

Thus, the anthropocentric paradigm, which is one of the factors causing humans to exploit natural resources less wisely, as revealed at the beginning, contradicts the concept of man as a *khalīfah* that entrusts a harmonious interaction between humans and the environment. Man is allowed to use natural resources, but he still preserves and respects the right to regenerate God's creatures. Man becoming God's caliph on earth means that man also replaces the role of God, who governs and cares for the earth to be the source of life for other created beings on earth, not just for man himself.

4.2 *Human as an ecology conserver*

Surah Hūd [11]: 61

"And to Thamūd (We sent) their brother Salih. He said: O my people! serve Allah, you have no god other than He; He brought you into being from the earth, ***and made you dwell in it****, therefore ask forgiveness of Him, then turn to Him; surely my Lord is Nigh, Answering."*

The word *ista'mara* is taken from the root of the word *'amara–ya'muru*, which means to prosper and nourish. The letters *sin* and *ta'* that accompany the word *ista'mara* can be understood with the meaning of the word command, so that it means: Allah commands to prosper the earth and its contents, or means reinforcement to make man truly capable of flourishing the earth (Departemen Agama RI 2007). Ibn Kathīr states that Allah made humans the builders who prospered on the earth and who worked on its use reasonably (Jabr 1989; Kathīr 2000). The word *ista'marakum* applies an active attitude, meaning that humans make a proportion of everything on earth but are also active in maintaining and conserving the earth, such as plowing the land, farming, caring for plants, digging trenches or rivers for irrigation, and maintaining the stability of the environment or nature.

The words *yastahzi'ūn* and *yastaskhirūna* and *wasta'marakum fīha* cannot be interpreted with the meaning of finding, believing, and incomprehensible by request. It is impossible for God, the Creator, to ask humans. This sentence uses an expression that means to ask to perform an act through the medium of pronunciation of the person who is lower or below it. If the demand comes from a higher one and is addressed to a lower person, then it is an order and doing something. Whereas if someone asks for a higher one or a person who is above him, then it is a desire.

Surah Al-Rūm [30]: 9

Have they not traveled in the earth and seen how was the end of those before them? They were stronger than these in prowess, ***and dug up the earth, and built on it in greater abundance than these have built on it****, and there came to them their messengers with clear arguments; so it was not beseeming for Allah that He should deal with them unjustly, but they dealt unjustly with their own souls.*

The word *athār al-arḍ* means they cultivate the earth. According to al-Rāghib al-Aṣfahānī, the root of the word *athāru* is *thāwara*, meaning the spread of something. In the verse in QS. Al-Baqarah [2]: 71 *tuthīr al-arḍ* means to cultivate the earth or plow the land. Thus the word *athār al- arḍ* means that they have spread throughout the earth to cultivate the earth and grow crops (Al-Aṣfahānī 1997; Al-Qurṭubī 2006).

The meaning of *'amarū* is linguistically maximized in building, constructing, and establishing. The opposite word of lafaz *'amarū* is *al-khurāb* (destroying/knocking down). The process of building is not only a discourse but also requires concrete efforts. Similarly, building or creating environmental sustainability requires effort and a long process. Even this is apparent in the term used. The use of *fi'il māḍi*, which was used in the past, is used in this context, which tends to be conceptual. This shows that environmental conservation processes and plans must be definitive and calculating.

This verse is a warning to all humans, wherever and whenever they are, that they may know and live the essence of life and know the purpose for which God created humans. Humans were created by God as a caliph on earth to worship Him. Whoever whose purpose in life is not in accordance with what God outlines have deviated from that purpose. For them apply the *sunnatullāh* above as the end of the 'Ād and Thamūds (Al-Qurṭubī 2006; al-Ṣābūnī 1981).

5 CONCLUSION

Conservation of nature or the environment requires a balance between human rights and obligations to nature. According to the Qur'an, both humans and nature have their own roles and purposes in creation as fellow creatures of Allah. The pattern of interaction between nature and humans in their ecosystem is mutually beneficial and framed for the common goal of obedience to God. Humans can use nature and the environment to meet their needs, but they must also protect and preserve the environment.

The Qur'an describes the pattern of harmonious interaction between nature and humans as fellow creatures of God. Only humans are entrusted with managing the world and its contents. In carrying out this belief, humans are given various potentials/abilities to prosper and conserve nature, rather than damaging nature. The Qur'an invites humans to take lessons from the damage to nature caused by human actions in the past and to conserve nature so as not to repeat the mistakes.

REFERENCES

'Atiyyah, M.H. 1998. *Innahum Yaqtulūn al-Bī'ah*. Kairo: Maktabah al-Usrah.

Al-Aṣfahānī, R. 1997. *Mufradāt Alfāzh al-Qur'ān*. Riyaḍ: Maktabah Naẓar Muṣṭafā al-Bāz.

Al-Bāqī, F.A. 1364 H. *Al-Mu'jam al-Mufahras li Alfāẓ al-Qur'ān al-Karīm*. Kairo: Dār al-Kutub al-Miṣriyyah.

Al-Jirah, 'A. al-Raḥman. 2000 M/1420 H. *Al-Islām wa al-Bī'ah*. Kairo: Dār al-Salām.

Al-Ṣābūnī, Muḥammad 'Ālī. 1981. *Ṣafwah al-Tafāsīr*. Bairūt: Dār al-Qur'ān al-Karīm.

Al-Suḥaibanī, 'A.I.'U.I.M. 2008 M/1429 H. *Aḥkām al-Bī'ah fī Fiqh al-Islāmī*. Saudi 'Arabia: Dār Ibn al-Jauziyyah.

Departemen Agama RI. 2007. *Al-Qur'an dan Tafsirnya*. Jakarta: Departemen Agama RI. Vol. 4.

Duwaidiri, R.W. 2004. *Al-Bī'ah Mafhūmihā al-'Ilm al-Mu'āṣir wa 'Umūqihā al-Fikr al-Turāthī*. Damshiq: Dār al-Fikr.

Hardjasoemantri, K. 1994. *Hukum Tata Lingkungan*. Yogyakarta: Gadjah Mada University Press.

Ḥasan, 'A. Al-Hadī. 2003. *Himāyah al-Bī'ah al-Tulūth bi al-Mubayyidāt al-Kimawiyyah wa Afdal al-Hulūl*. Suriah: Dār 'Alā al-Dīn. https://www.ipcc.ch/site/assets/uploads/sites/2/2019/06/SR15_Full_Report_High_Res.pdf. Accessed on 26 Mei 2020

Jabr, M.b. 1989. *Tafsīr al-Imām Mujāhid bin Jabr*. Ghiza: Dār al-Fikr al-Islāmī al-Hadīthah.

Kathīr, I. 2000. *Tafsīr Al-Qur'ān al-'Aẓīm*. Kairo: Mu'assasah Qurṭubah. Vol. 7.

Manẓūr, I. no date. *Lisān al-'Arab*. Kairo: Dār al-Ma'ārif.

Rice, S. A. 2009. *Green Planet: How Plants Keep The Earth Alive*. New Brunswick: Rutgers University Press.

Religion, Education, Science and Technology towards a More Inclusive and Sustainable Future – Rahiem (Ed.)

Gotong royong, an indigenous value for a more inclusive and sustainable future

R. Latifa* & N.F. Mahida
UIN Syarif Hidayatullah, Jakarta, Indonesia

ABSTRACT: Indonesia has a diversity of tribes, languages, religions, and belief systems. This diversity, on the one hand, has a great potential for the development of the Indonesian nation; on the other hand, there is also a sectarian danger where each person wants to prioritize the interests of their own group, resulting in conflicts between ethnic groups. If such conflicts are ignored, it will threaten the unity of the Indonesian nation. Therefore a strategy is required that will unite all ethnic groups, languages, religions, and belief systems. One of the cultures shared by all ethnic groups in Indonesia is *gotong royong. Gotong royong*, or mutual cooperation, can unite the entire Indonesia toward a just and prosperous nation. This paper wants to confirm the indicators of *gotong royong* and the sample of behavior that may explain the concept of *gotong royong.* Using confirmatory factor analysis, we found that *gotong royong* can be explained with the help of three indicators, namely: helping each other, making decisions together, and respecting the neighborhood. Furthermore, from the item difficulty analysis, the easiest item to answer as a sample of *gotong royong* is to *respect the neighborhood.* However, the most difficult item to answer comes from *making decisions together.*

Keywords: *gotong royong*, Indonesia, mutual cooperation, indigenous psychology

1 INTRODUCTION

Gotong royong, in particular, has been a major cultural operator in contemporary Indonesian writing about society. The term *gotong royong* might mean mutual help and reciprocity (Bowen 1986). The root of this expression probably comes from the Javanese verb *ngotongg* (which is cognate with Sundanese *ngagotongg*), meaning "several people bring something together." This gives rise to images of social relations in traditional Javanese villages working smoothly, harmoniously, and privately, where work is done through reciprocal exchange, and villagers are motivated by a general ethos of selflessness and concern for the common good.

Gotong royong can be interpreted as a collective social activity. However, the deepest meaning of *gotong royong* can be explained as a philosophy of life that prioritizes collective life. The philosophy of *gotong royong* is now part of Indonesian culture because *gotong royong* does not belong to any particular ethnicity.

Based on his field research in Central Java, the Indonesian anthropologist Koentjaraningrat (1961) categorizes *gotong royong* into two types, namely spontaneous assistance and mutual cooperation. Spontaneous assistance generally occurs in collective activities in agriculture, house building, celebrations, public works, and in the event of a disaster or death. However, *gotong royong* is usually based on the principle of individual reciprocity, either at the initiative of citizens or imposed as an expression of *gotong royong.*

*Corresponding Author: rena.latifa@uinjkt.ac.id

DOI: 10.1201/9781003322054-12

From a socio-cultural perspective, the value of *gotong royong* as a traditional institution is spiritual in nature, manifested in the form of individual behavior or action to do something together for the common interest of the community (Slikkerveer 2019). In the context of community development, Koentjaraningrat's (1961) observation is important that the Gotong Royong institution makes human life in Indonesia more empowered and prosperous: "With *gotong royong*, various problems of living together can be solved easily and cheaply, as well as community development activities."

Gotong royong connects everyone with many other communities, both for their activities and for their communities. The purpose of *gotong royong* is to strengthen solidarity and relationships, or togetherness, of citizens. The values of *gotong royong* are important for the implementation of sustainable development in Indonesian society (Rahmi *et al.* 2001). Many Indonesians see *gotong royong* as a hallmark of Indonesia's national identity.

According to Bowen (1986), there are several forms of *gotong royong* including helping each other. Hatta (in Feith and Castle 1970) referred to reciprocal assistance (*tolong-menolong*): all heavy work that could not be done by one individual person was performed by the system of *gotong royong*. Villagers, by virtue of their status as community members, are obliged to assist on occasions such as raising the roof of a house, a child's wedding, or the death of a relative. Generalized reciprocity involves general liability and the notion of ultimate return. The result is that within a certain circle of relatives or neighbors, a person feels a general obligation to help, but a person also remembers how much he has been helped when he needed help in the past.

Furthermore, as an additional explanation of the concept of helping each other, according to the World Giving Index (2018), Indonesia is the most generous country in the world. The dimensions of giving are related to (1) helping strangers, (2) donating money, and (3) volunteering time. Indonesians typically like to live in a group or collective. They would kindly help somebody else.

Besides *reciprocal assistance* or helping each other, this way of life, based on common ownership of land in Indonesian history, has created *mutual consultation*. All decisions concerning common interests are taken by mutual consent, or as the Minangkabau (West Sumatran people – a part of Indonesia) proverb goes: "Water becomes one by passing through a bamboo pipe; words become one by mutual agreement." When they are in trouble they would like to discuss it first. The diplomatic way is the best consideration for getting a solution to the problem. Group harmony and cohesiveness are highly valued.

In the sense of mutual consultation, *gotong royong* is described as a sign of the political autonomy of the rural community and the enduring nature of village democracy as opposed to a strong state. Thus the spirit of *gotong royong* grows from the spirit of resistance to domination.

In cross-cultural comparisons, Hofstede (2001) found that in collectivist cultures where people respect their 'ingroup' as a whole, individuals identify themselves as part of a group, valuing their group goals as more important than individuals. As such, he is more connected to his group and tends to care less about his personal goals as an individual, but more about the combined goals of the group as a whole. Hofstede (2001) concludes that in collectivist societies: "people respect their 'ingroup' as a whole, considering how their actions give positive or negative impressions to the 'outgroups' while remaining closely tied to their ingroup." When they live in one area, the neighborhood sometimes becomes their closest family. Thus, respecting the neighborhood also becomes important, especially in forming *gotong royong* around neighbors.

Sukarno, the first president of Indonesia, highlighted the conception of *gotong royong* as a dynamic character, describing it as *satu karyo*, *satu gawe*, a combination of two phrases in Javanese and Indonesian languages, meaning "one, united, task." *Gotong royong* is to unite Christians and Muslims, rich and poor, and indigenous Indonesians and naturalized citizens in a mutually tolerant struggle against enemies. "*Gotong royong* is necessary in the fight against imperialism and capitalism, in the present just as in the past. Without bringing together all our revolutionary forces to be thrown against imperialism and capitalism, we cannot hope to win!" (Soekarno 1965:413).

2 METHODS

2.1 *Participants*

The participants consist of 127 Indonesian people. In the total sample, 38.6% are females and 61.4% are males. The demographic data regarding age are dominated by respondents who have an age range of 18–24 years, which is 37.8% of the participants. The demography of participants can be seen in Table 1.

Table 1. Participants' information.

Demographic variable	N	Percentage
Gender		
Male	78	61.4%
Female	49	38.6%
Age		
Under 18	1	0.8%
18–24	48	37.8%
25–34	47	37.0%
35–44	23	18.1%
45–54	7	5.5%
55–64	1	0.8%
Total	127	100%

2.2 *Data collection*

The data is collected through Google Forms. Participants agreed to complete a questionnaire measuring *gotong royong*. The questionnaires were distributed to the participants through social media platforms, such as Facebook, WhatsApp group, etc. The checklist for Reporting Results of Internet E-Surveys (CHERRIES) was applied as follows: (1) the survey was anonymous and did not require any personal data; (2) the study has been approved by the IRB (UIN Syarif Hidayatullah Jakarta ethical committee); (3) participants have been informed regarding the length of time of the survey, the data storage, the investigator, and the purpose of the study; and (4) the participants were asked to fill out the informed consent form if they agreed to participate.

2.3 *Instruments*

Gotong royong indicators include: helping each other, making decisions together, and respecting the neighborhood. The instrument used is the *Gotong royong* scale, consisting of 6 items, using a 6-point Likert scale from "not like me at all" (score = 1) to "very much like me" (score = 6).

2.4 *Data analysis*

Confirmatory factor analysis (CFA) is used to validate the instruments using Lisrel 8.70. The demographic variables for gender were coded as 1 = male and 2 = female. We also used Rasch analysis to analyze which item answered was the easiest and which was the hardest.

3 RESULT

3.1 *Construct validity test result of gotong royong scale*

In this study, *gotong royong* is unidimensional, which means that it only measures *gotong royong*. CFA analysis was conducted to test the construct validity. In the initial calculation of the CFA data for the one-factor model, the fit model was obtained with chi-square = 11.2; df =

9; p-value = 0.22; RMSEA estimate = 0.049; RMSEA 90% CI = 0 to 0.12; probability RMSEA < 0.05 = 45; and CFI = 0.99. Furthermore, the factor loading coefficient and t-value of each item need to be checked to find out whether each item is valid in measuring what is being measured and determine whether there are items that need to be eliminated. According to Table 2, all items are valid because the factor loading coefficient is positive and the t-value > 1.96. Thus, all items only measure one *gotong royong* factor and can be included in the hypothesis analysis.

Table 2. CFA results for *gotong royong* scale.

Item	Coefficient	Standard error	T-Value	Item decision
1	0.67	0.08	8.08	Valid
2	0.63	0.08	7.40	Valid
3	0.73	0.08	9.18	Valid
4	0.66	0.08	8.00	Valid
5	0.81	0.08	10.44	Valid
6	0.83	0.08	10.97	Valid

Item analysis result

Table 3. Item statistics: Measure order.

Entry Number	Total score	Count	Measure	Model S.E.	INFIT		OUTFIT		PTMEA CORR.	EXACT OBS%	MATCH EXP%	Item	G
					MNSQ	ZTD	MNSQ	ZTD					
4	491	127	0.5	0.1	1.12	1	1.19	1.5	0.71	45.6	44.3	Q4	0
2	495	127	0.43	0.1	1.24	1.9	1.37	2.7	0.68	43.2	43.8	Q2	0
1	558	127	0.37	0.13	1.1	0.8	1.09	0.7	0.68	52.8	54.5	Q1	0
6	552	127	−0.24	0.12	0.75	−2.1	0.75	−2	0.78	57.6	51.4	Q6	0
3	552	127	−0.38	0.12	0.96	−0.3	0.95	−0.4	0.72	59.2	51.3	Q3	0
5	578	127	−0.68	0.13	0.77	−1.8	0.76	−2	0.76	60.8	53.3	Q5	0
MEAN	537.7	127	0	0.12	0.99	−0.1	1.02	0.1		53.2	49.8		
S.D.	32.8	0	0.45	0.01	0.18	1.5	0.22	1.7		6.7	4.2		

From Table 3 it can be seen that the item difficulty level ranges from 0.50 to −0.68. The higher the measure, the more difficult the item, and vice versa.

As for item fit, based on the INFIT MNSQ and OUTFIT MNSQ criteria, the item is said to be ideal if the value is between 0.5 and 1.5 (Linacre 2002). The results of the analysis show that all items meet the OUTFIT MNSQ criteria and the INFIT MNSQ criteria for all items on the *gotong royong* instrument. Thus, this instrument fits the Rasch model.

Furthermore, in the PTMEA CORR (Point Measure Correlation) column, the distinguishing power of the items can be seen. Based on the criteria of Alagumalai, Curtis, and Hungi (2005), an item is said to be very good if the point measure correlation value is more than 0.40; it is said to be good if it is in the range 0.30 to 0.39; it is sufficient if it is in the range of 0.20 to 0.29; it cannot discriminate if it is in the range of 0.00 to 0.19; and it needs to be revised if it is below 0. The analysis results show that all are able to discriminate, with the point measure correlation ranging from 0.78 to 0.68.

The characteristics of the item difficulty level (measurement) can be seen in Table 4. The results of the analysis show that the Q4 item, namely "When he is in a serious conflict with his neighbors, he thinks that he should ask the community leader (RT/RW)" has the highest measure of 0.50. Item Q5 is "It is important for him to always show respect for his neighbors by caring about the welfare of others." has the lowest measure of −0.68

Table 4. Item order based on item difficulty level.

No.	Indicators	Item (English)	Item (Indonesian)	Item difficulty
Q4	Make decisions together	When he got into a serious conflict with neighbors, he thought that he should ask the community leader (RT/RW)	*Ketika ia terlibat konflik berat dengan tetangganya, ia berpikir bahwa ia harus bertanya pada pemimpin warga (RT/RW)*	0.50
Q2	Respect neighborhood	It is very important for him to celebrate religious celebrations with his neighbors.	*Sangatlah penting baginya untuk merayakan perayaan agama bersama dengan tetangga.*	0.43
Q1	Help each other	He thinks it is very important to help neighbors who are in need, including helping with money.	*Ia berpikir sangatlah penting untuk membantu tetangga yang sedang membutuhkan, termasuk membantu dengan uang.*	0.37
Q6	Help each other	He thinks it is important to be involved in community service activities (cleaning with neighbors) in the home environment.	*Ia berpikir penting untuk terlibat pada aktivitas kerja bakti (bersih-bersih bersama tetangga) di lingkungan rumah.*	−0.24
Q3	Make decisions together	When there are problems in the neighborhood (e.g., flooding, environmental pollution, waste problems, and so on) it is important for him to discuss these problems with his neighbors and decide on a solution together.	*Ketika ada masalah di lingkungan tetangga (misal: banjir, polusi lingkungan, masalah limbah dan seterusnya) penting baginya membicarakan masalah ini dengan para tetangga dan memutuskan solusinya bersama.*	−0.38
Q5	Respect neighborhood	It is important for him to always show respect for his neighbors by caring about the welfare of others.	*Penting baginya untuk selalu menunjukkan hormat pada lingkungan tetangga dengan cara peduli pada kesejahteraan orang lain.*	−0.68

3 CONCLUSION AND RECOMMENDATION

From this research, we can conclude that the *gotong royong* scale is valid in measuring three factors: *helping each other, making decisions together, and respecting the neighborhood.* This conclusion is based on the factor loading coefficient that is positive and the t-value > 1.96.

Furthermore, from the item difficulty analysis, the easiest item to answer as a sample of *gotong royong* is to *respect the neighborhood.* However, the most difficult item to answer comes from *making decisions together.* The difficult item to answer is about *asking the community leader (RT/RW) if one has a serious conflict with their neighbor* (Q4).

From the item analysis, we can recommend increasing the intention of *gotong royong*; individuals can start by respecting the neighborhood. If this is already easy to do, *gotong royong* can be an indigenous value for a more inclusive and sustainable Indonesian future. As Sukarno said, it can unite Christians and Muslims, rich and poor, and indigenous Indonesians and naturalized citizens in a mutually tolerant struggle against enemies. *Gotong royong* is necessary in the fight against imperialism and capitalism, in the present just as in the past. We cannot hope to have a sustainable future, without bringing together the *gotong royong* values in our social lives.

The (re)application of this value of *gotong royong* into development policies and programs can indeed contribute to increasing local involvement and participation of all community members and is thus related to the realization of poverty alleviation and sustainable development throughout Indonesia. In the context of community development, Koentjaraningrat's (1961) observation is important that the Gotong Royong institution makes human life in Indonesia more empowered and prosperous: "With gotong royong, various problems of living together can be solved easily and cheaply, as well as community development activities."

REFERENCES

Bowen, J. R. (1986). On the political construction of tradition: Gotong Royong in Indonesia. *The Journal of Asian Studies*, *45*(3), 545–561.

CAF World Giving Index 2018; A Global View of Giving Trends. (2018). https://www.cafonline.org/docs/default-source/about-us publications/caf_wgi2018_report_webnopw_2379a_261018.pdf

Feith, Herbert, and Lance Castles, eds. 1970. *Indonesian Political Thinking, 1945–1965*. Ithaca, N.Y.: Cornell University Press.

Hofstede, G. (2001). *Culture's Consequences: Comparing Values, Behaviors, Institutions and Organizations Across Nations* (2nd ed.). Thousand Oaks, CA: Sage Publications Inc.

Koentjaraningrat. (1961). *Some Sociological-Anthropological Observations on Gotong Royong Practices in two Villages in Central Java*. Ithaca, New York: Cornell University Modern Indonesia Project.

Rahmi, D.H., Wibisono, B.H., Setiawan, B. (2001). *Rukun* and *Gotong Royong*: Managing public places in an Indonesian *Kampung*. In: Miao, P. (eds) *Public Places in Asia Pacific Cities. The GeoJournal Library*, vol 60. Springer, Dordrecht. https://doi.org/10.1007/978-94-017-2815-7_6

Slikkerveer, L. J. (2019). Gotong royong: An indigenous institution of communality and mutual assistance in Indonesia. In *Integrated Community-Managed Development* (pp. 307–320). Springer, Cham.

Soekarno, Ir. 1965. *Dibawah Bendera Revolusi [Under the Flag of Revolution]*. Vol. 2. Jakarta: Panitya Penerbit.

Religion, Education, Science and Technology towards a More Inclusive and Sustainable Future – Rahiem (Ed.)

The implementation of multicultural education in Indonesia: Systematic literature review

M. Huda*, A. Saeful, H. Rahim, D. Rosyada & M. Zuhdi
UIN Syarif Hidayatullah, Jakarta, Indonesia

S. Muttaqin
STAI Alhikmah, Jakarta, Indonesia

ABSTRACT: Indonesia is a country with a diverse society that potentially causes conflicts to occur between cultures, races, ethnicities, and religions. The theory used in this study is the theory of multicultural education from James A. Banks. The purpose of this study is to analyze the implementation of multicultural education in Indonesia. The method used is qualitative. The researchers identified various concepts in the literature and checked for patterns, parallels, and regularities. A systematic analysis of the literature was used to analyze the implementation of multicultural education. Data were collected online and 11 related studies were selected for further identification. The data shows that multicultural education in Indonesia is carried out by (1) integrating subjects with multicultural education; (2) through school culture; and (3) through local wisdom. Researchers conclude that multicultural education in Indonesia is not well organized, but can still be integrated into subjects with support from teachers, principals, education staff, and school management. Further research can be undertaken by looking at schools that claim to be multicultural and are applying multicultural theories appropriately.

Keywords: multicultural education, Indonesia, school

1 INTRODUCTION

Indonesia is a multicultural country that has diverse religions, ethnicities, cultures, languages, and races (Huda 2021). This multicultural reality is the nation's advantage (Nanggala 2019). However, the multicultural reality is also prone to conflict (Suryaman & Juharyanto 2020). On January 8, 2022, Indonesian history recorded a clash between two tribes in Papua, namely the Nduga tribe and the Lany Jaya tribe (MJO 2022). To minimize such occurrences, multicultural education can be a solution to nullify the conflicts that arise in a multicultural society (Chase & Morrison 2018).

The implementation of multicultural education in Indonesia is far from successful (Harjatanaya & Hoon 2018) because inconsistencies are occurring between policies and multicultural education practices. Ironically, the 2003 Education Law provides and mandates a sufficient basis for a national education policy standard that cares for and respects diversity. To optimally implement multicultural education in Indonesia, the government should direct educational policies and practices toward a more holistic approach to multicultural education by involving all school elements to create tolerant, peaceful, and harmonious citizens contributing positively to a pluralist and multicultural society (Raihani 2018).

*Corresponding Author: miftahulhuda20@mhs.uinjkt.ac.id

DOI: 10.1201/9781003322054-13

Multicultural education is not only implemented in Indonesia, it is implemented in other countries as well (Bakhov 2015). The policy of multiculturalism can be seen in the state's guarantee to recognize and respect the rights of ethnic communities in decision-making and public funding that is socially significant.

Limited studies explain the implementation of multicultural education in schools (Efendi & Lie 2021; Huda *et al.* 2021; Purba *et al.* 2019). The studies convey that multicultural education is implemented through two primary methods: school culture and classroom culture. Furthermore, research (Ozbilgehan & Celenk 2021) has found some problems related to the implementation of multicultural education in Turkey. The problems were grouped as educational, communication, curriculum, family-related, and general. Furthermore, studies (Jiyanto & Efendi 2016) show that multicultural education is not only achieved through theory or by adding it to an existing curriculum, but also through teaching practices such as integrating it into materials that discuss diversity issues, such as science, social knowledge, religious education, and citizenship education.

This research specifically delves into the implementation of multicultural schools in the Indonesian context. It aims at providing an overview as well as the possibility of adopting multicultural education in Indonesia. This study is also a preliminary or background study investigating approaches and strategies for the implementation of multicultural education in Indonesia.

2 METHODS

This article seeks to answer the question: How is the implementation of multicultural education in Indonesia? A qualitative method with a library research type is used in this research. Researchers explore this question and build hypothetical concepts through literature systematic review and critical thinking (Rahiem & Rahim 2020). This article is a powerful vehicle for theory building that uses one of the four conceptual article types proposed by Jaakkola (2020) namely synthesis theory. Below is the flow of the synthesis theory used in this article:

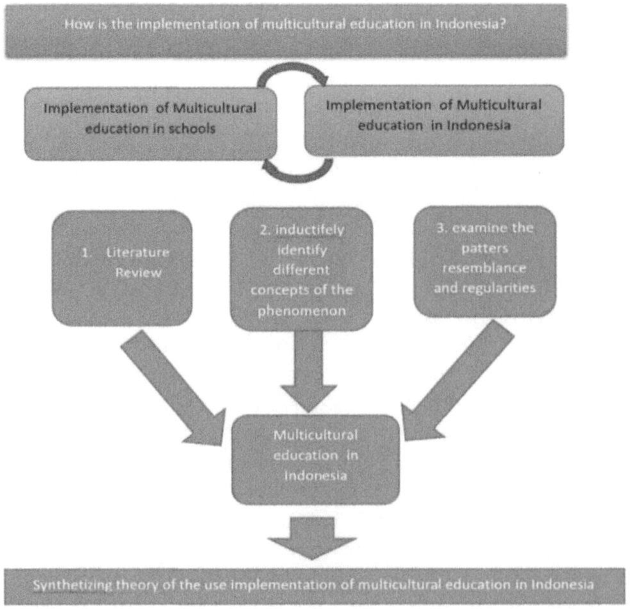

Figure 1. Methodology.

The researcher starts by looking at the idea of multicultural education before looking into how it is implemented in schools. Further study will concentrate on these two ideas. The researcher then looks at the focused phenomena, which isn't covered in enough detail in the earlier studies. The implementation of multicultural education in Indonesia is a phenomenon that is in the spotlight. By examining the literature gathered from Google Scholar using the terms "multicultural education in Indonesia," "implementation of multicultural education in Indonesia," and "multicultural education in Indonesia," the researchers were able to identify various conceptualizations of this phenomenon. The observed study made no explicit mention of Indonesian education. The researcher looks for regularities, parallels, and patterns in the observed premises, finds explanations from the theoretical framework, and then draws a conclusion.

3 RESULTS

The systematic literature review data reveals that the implementation of multicultural education in Indonesia is carried out through (1) integrating subjects and multicultural education content; (2) school culture; and (3) local wisdom. (Figure 2). The conclusion was obtained from a systematic literature review of 11 articles found in the Google Scholar database using the following keywords: multicultural education in Indonesia, implementation of multicultural education in Indonesia, and multicultural education in Indonesia.

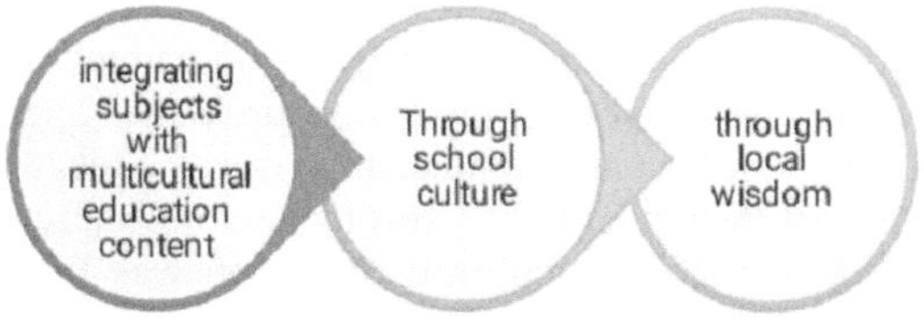

Figure 2. Research finding.

3.1 *Integrating subjects with multicultural education content*

Five articles discuss the integration of multicultural education content with subjects:

Primasari and Marini (2021) found that the implementation of multicultural values at the Jakarta Multicultural School (JMS) through learning is carried out in an integrated manner with intracurricular and extracurricular activities; the role of principals, teachers, and students who have an understanding of developing multicultural values has a positive effect in building the character of students. Documentation analysis, interviews, and observations were used to determine the language of instruction in schools, implementation of international curriculum programs, school culture, school co-curricular activities, and school hidden curriculum.

Januarti and Zakso (2019) found that the content integration factor was used to assess the implementation of multicultural education in schools. The integration is carried out through self-development activities that are both programmed and nonprogrammed. Extracurricular activities are examples of programmed self-development activities, while routine activities carried out regularly, spontaneous activities, and exemplary activities are examples of nonprogrammed self-development activities. In this regard, the school atmosphere, curriculum, buildings and infrastructure, instructors' roles, school programs/activities, and students are all supportive aspects.

Supriatin and Nasution (2017) found that multicultural education is being implemented in schools by designing the learning process, preparing the curriculum, designing evaluations, as well as educating instructors with multicultural perspectives, attitudes, and behaviors to foster the multicultural attitudes of their students. Multicultural education is being incorporated into the school curricula and can be carried out comprehensively through civic education, religious education, or integrating with other subjects.

Parinduri (2017) explains that multicultural education is implemented in religion-based schools in three ways. The first is through school rules that are open to accepting pupils from other religious backgrounds. The second is by incorporating multicultural values into curriculum creation, teaching, and learning activities. The third is by providing pupils with the same level of service (equality) in their social interactions and providing the same reward for differences in cultural values that have been developed.

Jiyanto and Efendi (2016) found that the implementation of multicultural education does not have to stand alone but can be integrated into subjects and educational processes in schools, including role modeling in schools by teachers and adults. The content of multicultural education must be implemented in the form of actions, both in schools and in the community. The application of multicultural education is an effort to build awareness and understanding of a multicultural society. This is because multicultural education is a method of instilling a polite, sincere, and tolerant attitude toward cultural variety in a plural society.

3.2 *Through school culture*

Four more articles discuss multicultural education through school culture:

Zamroni *et al.* (2021) reveal that exposure to social and cultural diversity is relatively high because Indonesia has been built on ethnic, religious, and linguistic diversity. So, multicultural education has become a more complicated problem. The goal of multicultural education is to provide equal and nondiscriminatory education for all students. Multicultural education can be implemented through school culture by greeting the five religions. Adib (2020) found that multicultural education was built by students through school culture because there was an agency when students took part in extracurricular activities. Sulistiyo and Alfian (2019) investigated how multicultural education is implemented by empowering the school environment as a hidden curriculum, in the provision of learning facilities, worship facilities, administrative services, and numerous other services.

Purba *et al.* (2019) found that multicultural education was implemented using two primary methods: school culture and classroom culture. School culture refers to the foundation's efforts to promote intercultural values through a variety of programs and policies. The foundation establishes numerous regulations in which instructors take a greater role in imparting knowledge of diversity in the classroom through their particular subject matter.

3.3 *Through local wisdom*

Two articles were found on the implementation of multicultural education through local wisdom. Sopiansyah (2021) investigated that the integration of multicultural education with national identity can be achieved through (1) curriculum design that incorporates local wisdom; (2) the optimization of civic education in its efforts to strengthen national identity through multicultural and local wisdom possessed by the Indonesian people; and (3) the placement of multicultural education as an educational philosophy, educational approach, and field of study. Noor and Sugito (2019) found that children's knowledge of the significance of respecting others and diverse cultures will grow as a result of multicultural education based on local wisdom. Every community's wisdom is a reflection of a unified (holistic) philosophy of existence. Teachers must integrate context with local wisdom values, such as using Java and Prambanan Temple as teaching-learning materials. Learning models based on local wisdom can help increase harmony, national identity, and learning mastery.

4 DISCUSSION

Learning is the most effective way of conducting multicultural education. Therefore, enhancing teachers' competence in integrating multicultural values into learning is considered a key component (Setyowati *et al.* 2019). The dimensions of multicultural education applied in schools are evidence that some dimensions are less related to practical applications, such as providing more detailed examples of issues related to multiculturalism and diversity that are relevant to different subjects (Erbaş 2019). Multicultural education is implemented in an integrated manner in several subjects, such as civics and the Indonesian language. This study recommends that multicultural education should be a part of the education curriculum. Teachers need to be given training on the implementation of multicultural education to provide an understanding of the practicality of multicultural education in religion-based schools (Arifin & Hermino 2017).

As a practical example of multicultural education implementation in Turkey, teachers do not separate foreign students from local students. They do not discriminate against them in grading because teachers are not biased against foreign students and are oriented toward the happiness and success of the students (Ozbilgehan & Celenk 2021). With students coming from various cultural backgrounds placed in one room, it can be a medium for them to directly apply multicultural education in their daily lives (Dover 2013). The implementation of multicultural education from the perspective of teachers, principals, and student activities in elementary schools is based on transformative pedagogy (Iulia 2015).

In the implementation of multicultural education, not only local wisdom but also madrasa culture has developed to appreciate the variety of processes in learning inside and outside the classroom (Abisdin & Murtadlo 2020). Furthermore, two models of multicultural education that can be implemented in early childhood education are the contribution approach and the additive approach (Alghamdi 2017). The application of multicultural education through cultural arts education has an important role in shaping students' character. Therefore multicultural-based art and culture learning is needed as a solution to tackle problems in the character of adult learners (Perdana *et al.* 2018; Rahmawati *et al.* 2018).

The 11 papers, described above, contain three important points: (1) Multicultural education is an activity that provides equal opportunities for justice for every individual to get an education regardless of ethnicity, culture, ethnicity, race, and religion so that they gain both knowledge and skills; (2) Multicultural education in schools does not stand alone but is integrated with subjects; and (3) It is possible to undertake multicultural education in Indonesia in three ways, namely integrating the content of multicultural education with subjects, school culture, and local wisdom.

5 CONCLUSION

The implementation of multicultural education in Indonesia can be carried out by integrating multicultural education content with subjects, through school culture, and by acknowledging and respecting cultural diversity. This research does not cover the whole problem comprehensively because it is solely based on a literature review. In addition, the number of articles found is also very limited. However, this research serves as a solid basis for future investigations on how to implement multicultural education in Indonesia. Researchers suggest further studies with a larger number of articles and databases. The implication of this research is based on the theory that multicultural education can contribute to implement multiculturalism in schools to minimize or prevent intolerance and radicalism from entering schools. The recommendation from this study is that there is no research on the multicultural education model in Indonesia, so further research is needed to find schools that claim to be multicultural to get new theories from the school model.

REFERENCES

Abisdin, A.A. and Murtadlo, M.A. (2020) 'Curriculum development of multicultural-Based islamic education as an effort to weaver', *IJIERM: International Journal of Islamic Education, Research and Multiculturalism*, 2(1), pp. 29–46. DOI:https://doi.org/10.47006/ijierm.v2i1.30.

Alghamdi, Y. (2017) 'Multicultural education in the us: current issues and suggestions for practical implementations', *International Journal of Education*, 9(2), p. 44. doi:10.5296/ije.v9i2.11316.

Arifin, I. and Hermino, A. (2017) 'The importance of multicultural education in schools in the era of ASEAN economic community, *Asian Social Science*, 13(4), pp. 78–92. doi:10.5539/ass.v13n4p78.

Awang, J. *et al.* (2018) 'Interreligious peace in multicultural society: A critique to the idea of peace multiculturalism', *International Journal of Civil Engineering and Technology (IJCIET)*, 9(9), pp. 1485–1493.

Bakhov, I.S. (2015) 'Historical dimension of the formation of multicultural education in Canada', *Pedagogika*, 117(1), pp. 7–15. doi:10.15823/p.2015.063.

Chase, S. and Morrison, K. (2018) 'Implementation of multicultural education in unschooling and its potential', *International Journal of Multicultural Education*, 20(3), pp. 39–59. doi:10.18251/ijme.v20i3.1632.

Deni Sopiansyah, M.M.E. (2021) 'Model pembelajaran dan implementasi pendidikan multikultural dalam pendidikan Islam dan Nasional', *Mimbar Kampius: Jurnal Pendidikan dan Agama Islam*, 20(2), pp. 88–98. doi:10.17467/mk.v20i2.467.

Dover, A.G. (2013) 'Teaching for Social Justice: From Conceptual Frameworks to Classroom Practices to Classroom Practices', *Multicultural Perspectives*, 15(1), pp. 3–11. doi:10.1080/15210960.2013.754285.

Efendi, M.Y. and Lie, H. (2021) 'Implementation of multicultural education cooperative learning to develop character, nationalism and religious', *Journal of Teaching and Learning in Elementary Education*, 4(1), pp. 20–38. DOI:http://dx.doi.org/10.33578/jtlee.v4i1.7817.

Erbaş, Y.H. (2019) 'Dimensions of multicultural education: Pedagogical practices knowledge of graduate students towards multicultural education in Turkey', *Mediterranean Journal of Educational Research*, 13 (27), pp. 142–181. doi:10.29329/mjer.2019.185.7.

Harjatanaya, T.Y., and Hoon, C. (2018) 'Politics of multicultural education in post-Suharto Indonesia: A study of the Chinese minority, *Compare*, 50(1), pp. 18–35. doi:10.1080/03057925.2018.1493573.

Huda, Mi., Nubatonis, T. and Uus Ruswandi (2021) 'Implementation of multicultural education in education practice in Indonesia', *Edukasi: Jurnal Pendidikan Islam (e-journal)*, 9(1), pp. 73–81. Available at: https://staim-tulungagung.ac.id/ejournal/index.php/edukasi/article/view/582.

Ika Firma Ningsih Dian Primasari, Arita Marini, A.M. (2021) 'Implementasi Nilai-nilai multikultural di Sekolah dasar', *Syntax Literate: Jurnal Ilmiah Indonesia*, 6(11), pp. 5680–5694. doi: http://dx.doi.org/10.36418/ Syntax-Literate.v6i11.1793.

Iulia, A. (2015) 'Integrating multicultural education In Pre-service teacher training courses Ioana Todor', *Journal of Linguistic Intercultural Education*, 8, pp. 213–224. Available at: https://e-resources.perpusnas.go.id:2227/ehost/detail/detail?vid=15&sid=04622602-a83d-4f2e-8e59-fc2546c2e4fc%40redis&bdata=JnNpdGU9ZWhvc3QtbGl2ZQ%3D%3D#db=ufh&AN=115394120.

Jaakkola, E. (2020) 'Designing conceptual articles: Four approaches', *AMS Review*, 10(1–2), pp. 18–26. DOI: https://doi.org/10.1007/s13162-020-00161-0.

Jiyanto; Amirul Eko Efendi (2016) 'Implementasi pendidikan multikultural di madrasah inklusi madrasah aliyah negeri maguwoharjo yogyakarta', *Jurnal Penelitian*, 10(1), pp. 25–44.

Miftahul Huda (2021) 'Sociological aspects of multicultural Islamic religious education', *Al-Insyiroh: Jurnal Studi Keislaman*, 7(2), pp. 122–143. DOI:https://doi.org/10.35309/alinsyiroh.v7i2.4990.

MJO (2022) *Bentrok Suku Nduga dan Lani Jaya di Papua, Kedua Pihak Berdamai, CNN Indonesia*. Available at: https://www.cnnindonesia.com/nasional/20220116072405-20-747175/bentrok-suku-nduga-dan-lani-jaya-di-papua-kedua-pihak-berdamai (Accessed: 12 April 2022).

Muhammad Abrar Parinduri (2017) *Pendidikan di Sekolah Berbasis Agama dalam Perspektif Multikultural (Studi Kasus pada Sekolah Islam dan Sekolah Kristen di Sumatera Utara)*. UIN Syarif Hidayatullah Jakarta.

Nanggala, A. (2019) '*Pendidikan Kewarganegaraan Sebagai Pendidikan Multikultural*', 3(2), pp. 197–210. doi: https://doi.org/10.36787/jsi.v3i2.354.

Noor, A.F., and Sugito (2019) 'Multicultural education based in the local wisdom of Indonesia for elementary schools in the 21st century', *Journal of International Social Studies*, 9(2), pp. 94–106. Available at: https://iajiss.org/index.php/iajiss/article/view/408/325.

Ozbilgehan, M. and Celenk, S. (2021) 'A review of multicultural education in Northern Cyprus and Turkish learning levels of students from different cultural backgrounds', *Revista de Cercetare si Interventie Sociala*, 73, pp. 114–132. doi:10.33788/rcis.73.8.

Perdana, Y., Djono and Ediyono, S. (2018) 'The implementation of multicultural education in history learning at SMAN 3 surakarta', *International Journal of Multicultural and Multireligious Understanding*, 5(3), pp. 11–18. doi:http://dx.doi.org/10.18415/ijmmu.v5i3.135.
Purba, A.S., Malihah, E. and Hufad, A. (2019) 'The Implementation of multicultural education in senior high schools in medan', *Budapest International Research and Critics Institute (BIRCI-Journal): Humanities and Social Sciences*, 2(3), pp. 226–233. doi:10.33258/birci.v2i3.411.
Rahiem, M.D.H. and Rahim, H. (2020) 'The dragon, the knight and the princess: Folklore in early childhood disaster education', *International Journal of Learning, Teaching and Educational Research*, 19(8), pp. 60–80. doi:10.26803/IJLTER.19.8.4.
Rahmawati, N.N., Kumbara, A. and Suda, I.K. (2018) 'Multiculturalism towards religious life in Tewang Tampang Village: Moral education implementation based local wisdom', *International Journal of Linguistics, Literature and Culture*, 4(6), pp. 63–71.
Raihani, R. (2018) 'Education for multicultural citizens in Indonesia: Policies and practices', *Compare: A Journal of Comparative and International Education*, 48(6), pp. 992–1009. doi:10.1080/03057925.2017.1399250.
Setyowati, R., Sarmini and Amaliya, N. (2019) 'From multicultural towards national identity: Teacher construction on strategies for implementing multicultural education in schools', *Advances in Social Science, Education and Humanities Research*, 383(Icss), pp. 405–410. doi:10.2991/icss-19.2019.177.
Sulistiyo and Alfian (2019) 'The Implementation of multicultural education of Sosial Studies in Indonesia', *Business & Social Sciences Journal*, 4(2), pp. 1–11. doi:10.35940/ijmh.L0335.0831219.
Supriatin, A. and Nasution, A.R. (2017) 'Implementasi pendidikan multikultural dalam praktik pendidikan di indonesia', *Elementary: Jurnal Ilmiah Pendidikan Dasar*, 3(3), pp. 1–13. Available at: https://e-journal.metrouniv.ac.id/index.php/elementary/article/view/785.
Suryaman and Juharyanto (2020) 'The role of teachers in implementing multicultural education values in the curriculum 2013 implementation in Indonesia', *Journal of Education and Practice*, 11(3), pp. 152–156. doi:10.7176/JEP/11-3-16.
Zamroni *et al.* (2021) 'Cross-cultural competence in multicultural education in Indonesian and New Zealand high schools', *International Journal of Instruction*, 14(3), pp. 597–612. doi:10.29333/iji.2021.14335a.

Religion, Education, Science and Technology towards a More Inclusive and Sustainable Future – Rahiem (Ed.)

Strengthening emotional intelligence, spiritual intelligence, and school culture on student character

K.I. Rosadi, A.A. Musyaffa*, K. Anwar, M. El Widdah & M. Fadhil Idarianty
UIN Sulthan Thaha Saifuddin, Jambi

ABSTRACT: School is one of the institutions and places of effort to foster the character of a student. The purpose of the research is to find out the effect of emotional intelligence, spiritual intelligence, and school culture on student character strengthening. While the research was conducted on high school students in Jambi province, the method used is a survey method with a quantitative approach using path analysis. The population of this study was 855 Islamic high school teachers in Jambi Province, with a random sampling technique. Data were collected by administering a questionnaire. The results showed (1) a positive and significant direct influence of emotional intelligence on school culture; (2) a positive and significant direct effect on spiritual intelligence; (3) a direct positive and significant effect of emotional intelligence on student character strengthening; (4) a direct positive and significant influence of spiritual intelligence on strengthening student character; (5) a direct positive and significant effect of work culture on student character strengthening; (6) an indirect effect of emotional intelligence on student character strengthening through school culture; and (7) a positive and significant indirect effect of spiritual intelligence on student character strengthening through school culture in senior high schools in Jambi Province. To conclude, the higher the application of emotional intelligence, spiritual intelligence, and work culture, the higher the student character strengthening.

Keywords: emotional intelligence, spiritual intelligence, school culture, character strengthening

1 INTRODUCTION

Student character formation takes place through the educational process. According to British scholars Bridges and Bridges (2017) and Menzies and Baron (2014), "transnational education refers to all types of higher education programs, complexes of educational courses or educational services, dual enrolment opportunities, summer bridge programs and team projects." Such programs can be provided by a country's educational system or may be independent of any system. International student exchange may be studied theoretically and applied practically. According to the meta-analysis of different research, the main problems of student exchange are the institutionalization of international education and management of student needs, psychological and social support shapes and structures, and the transition from national to international. Thus, the "research focuses on the methods of internationalization at the individual, institutional, and national levels." In this context, character education is directed at educating students to become human beings who do good with good actions based on righteousness toward God. (Bedenlier et al. 2018, p. 108; Makarova & Egorova 2021). In the concept of *ulul albab*, Fahim Tharaba et al. (2021) state that education aims to encourage students to become active learners who convey knowledge to others, give warnings, and solve problems in society. The sixth President of

*Corresponding Author: musyaffa@uinjambi.ac.id

DOI: 10.1201/9781003322054-14

Indonesia, Susilo Bambang Yudhoyono, hopes that this character education will make Indonesian people excel in science and technology. It was further emphasized that there are five fundamental goals of the national movement for character education: (1) people must have good morals, noble character, and good behavior. (2) the nation should become a smart and rational nation; (3) the nation should be innovative, moving forward, and willing to work hard; (4) building optimism; and (5) becoming a true patriot by showing love for the nation, the state, and the homeland of Indonesia. Strengthening student character in schools cannot be separated from emotional intelligence, spiritual intelligence, and school culture. The essence of emotional intelligence that has been introduced so far is our ability to build emotions well concerning ourselves and others. One of the characteristics of people who have emotional intelligence is empathy. Empathy is the ability to understand the feelings of others or the ability to feel what other people think. Generally, teachers' emotional intelligence in senior high schools in Jambi Province is standard or mediocre. The guidance of students on the development of the spiritual aspect is left entirely to religious teachers, and other teachers only assist in implementing spiritual activities. Furthermore, the factor that influences the strengthening of student character is school culture. School culture is a set of values that underlies the behavior, traditions, habits, and daily lives of students. The principal, teachers, students, and school employees should strive to achieve the formation of these characteristics so as to provide an example for the wider community. This statement is in accordance with the results of the study conducted by Ma and Wang (2022). The study enumerated the existing gaps and offered future directions and implications for educational practitioners and researchers whose awareness of spiritual intelligence and its impact on education and learner-psychology variables can improve based on researchers' observations in senior high schools in Jambi Province. The school culture began to change due to the development of community culture in general and the development of information technology. As a result, the development affected the school culture, especially the character of students. However, the results showed that in senior high schools in Jambi Province, the strengthening of student character was not carried out properly. There are still students who (1) do not respect teachers and their friends; (2) speak inappropriately; (3) lack discipline in the learning process; (4) have not shown high learning motivation; (5) do not keep the school clean; and (6) annoy other students.

2 RESEARCH METHODS

This research was conducted in Jambi Provincial High School. The research activity was carried out for 5 months from January to May 2022. This research used a quantitative approach, survey method, and path analysis techniques (Barreiro & Justo 2001). This research consisted of exogenous variables, intervening variables, and endogenous variables. The exogenous variables are emotional intelligence (X1) and spiritual intelligence (X2); the intervening variable is school culture (Y). The endogenous variable was strengthening student character (Z). After the data was collected, it was analyzed using data management techniques. The data analysis used by the researcher in this study aims to answer the questions listed in the identification of the problem. For this reason, the researcher described the results of data processing as the influence of variables (emotional intelligence), (spiritual intelligence), Y (school culture), and Z (student character strengthening) as endogenous variables. The analytical method used in this study is path analysis.

3 RESEARCH RESULTS AND DISCUSSION

3.1 *Effect of emotional intelligence on school culture*

The first research hypothesis was: *There is a direct influence of emotional intelligence on school culture.*

This hypothesis is statistically formulated as follows:

$$\text{Ho}: P_{y1} = 0 \ \text{H1}: \text{Py1} > 0.$$

Table 1. Hypothesis testing the effect of emotional intelligence on school culture.

Model		Unstandardized Coefficients B	Std. Error	R	R2	FCount	TCount	Sig.
1	(Constant)	7.253	2.847	0.844	0.712	0.844	2,548	0.000
	X1	.842	,033				25,437	0.000

Based on coefficients in the Unstandardized Coefficients column in Table 1, the regression equation obtained is as follows:

$$\widehat{Y} = 7.253 + 0.842X1 \text{ Y} = \text{School culture X1} = \text{Emotional intelligence.}$$

The constant of 7,253 states that when there is no emotional intelligence, the school culture is 7,253. The regression coefficient of 0.842 states that each addition of 1 value of emotional intelligence would increase school culture by 0.842.

The table shows that the value of R = 0.844 with a significance of 0.00, which is smaller than the value of = 0.05, so Ho is rejected. This means that there is an influence of emotional intelligence on school culture.

The R-Squared obtained from the model summary table is 0.712, or 71.2%, which means that as much as 71.2% of the regression model of the Y function (school culture) could be explained by the emotional intelligence factor (X1).

3.2 *Influence of spiritual intelligence on school culture*

The second research hypothesis is: *There is an influence of spiritual intelligence on school culture.*

This hypothesis is statistically formulated as follows:

$$\text{Ho} : P_{y2} = 0 \text{ H1} : \text{Py2} > 0.$$

Table 2. Hypothesis testing the influence of spiritual intelligence on school culture.

Model		Unstandardized Coefficients B	Std. Error	R	R2	FCount	TCount	Sig.
1	(Constant)	11.166	3,595	0.844	0.712	362.54	3.106	0.000
	X2	,746	,039				19,061	0.000

Based on coefficients in the Unstandardized Coefficient column in Table 2, the regression equation obtained is as follows:

$$\widehat{Y} = 11.166 + 0.746X2 \text{ Y} = \text{School culture X2} = \text{Spiritual intelligence.}$$

The constant of 11,166 states that when there is no spiritual intelligence, the school culture is 11,166. The regression coefficient of 0.746 states that each additional 1 value of spiritual intelligence would increase school culture by 0.746.

The value of R = 0.762 with a significance of 0.00 is obtained based on Table 2. Because the significance value was smaller than the value of = 0.05, then Ho is rejected.

3.3 *Effect of emotional intelligence on character strengthening*

The third research hypothesis is: *There is an influence of emotional intelligence on strengthening student character.*

This hypothesis is statistically formulated as follows:

$$Ho : P_{z1} = 0 \text{ H1 : Pz1} > 0.$$

Table 3. Hypothesis testing the effect of emotional intelligence on student character strengthening.

Model		Unstandardized Coefficients						
		B	Std. Error	R	R2	FCount	TCount	Sig.
1	(Constant)	55,620	3.069	0.399	0.159	49,517	18,123	0.000
	X2	,251	0.036				7,037	0.000

Based on coefficients in the Unstandardized Coefficients column in Table 3, the regression equation obtained is as follows:

$$\widehat{Z} = 55.620 + 0.251\text{X1 Z} = \text{Character gain X1} = \text{Emotional intelligence.}$$

The constant of 55.620 states that when there is no emotional intelligence, the student character strengthening is 55,620. The regression coefficient of 0.251 states that each addition of 1 value of emotional intelligence would increase student character strengthening by 0.251.

Based on the SPSS program analysis results, the value of R = 0.399 with a significance of 0.00, because the significance value was smaller than the value of = 0.05, then Ho is rejected. This means there is a relationship between emotional intelligence and character strengthening of students.

The R-Squared obtained from the model summary table is 0.159 or 15.9%, which means that as much as 15.9% of the regression model of the Z function (character strengthening of students) could be explained by the emotional intelligence factor (X1). Testing the significance of the influence of emotional intelligence (X1) on character strengthening of students (Z) the value of F = 49.517 is obtained with a significance value of 0.00, because the significance value is smaller in which 0.00 < 0.05, then Ho is rejected.

3.4 *Influence of spiritual intelligence on student character strengthening*

The fourth research hypothesis is: *There is an influence of spiritual intelligence on student character strengthening.*

This hypothesis is statistically formulated as follows:

$$H_o : P_{z2} = 0 \text{ H1 : Pz2} > 0.$$

Table 4. Hypothesis testing the influence of spiritual intelligence on student character strengthening.

Model		Unstandardized Coefficients						
		B	Std. Error	R	R2	FCount	TCount	Sig.
1	(Constant)	57,757	3,290	0.343	0.118	34,899	17,555	0.000
	X2	,212	0.036				5,907	0.000

Based on coefficients in the Unstandardized Coefficients column in Table 4, the regression equation obtained is as follows:

$$\widehat{Z} = 57.757 + 0.212\text{X2 Z} = \text{Character gain X2} = \text{Spiritual intelligence.}$$

The constant of 57.757 states that when there is no spiritual intelligence, the student character strengthening is 57.757. The regression coefficient of 0.212 states that adding 1 value of spiritual intelligence would increase student character strengthening by 0.212.

Based on the SPSS program analysis results, the correlations table (attached) obtained a value of R = 0.343 with a significance of 0.00. Because the significance value was smaller than the value of = 0.05, then Ho is rejected. This means there was an influence of spiritual intelligence on student character strengthening.

The R-Squared obtained from the model summary table is 0.118 or 11.8%, which means that as much as 11.8% regression model of the Z function (student character strengthening) could be explained by the spiritual intelligence factor (X2). Testing the significance of the influence of spiritual intelligence (X2) on student character strengthening (Z) a value of F = 34,899 is obtained with a significance value of 0.00, because the significance value is smaller in which 0.00 < 0.05, then Ho is rejected.

3.5 *The influence of school culture on strengthening student character*

The fifth research hypothesis is: *There is an influence of school culture on character strengthening.*

This hypothesis is statistically formulated as follows:

$$\text{Ho} : P_{zy} = 0 \text{ H1} : \text{Pzy} > 0.$$

Table 5. Hypothesis testing the influence of school culture on strengthening student character.

		Unstandardized Coefficients						
Model		B	Std. Error	R	R2	FCount	TCount	Sig.
1	(Constant)	59.096	2,903	0.360	0.129	38,958	20,359	0.000
	Y	,227	0.036				6,242	0.000

Based on coefficients in the Unstandardized Coefficients column in Table 5, the regression equation obtained is as follows:

$$\widehat{Z} = 59.096 + 0.227\text{Y Z} = \text{Character gain student Y} = \text{School culture.}$$

The constant of 59,096 stated that whether there was no school culture, then the strengthening of student character was 59.096. The regression coefficient 0.227 states that every addition of 1 school culture value would increase student character strengthening by 0.227.

Based on the results of the SPSS program analysis, the correlations table (attached) obtained a value of R = 0.360 with a significance of 0.00. Because the significance value was smaller than the value of = 0.05, then Ho is rejected.

3.6 *Effect of emotional intelligence on student character strengthening through school culture*

The sixth research hypothesis is: *There is an influence of emotional intelligence on student character strengthening through school culture.*

This hypothesis is statistically formulated as follows:

$$\text{Ho} : P_{z1y} = 0 \text{ H1} : \text{Pz1y} > 0.$$

Table 6. Coefficient of X1 for Z through Y.

Coefficients[a]

		Unstandardized Coefficients		Standardized Coefficients		
Model		B	Std. Error	Beta	t	Sig.
1	(Constant)	56,369	3.273		17,220	,000
	Y over X1	.307	.112	.487	2,732	,007
	Z over X1	,946	,086	,948	11.018	,000
	Z over Y	0.045	,067	,072	,679	,000

Based on Table 6, the Standardized Coefficient (beta) value of each variable is known. The direct effect of emotional intelligence on student character was 0.487, while the indirect effect was 0.948 x 0.072 = 0.068. These results show that the total influence of emotional intelligence on student character strengthening through school culture was 0.487 + 0.068 = 0.555. This means that the total influence of emotional intelligence on student character strengthening through school culture was 55.5%.

The results of the analysis of this study indicated that there was a positive and significant indirect effect of emotional intelligence on student character strengthening through school culture. This illustrated that the higher the emotional intelligence, the better the school culture so that the student character strengthening would be as good as possible. When referring to the path coefficient value, it could be understood that 6.8% of the variation in student character strengthening could be explained by emotional intelligence through school culture. While the total influence of emotional intelligence on student character strengthening through school culture is 55.5%.

3.7 *Influence of spiritual intelligence on student character strengthening through school culture*

The seventh research hypothesis is: *There is an influence of spiritual intelligence on student character strengthening through school culture.*

This hypothesis is statistically formulated as follows:

$$\text{Ho} : P_{z2y} = 0 \ \text{H1} : \text{Pz2y} > 0.$$

Table 7. Coefficient of X2 with respect to Z through Y.

Coefficients[a]

Model		Unstandardized Coefficients		Standardized Coefficients		
		B	Std. Error	Beta	t	Sig.
1	(Constant)	56,369	3,273		17,220	,000
	Y over X2	,100	,091	,161	1.091	,000
	Z over X2	,110	,084	,113	1.310	,000
	Z over Y	0.045	,067	,072	,679	,000

The Standardized Coefficient (beta) value of each variable is known based on Table 7. The direct effect of a teacher's spiritual intelligence on student character strengthening is 0.161, while the indirect effect is 0.113 x 0.072 = 0.008.

From these results, the total influence of emotional intelligence on strengthening student character through school culture is 0.113 + 0.008 = 0.121. This means that the total influence of the teacher's spiritual intelligence on student character strengthening through school culture is 12.1%.

The results of the analysis of this study indicate that there is a positive and significant indirect effect of spiritual intelligence on school culture. This illustrates that the higher the spiritual intelligence, the better the school culture so that character strengthening would be even better. When referring to the path coefficient value obtained through SPSS, it could be understood that 55% of the variation in character strengthening could be explained by spiritual intelligence through school culture. At the same time, the total influence of spiritual intelligence on character strengthening through school culture is 65%.

4 CONCLUSION

Based on the results of the study, it could be concluded as follows: (1) There is a direct and significant positive effect of emotional intelligence on school culture in senior high schools in

Jambi Province. (2) There is a direct and significant positive influence of spiritual intelligence on school culture in senior high schools in Jambi Province. (3) There is a direct and significant positive effect of emotional intelligence on student character strengthening in high schools in Jambi Province. (4) There is a direct and significant positive effect of spiritual intelligence on strengthening the character of students in high schools in Jambi Province. (5) There is a direct and significant positive influence of school culture on strengthening the character of students in high school in Jambi Province. (6) There is an indirect influence of emotional intelligence on character strengthening through school culture in senior high schools in Jambi Province. (7) There is an indirect influence of spiritual intelligence on character strengthening through school culture in senior high schools in Jambi Province.

REFERENCES

Abbas A. (2020) Strengthening character Education in vocational high schools. *Novateur Publications JournalNX- A Multidisciplinary Peer Reviewed Journal* ISSN No: 2581 - 4230 VOLUME 6, ISSUE 10, Oct. -2020.

Barreiro, P. L., & Justo, P. A. (2001). *Population and Sample: Sampling Techniques*. Spain: MaMaEuSch. http://www.mathematik.unikl.de/mamaeusc

Bedenlier, S., Kondakci, Y., & Zawacki-Richter, O. (2018). Two decades of research into the internationalization of higher education: Major themes in the journal of studies in international education (1997–2016). *Journal of Studies in International Education*, 22(2), 108–135. https://doi.org/10.1177/1028315317710093

Fahim Tharaba, M., *et al.* (2021) The science integration of Ulu al-albab perspective (campus development towards world class university), *Psychology and Education*, 58(1): 1284–1291. https://doi.org/10.17762/pae.v58i1.877

Makarova A. & Egorova 1., (2021) International student exchange management as factor of educational services development, *International Journal of Cognitive Research in Science, Engineering and Education (2021)*, https://doi.org/10.23947/2334-8496-2021-9-1-75-90

Qiangqiang M. and Fujun W. (2022), The Role of Students' spiritual intelligence in enhancing their academic engagement: A theoretical review, MINI REVIEW article. *Psychol.*, 06 May 2022 *Sec. Educational Psychology*. https://doi.org/10.33adiH89/fpsyg.2022.857842

Religion, Education, Science and Technology towards a More Inclusive and Sustainable Future – Rahiem (Ed.)

Problems facing digital learning in Jakarta (secondary schools as a case study)

A.A. Sihombing*
The National Agency for Research and Innovation (BRIN), Jakarta

M. Fatra & H. Rahim
UIN Syarif Hidayatullah, Jakarta, Indonesia

F.D. Maigahoaku
St. Cyril Ruteng Pastoral High School Ruteng, Indonesia

L.R. Octaberlina
UIN Maulana Malik Ibrahim, Malang, Indonesia

R.M. Jehaut
St. Cyril School of Higher Postoral Education of Ruteng (STIPAS), Indonesia

ABSTRACT: The restriction of digital learning is a controversial policy in which the government promotes "Freedom of Learning," but eliminates it. This qualitative study aims to demonstrate the paradox of education by highlighting the risks that must be taken. Primary data were collected through observation, Google Forms, and in-depth interviews via WhatsApp. Respondents consisted of junior and senior high school students, parents, and teachers in DKI Jakarta. Secondary data came from previous research and online news. This finding shows three facts, namely students who are low motivated or undisciplined, unskilled teachers, and parents who do not support students' digital learning processes. Thus, the basic needs of students are sacrificed by forcing face-to-face learning. The development and improvement of teacher skills in digital pedagogy is very urgent so that education can answer the demands of the times: technology-friendly, collaborative, creative, competent, innovative, and able to create participatory opportunities for future nation-building.

Keywords: Merdeka Belajar, digital pedagogy, face-to-face education, digital native

1 INTRODUCTION

Digital pedagogy (Jooston *et al.* 2020) has promoted online learning, which is in line with the existence of digital communication in this era. However, this has not been viewed as an urgent need in the learning process and has even been considered to bring a less positive impact on the learners' development (Anjani 2021). The government thus imposes the policy to return to school, and implement hundred percent face-to-face learning during the pandemic (Prastiwi 2021). The policy was found in the joint decree (*Surat Keputusan Bersama)* of four ministers, the Minister of Education, Culture, Research, and Technology; the Minister of Religious Affairs; the Minister of Health; and the Minister of Home Affairs (Kemendikbud 2021). Since 2022, a hundred percent offline learning has been implemented.

*Corresponding Author: sonadi2017@gmail.com

DOI: 10.1201/9781003322054-15

All students are required to enter school and online education services are removed. The attitude is counter-productive with the program of *Merdeka Belajar*, which gives freedom to both teachers and students to be flexible in carrying out the learning process by maximizing the use of digital technology. Instead of giving them flexibility, the government returns the students and teachers to the conventional class with the same time, place, materials, and method in the learning process. This gap becomes the main reason for conducting this research.

To overcome this gap, this research has come up with three research questions stated as follows. First, how did the teachers, parents, and students express their views on the digital learning experience? Second, to what extent was the infrastructure and technical knowledge sufficient for the introduction of digital learning? Third, what solutions did the interviewees propose to improve digital learning?

The educational literature of the last decade focused on formal classrooms and highlighted technological inequality that could be mapped into three categories. The first is the trend of digital technology integration in the educational process that required a change in the perception of educational culture (Casey & Hallissy 2016; Pokhrel & Chhetri 2021; Watermeyer *et al.* 2021). The second examines the influence, local and global effects, of digital media on the educational environment. The dominance of digital technology became significant because it required a paradigm shift in facilities, infrastructure, and the way teachers worked (El-Sofany & El-Haggar 2020; Núñez *et al.* 2020). The third is the study related to the challenges and consequences of digital education on society in the digital age (Flavin & Quintero 2018; Hanim *et al.* 2020). The current study was different as it focused on misperceptions about digital pedagogy that led to the controversial attitudes of education providers. On the one hand, the government echoed and implemented the program of *Merdeka Belajar.* On the other hand, policymakers happened to close the space and the opportunity for freedom to learn.

This study aims to explore the difficulties of Indonesian education in adapting the opportunities for educational advancement in digital pedagogy. The issue was studied from the perspective of the experiences of educators, students, and parents who had undergone digital learning over the past two years. Correspondingly, three research questions are formulated: First, how did the teachers, parents, and students express their views on the digital learning experience? Second, to what extent were the infrastructure and technical knowledge sufficient for the introduction of digital learning? Third, what solutions did the interviewees propose to improve digital learning? The answers to the three questions are the core and focus of the discussion in the next sections.

The grounded argument upon compiling this study was that the hardship encountered by educational institutions, educators, and parents in digital pedagogy came from the limited knowledge, understanding, and skills of digital pedagogy. It led to the misperception that education through digital platforms was ineffective, even resulting in unfavorable academic development of the learners. Therefore, a strong urge to immediately open a school, to apply hundred percent on-site learning became inevitable. This attitude was paradoxical because, at the same time, the government echoed *Merdeka Belajar*, which gave students a space to choose.

2 METHODS

This study is a qualitative research since it involves the perspectives of the participants. The participants of the research were chosen by using simple random techniques where three parties were put in data collection. They were junior and senior high school students, teachers, and parents. Each of the parties was twenty in number. They were involved in examining how they exerted a significant influence on digital pedagogical practices. The research process took place in mid-2021 until early 2022, and it began with desk review and field observation. The instrument for collecting the data is open interview questions through

Google Forms and observations, and the data analysis was done by analyzing the content of the interview results. In case the authors found some answers that required broader information, they conducted an in-depth interview by phone. Observations were made around school buildings, classroom infrastructure facilities, teacher rooms, and health rooms.

This study chose the case of the 9th-grade students of Junior High School (SMP) and 12th-grade of State Madrasah (MAN) in DKI-Jakarta over two considerations: (a) Jakarta was a metropolitan city and the capital of Indonesia, which should have fine digital technology infrastructure; (b) 9th and 12th-grade students were not classes of children anymore, and they have undergone digital education for two years; (c) since SMA (*Sekolah Menengah Atas* or Public Senior High School) and MAN (*Madrasah Aliyah Negeri* or Public Islamic Senior High School) had exactly similar curriculum and learning process, the chosen 12th-graders of MAN also represented the students of SMA. Based on this, it was assumed that they had been able to use digital technology well, and able to assess and voice up views regarding digital pedagogy so far.

The total answers of the students in Google form were 40; 18 of which were from senior high school students and 22 were from junior high school. Only 20 answers were taken on the ground that 5 students did not provide complete answers and 5 others gave ambiguous answers. The selection criteria of the 20 answers was even numbers because the even answers were complete, clear, and convincing. However, 24 answers were gained by the teachers in total, yet only 20 of them were taken according to the number of students because their answers were complete and clear. The answers from parents were 62 in total, but only 20 were taken from numbers 1 to 20 because their answers were complete and clear, besides the consideration of matching the number of students' and teachers' answers.

3 FINDINGS

The study of digital pedagogy in the frame of *Merdeka Belajar* showed the difference in experience and perspective of teachers, students, and parents on digital technology. These inhomogeneous experiences and views exerted an impact on their attitude in providing responses, as denoted in the following interview excerpts:

Table 1. Teachers' experience.

Informant	Teachers' response to digital education restriction policy	Code
1, 3, 4, 5, 6, 7, 9, 10, 11, 12, 13, 14, 19, 20	There are many barriers to online learning. The effectiveness of online learning with face-to-face learning is 1:6, which means that learning 1 year online equals face-to-face learning for 2 months.	Agree
2, 8, 13, 15, 16, 17, 18	Online/blended learning amid the pandemic is still highly effective in learning. Loss of learning does not happen in our school.	Disagree
Informant	**Teachers' advice on digital education**	**Code**
2, 9, 10, 11, 12, 13, 18	*Model baru, menambah wawasan, inovasi, respons terhadap kemajuan* [New models, increasing insight, innovation, and response to progress]	New model, responding to the progress of the times
1, 3, 4, 5, 6, 7, 8	It should be applied for effective sustainability, enhanced digital capabilities, and technological facilities.	Digital literacy
14, 15, 16, 17, 19, 20	There is practical, intensive training at schools, and human resources need to be upgraded. The government should provide facilities to provide learning quotas for teachers and students for free.	Practical and intensive innovation

Table 2. Parents' experience.

Informant	Teachers' response to digital education restriction policy	Code
2, 4, 5, 6, 7, 8, 10, 11, 12,13, 14, 15, 16, 17, 18, 19, 20	Digital education has many shortcomings: making children less interactive with teachers, less understanding of the subject, urgent need for stable internet access, more sensitive children, less socialization, less discipline, and less supervision in learning.	Agree
1, 3, 9	The use of e-learning and blended learning is good, providing flexibility in choosing the time and place to access lessons.	Disagree
Informant	**Related parents' advice digital education model**	**Code**
1, 2, 3, 7, 8, 9, 12, 13, 14, 15	For children to not be outdated with technology, and to master the digital world demand, communication between teachers and students, such as learning via Zoom or Google Meet, needs to be promoted. Applications should adapt to the development of the times, be interesting, and be updated.	update technology and communication
4, 5, 6, 10, 11, 16, 17, 18, 19, 20	Teachers prepare materials and provide digital-based assignments so that online learning can still be conducted.	Digital-based materials and tasks

Table 3. Students' experience.

Informant	Students' response to digital education restriction policy	Code
12, 16, 18, 20, 22, 24, 26, 38, 40	Digital education makes students less productive in learning and understanding the material, but face-to-face can make them understand the materials provided.	Agree
2, 4, 6, 8, 10, 14, 28, 30, 32, 34, 36	Digital pedagogy encourages the ability of all parties to be more creative in carrying out our daily life that is different than ever.	Disagree
Informant	**Related parents' advice digital education model.**	**Code**
4, 8, 12, 20, 24, 28, 40	Continuing digital education for students' creativity experience. Blended learning creates more effective learning.	Extend digital education
8, 16, 18, 34	Materials, practices, and tasks need to be balanced. Teachers' ability to explain the material digitally needs to be improved. Parents supervise children in using gadgets, and students are trained to maximize the use of technology by looking for new things, such as how to create a masterpiece or read books online.	Balance of materials and learning techniques
22, 24, 26, 28, 30, 32, 36, 38, 40	The materials cannot be understood because the teachers do not provide explanations. The teacher only gives continuous tasks. Teachers are expected to be more creative, and digitally communicative, leaving the boring old methods.	Media and communicative learning model development

The development and improvement of teachers' skills in digital pedagogy were urgent so that education could respond to the demands of the time: to be technology-friendly, collaborative, creative, competent, innovative, and capable of creating participatory opportunities for nation-building in the future. The results of the school environment observations proved significant changes: renovation of toilets; available places; hand washing soap in front of the school; a clean and tidy school environment; and spaced student benches. Likewise, the teachers' room was set in a distant seat, a health room was available, and the school cafeteria was closed so as not to cause crowds.

4 DISCUSSION

The controversy of Indonesian education was indicated by the absence of educational services on digital platforms. Even though the government socialized *Merdeka Belajar*, at the same time, educational institutions returned to the conventional learning process. This reality showed that *Merdeka Belajar* was still limited to a theoretical level. Instead of giving a space for the independence to learn, the government was colonizing the digital native generation. Students lost the freedom to choose and the opportunity to explore and produce potential digital natives who only work when they are connected to the internet, enjoying the use of digital technology in everyday life. It was a great and valuable opportunity that can be used to improve students' academic performance. It requires that digital pedagogy be viewed positively, that students are motivated, and that learning outcomes are easy to achieve (Jooston *et al.* 2020; Mundir 2020; Okoye *et al.* 2021).

Transition in the teacher's role, from being the expert in the classroom to a collaborator and facilitator, was a precondition of digital pedagogy as technology has impacted the teaching and learning process. Students were no longer dependent on teachers. Teachers acted as collaborators or moderators (Mahmud 2021; Prawitasari & Suharto 2020; Rahmawati & Suryadi 2019; Susanto, 2020). The 21st-century learning requires different intelligences, creativity, skills, and attitudes (Cope 2014). Therefore, the significance of digital pedagogy adaptation in the learning process corresponds with the learners' experiences and social life.

Schools, educators, and parents were unable yet to take innovative steps so that digital pedagogy becomes interesting for students. As shown in Table 3, students hoped that, amid limited infrastructure, digital education would not be abandoned altogether. In addition to their distinctive character, they reflected the needs of the times and the needs of 21st-century education. Parents needed to be given understanding and knowledge for their insight on the role of digital technology to be positive. As shown in Table 2, parents mostly perceive that digital technology brings about a negative impact on the development of their children. A good understanding of digital technology would provide wider space and opportunities for their positive development. It was obvious that the use of digital technology by students should be under the control of parents and teachers. This could be done by having intensive communication between parents and teachers for monitoring, coordinating, and discussing the progress and obstacles of parents in accompanying the students in learning from home (Ismail *et al.* 2022; Manca & Delfino 2021). The current study of digital pedagogy reflected that the important aspects that become preconditions for educational success were less noticed by the government: the elements of educators and the competence of digital pedagogy – mastery, knowledge, and skills – tend to be neglected. The knowledge, insight, mastery, skills, and competencies of teachers in digital pedagogy were limited. The phenomenon proved that the problem of teachers and digital pedagogy was not yet a concern of the government. Learning was still perceived to go a long way with the teachers in the classroom. Learning in virtual space, or online learning, was considered less effective due to the strong influence of the old paradigm.

5 CONCLUSIONS

It has been identified how digital learning has created some challenges for students, teachers, and parents. This was because children, teachers, and parents were not ready for the rapid innovation of digital learning. This started with the lack of digital knowledge for students, teachers, and parents and ended up with difficulties in conducting the learning process either at school or at home, as well as the lack of motivation to perform better. However, this could be overcome by having good coordination among teachers, parents, and students in discussing the progress and problems of digital learning and also finding solutions to overcome the problems. By having intensive coordination, digital learning will not only become an emergency help for education during the pandemic but also the best education breakthrough that can be applied at any time and place. However, there must be limitations to the results of this research due to the limited time, place, and subjects involved in carrying out this research. Thus, further research would be suggested to be conducted over a longer time period and involve more potential participants to broaden the insight and feasibility of the research suggestions.

REFERENCES

Anjani, A., 2021. *Nadiem Dorong Sekolah Tatap Muka: PJJ Tidak Lagi Efektif [online]*. DetikEdu.

Casey, L. and Hallissy, M., 2016. Live learning: Online teaching, digital literacy and the practice of inquiry. *Irish Journal of Technology Enhanced Learning*, 1 (1).

Cope, M.K. and B., 2014. 'Education is the new philosophy', to make a metadisciplinary claim for the learning sciences. *A Companion to Research in Education*, 9789400768 (October), 115.

El-Sofany, H.F. and El-Haggar, N., 2020. The effectiveness of using mobile learning techniques to improve learning outcomes in higher education. *International Journal of Interactive Mobile Technologies*, 14 (8), 4–18.

Flavin, M. and Quintero, V., 2018. UK higher education institutions' technology-enhanced learning strategies from the perspective of disruptive innovation. *Research in Learning Technology*, 26 (1063519), 1–12.

Hanim, Zaenab, Masyni, R.S.S., 2020. Learning innovation management on effective classes at SMPIT Cordova Samarinda. *Frontiers in Education*, 6 (1), 1–12.

Ismail, S.N., Omar, M.N., Don, Y., Purnomo, Y.W., and Kasa, M.D., 2022. Teachers' acceptance of mobile technology use towards innovative teaching in Malaysian secondary schools. *International Journal of Evaluation and Research in Education*, 11 (1), 120–127.

Jooston, T. (University of W.-M., Lee-McCarty, K. (Online L.C., Harness, Li. (Alverno C., and Paulus, R. (National R.C. for D.E. and T.A., 2020. *Digital Learning Innovation Trends*, 33.

Kemendikbud, 2021. *Keputusan Bersama 4 Menteri Tentang Panduan Penyelenggaraan Pembelajaran di Masa Pandemi COVID-19 [online]*. Kementerian Pendidikan dan Kebudayaan.

Mahmud, R., 2021. Blended learning model implementation in the normal, pandemic, and new normal era. *Proceedings of the 5th Progressive and Fun Education International Conference (PFEIC 2020)*, 479 (Pfeic), 130–139.

Manca, S. and Delfino, M., 2021. Adapting educational practices in emergency remote education: Continuity and change from a student perspective. *British Journal of Educational Technology*, 52 (4), 1394–1413.

Mundir, 2020. Digital literation based learning strategy in improving student learning participation in education in pandemi era. *International Journal of Supply Chain Management*, 9 (5), 1702–1708.

Núñez, J.A.L., Belmonte, J.L., Guerrero, A.J.M., and Sánchez, S.P., 2020. Effectiveness of innovate educational practices with flipped learning and remote sensing in earth and environmental sciences-An exploratory case study. *Remote Sensing*, 12 (5).

Okoye, K., Rodriguez-Tort, J.A., Escamilla, J., and Hosseini, S., 2021. Technology-mediated teaching and learning process: A conceptual study of educators' response amidst the Covid-19 pandemic. *Education and Information Technologies*.

Pokhrel, S. and Chhetri, R., 2021. A Literature review on impact of COVID-19 pandemic on teaching and learning.

Prastiwi, M., 2021. *SKB 4 Menteri Terbaru, Januari 2022 Satuan Pendidikan Wajib Gelar PTM [online]*. Kompas.com.

Prawitasari, B. and Suharto, N., 2020. *The Role of Guru Penggerak (Organizer Teacher) in Komunitas Guru Belajar (Teacher Learning Community)*, 400 (Icream 2019), 86–89.

Rahmawati, M. and Suryadi, E., 2019. Guru sebagai fasilitator dan efektivitas belajar siswa. *Jurnal Pendidikan Manajemen Perkantoran*, 4 (1), 49.

Susanto, H., 2020. Profesi Keguruan. *In*: H. Bambang Subiyakto; Akmal, ed. *Profesi Keguruan*. Banjarmasin: Fakultus Keguruan dan Ilmu Pendidikan Universitas Lambung Mangkurat, 9–98.

Watermeyer, R., Crick, T., Knight, C., and Goodall, J., 2021. COVID-19 and digital disruption in UK universities: Afflictions and affordances of emergency online migration. *Higher Education*, 81 (3), 623–641.

Religion, Education, Science and Technology towards a More Inclusive and Sustainable Future – Rahiem (Ed.)

Muslim education in Kenya: Challenges and opportunities

A.A. Ali*
School of Education and Social Sciences, Umma University, Kenya

ABSTRACT: The word knowledge is associated with one of the beautiful names and attributes of Allah, and Islam is the religion of mercy that champions intelligent and emotional intelligence for the spiritual, cultural, social, economic, and political well-being of people. It was for this reason that Adam, the father of humanity, was taught knowledge of all things, and the first revelation (Q 96:1–5) to the Prophet Muhammad (peace be upon him) was "Read." Advancing Muslim education is critical, as it makes Muslim societies functional, productive, innovative, and sustainable. Despite such religious encouragement and motivation, Muslim societies in Kenya are not doing well in terms of education, as Muslim education suffers from historical, developmental, and structural crises. Education has always been used by successive governments as a weapon of manipulation and marginalization, and it has been largely controlled by Christian missionaries since independence. The purpose of this paper is to diagnose challenges facing Muslim education with the objective of exploring sustainable solutions to improve it. The research methods used in this study are philosophical approaches, observations, and document analysis for enhancing understanding and explaining, as well as exposing and evaluating underlying assumptions and connectedness. The study findings revealed the significant impact of the Christianization of education during and after colonization, which has traditionally had an impact on the Muslim population of learners, teachers, and institutions. According to the literature, the Muslim education problem is centered on a problem of inequality, which is attributed to the historical and structural features of colonial education systems. Qur'anic schools are relegated to complementary nonformal education and have remained unrecognized as an education system. Other findings from this study include problems of understaffing, poverty in Muslim families, illiterate parents, leadership and administrative weakness, early marriages, drug and substance abuse, rampant cheating in examinations, competition between Islamic and secular education for Muslim students, and the environmental and climate change crisis. The study appreciated efforts made by Muslims and Muslim organizations to establish their own educational institutions ranging from basic to tertiary in the last three decades, prompting the government to recognize these efforts. The paper recommends urgent reforms of the Muslim Education Council in terms of policy, institutional, curriculum, and quality assurance aspects so that it drives the culture of academic excellence and invests in pedagogical capacity building and skills for the delivery of education.

Keywords: knowledge, Muslim education, Christianization, challenges, Kenya

1 INTRODUCTION

Islam reached East Africa as early as the 7th century (Holway 1971; Schacht 1965) from Arabia and India. Some Muslims from Arabia were fleeing the persecution of the Umayyad

*Corresponding Author: aadan@umma.ac.ke

DOI: 10.1201/9781003322054-16

governor of Iraq, Hajjaj ibn Yusuf[2], and the inflow of traders to the East African Coast. Islam spread through the interactions of individuals and communities. On the Kenyan coast, the Arabs intermarried with local Bantu people, resulting in the Swahili people, most of whom converted to Islam, and the Swahili language, a structurally Bantu language with a rich vocabulary from the Arabic language, was born.

The Somalis were the earliest non-Arabs that converted to Islam [5] due to the influence of the Arab traders that had settled at the coast of present-day Somalia. Key factors that favored the spread of Islam in Kenya include intermarriages, Swahili language communication, Quran translation to Swahili, trade and commerce, and emergencies at Muslim institutions such as mosques and madrasahs. The colonization of Kenya by the British brought missionaries, whose work was encouraged and promoted. The colonialists suppressed the study of Arabic and discouraged the use of Swahili. Moreover, the missionaries established educational institutions with the aim of preparing the students for jobs to assist colonial administrations and help them convert to Christianity. Muslims were afraid that their children would lose their faith. Therefore, they avoided the missionary schools and resorted to an alternate education system based on the madrasah and Qur'anic schools. The Qur'anic schools were relegated to complementary nonformal education or preschool institutions by the British.

According to Musisi (2018), the literacy and educational levels of Muslims in Uganda are far below those of their Christian counterparts, and the situation for Kenya's Muslims is the same. Educational marginalization affects the social, cultural, economic, and political orientation of any community. To date, madrasah-based education suffers from policy recognition, quality, and sustainability problems in Kenya. The graduates of the missionary schools ended up finding good jobs in the colonial administrations, while those of the madrasahs couldn't. When colonization ended and Kenya became independent, it was the Christians who were in control of the social, economic, and political administration of the country, with Muslims having a marginal presence. The discrepancy in education introduced an element of tension, mistrust, and fear between Muslims and Christians. This left the Muslims to lag behind socioeconomically. It took the Muslim communities more than three decades to establish their own schools. Between 1963 and 2000, the number of Muslim students qualifying for public universities was marginally small. However, this ended after the liberalization of the education sector from 1991–2000. The Supreme Council of Kenya Muslims (SUPKEM), Jamia Mosque Committee, Young Muslim Association, Africa Muslim Agency, Islamic Foundation, and Muslim Education Welfare Association, among others, have played a critical role in promoting Muslim education and related infrastructure, including teachers' training colleges and secondary and primary schools (Ndzovu 2014). Some of the established Muslim educational institutions were affected by the war on terror, which denied them financial flow. In this paper, the key questions that will be investigated will include: (1) How did colonial machinations impact Muslim educational growth and development in Kenya? What role did Muslim organizations play in the educational advancement of the country? What are the experiences and encounters of Muslim education in Kenya, in relation to lessons and experiences?

2 METHODS

The study utilized philosophical research methods described by Žukauskas, Vveinhardt, and Andriukaitienė (2018); observational data collection methods described by Ciesielska and Dariusz Jemielniak (2018) and Ewing and Griffiths (2011); and document analysis as illustrated by Bowen (2009) with the objective of enhancing understanding, explanation, as well as exposing and evaluating underlying assumptions and connectedness.

3 RESULTS AND DISCUSSION

The study findings indicated that colonization marginalized Muslim education by pushing out the Islamic education offered by Quranic schools and madrasahs. According to Hussein (2016), dualism in education has also become a big burden for many Muslim learners due to competition for their time and minds.

Other challenges encountered by Muslim education are staffing and poverty problems, as reported by Shaaban (2012) and Issack (2018). Due to the marginalization problem, the illiteracy and education level of Muslim parents are far below that of Christians, as indicated by Sheikh (2013). According to Shaaban (2012) and Omari (2014), Muslim institutions are personalized as family affairs and thus suffer from nepotism and poor administrative malpractices. Since Muslims dominate the coastal and northern parts of the country, girls from these regions are subjected to early marriages (Sakwa 2020), forcing them to drop out of education. Although now eradicated, female genital mutilation was also a problem for a Muslim girl child (Njogu 2015). Further, environmental and climate change crises have also contributed to the Muslim education problem of poor performance (Ndichu 2009). NACADA (2019) has reported the problem of drug and substance abuse affecting Muslim education performances, mainly due to the chewing of Khat (*miraa*) and smoking of tobacco and other drugs, which has forced the students to perform poorly or resort to examination malpractices. Nowadays, social media is overtaking these problems.

(Svensson 2018) has reported in his research how the colonial government relegated Qur'anic schools to complementary nonformal education, causing the uncertainty of unrecognition. Sometimes, Muslim educational institutions are wrongly associated with radicalization problems, impacting the learners' and parents' psychology. Muslim parents are always confronted with the problem of dualism in education. Preferring one will cause them to lose the other, resulting in the difficult compartmentalization of education into religious and secular, with Muslim children bearing the heavy burden of pursuing both education under resource and time constraints (Abdi 2017; Hussein 2016). Although efforts have been made to integrate the two systems of education, the Islamic component has always remained unrecognized and marginalized. The problem is that Kenya lacks a unified Islamic education curriculum, standard, quality assurance, and examination system. According to Izama (2014), because of the problem of inequality, Muslim learners have fewer years of schooling, making them less literate compared with Christian learners.

The Muslim-dominated regions of northern Kenya and the coast region are generally arid and semi-arid with harsh and fragile environments. With threats of climate change increasing, learners are often exposed to food and nutritional scarcity and crisis. Children from pastoral communities frequently migrate with the animals, skipping school. When all these factors prevail, the quality of learning of pupils in schools is significantly affected (Ndichu 2013). The poverty rate is very high in Kenya, where 30% of the population live on less than US$1.25 and 40% of the rest live on less than US$2.00 a day as of 2013. The majority of these poor people are Muslims and live in remote and rural areas such as Northern Kenya (Ali 2017). Some Muslim girls drop out of school due to a lack of fees. A study by NACADA (2019) indicated that drug and substance abuse among learners in schools is on the increase and is becoming a social and public health concern. Muslim communities are significantly affected by miraa chewing (*Catha edulis*), Bhangi (*Cannabis Sativa*), and the consumption of hard drugs such as cocaine and heroin. According to the above report, the drivers for the uptake of drugs are significantly associated with class repetition and a decline in academic performance.

Cheating in the national examination by schools in the Muslim-dominated region is a growing phenomenon. This problem was also confirmed by two principals of local secondary schools from Mandera, whom I managed to interview on July 12th, 2022. Another emerging concern facing Muslim education in Kenya is that young learners are dropping out to pursue the business of riding a motorbike, *Tuk Tuk,* or hawking.

Both peer pressure and poverty are contributing to these problems. Muslim education in Kenya also suffers from corruption, where bursaries and scholarships are distributed based on nepotism and political correctness. Children, families of politicians, and the rich benefit most. Muslim minds are greatly confronted and bombarded by social media, causing both knowledge and thought crises. These challenges are impacting Muslim parents' and children's quality of life and faith. Muslim-owned primary and secondary schools always perform poorly in the national examinations, blamed on the above-described menace. Muslim educational organizations lack dynamic leadership, strategic planning, and funding to help them move forward sustainably. This also represents a social risk for the country and Muslims because, through education, humanity is transformed to be cultured, civilized, and wise (Prof. Rosanani 2017). Aristotle added that public education should cultivate virtues to sustain a democratic society. Muslim education is under siege due to the tussle between the radicals, the liberals, and the conservatives.

The secular system of education aims to produce workers for employment, while Islamic education often aims to produce global citizenship with the meaning of life, ethics, behaviors, morality, and existence. Through the joint efforts of several mosques in Nairobi, the Muslim Education Council was established in 2011. However, the Council is understaffed, lacks dynamic leadership, and is poorly funded despite being the only legal entity recognized by the Government of Kenya under the Basic Education Act of 2013. All the Muslim educational institutions are uncoordinated. Allah SWT said in the Quran, Chapter 13 Verse 11: *Indeed, Allah will not change the condition of a people until they change what is in themselves.* Muslim education as the basis for social transformation and resilience building calls for *ummatic* action, reform, and Q*alb* leadership. Efforts are also required to pursue the promotion of knowledge integration and pedagogy training to skill teachers professionally and strategically.

Muslim education should cultivate knowledge, skills, and wisdom to lead to justice, peaceful coexistence, open-mindedness, social responsibility, and a sense of belongingness (AEMS 2021). Education is a service to humanity and should be free from obstacles so that it becomes the catalyst to lift families and communities out of the vicious cycle of poverty. It should not be for profit-making in societies. This article strongly advocates for accountable leadership so that they can help communities overcome all educational obstacles by walking the path of dialogue, increasing access to quality education, and helping people in their development. This is possible when inclusive and quality education is accessed justly to eliminate poverty, ignorance, and social fragmentation with the flame of love and passion for learning (Adan 2021). The Kenya Constitution (2010), the Devolution Act (2012), the emerging class of Muslim entrepreneurs, increasing levels of Muslim communities' literacy, and the value of education are all pointers to the future seeds of Muslim education in Kenya. The paper may not have done complete justice to the subject due to the limited financial, human, and time resources to engage in a detailed study. There is a need to conduct more quantitative and qualitative research on Muslim education in terms of national examination performances, mentorship, and learners' perceptions.

4 CONCLUSION

According to this article, Muslim education is at a crossroads as a result of the clash of many global ideologies, including the significant impact left by the historical process and structure of colonial education systems. The impact of education dualism on Muslim learners' minds cannot be underestimated, thus the demand for urgent attention. Efforts made by various Muslim educational entities and the liberalization of education by the government need to be appreciated.

Weaknesses associated with the Muslim Education Council—the overall body legally recognized by the government—call for urgent action in terms of its restructuring, staffing, and funding. Due to these difficulties, planning, coordination, and partnership building between Muslim educational institutions have largely remained the biggest headache. There is a need to build successfully on the lessons and experience gained through the integration of Islamic education in Kenya. However, the modes of integration currently existing are not unified or properly done and suffer from mixing of subjects in timetabling, separate timings, overloaded subjects, irrelevant units for the levels, quality assurance problems, examination problems, and teaching aid scarcity. Further, there is a need to reject the issue of classifying education as secular and religious since all knowledge is from Allah for humanity's benefit.

REFERENCES

Abdi, A., 2017. Integration of Islamic and secular education in Kenya: A synthesis of the literature. *International Journal of Social Science and Humanities Research*, 5 (3), 67–75.

Adan, A., 2021. Education system and inclusive society. Religious radicalism-christian and muslim understanding and responses. In: *Pontificium Consilium Pro Dialogo Inter Religious*. Vatican.

Ali, A.E.E.S., 2017. The challenges facing poverty alleviation and financial inclusion in North-East Kenya Province (NEKP). *International Journal of Social Economics*, 44 (12), 2208–2223.

Bowen, G.A., 2009. Document analysis as a qualitative research method. *Qualitative Research Journal*.

Ciesielska, M. and Dariusz Jemielniak, 2018. *Qualitative Methodologies in Organization Studies*. Palgrave Macmillan.

Izama, M.P., 2014. Muslim education in Africa: Trends and attitudes toward faith-based schools. *The Review of Faith & International Affairs*.

NACADA, 2019. Status of drugs and substance abuse among primary school pupils in Kenya. *National Authority for The Campaign Against Alcohol and Drug Abuse (NACADA) and Kenya Institute for Public Policy Research and Analysis (KIPPRA)*.

Ndichu, G.D., 2009. *Impact of Drought on Primary Schools Learning in Laikipia West District of Laikipia County, Kenya*. Doctoral dissertation, thesis, master of environmental studies. Kenyatta University.

Ndzovu, H., 2014. *Muslims in Kenyan politics: Political involvement, marginalization, and minority status*. Northwestern University Press.

Omari, H.K., 2014. Islamic leadership in Kenya: A case study of the supreme council of Kenya muslims (Supkem). Doctoral Dissertation. University of Nairobi.

Sakwa, H.N., 2020. *Effects of Early Marriages on the Education of Primary School Girls in Buna Sub-county*, Wajir County, Kenya. Doctoral dissertation.

Schacht, J., 1965. Notes on Islam in East Africa. *Notes on Islam in East Africa*, 23, 91–136.

Shaaban, J.M., 2012. *The Challenges of Teaching Islamic Religious Education on Spiritual and Academic Formation of Secondary School Students in Nairobi, Kenya*. MA Thesis. Kenyatta University.

Sheikh, A.S., 2013. *Islamic Education in Kenya: A Case Study of Islamic Integrated Schools in Garissa County*. University of Nairobi.

*Religion, Education, Science and Technology towards a More Inclusive and Sustainable Future –
Rahiem (Ed.)*
© 2024 the Author(s), ISBN: 978-1-032-56461-6
Open Access: www.taylorfrancis.com, CC BY-NC-ND 4.0 license

Child-friendly schools in Indonesia: Validity and reliability of evaluation questionnaire

T. Wulandari*, A. Mursalin, Sya'roni, F.K. Dewi, Atika & Baharudin
UIN Sulthan Thaha Saifuddin Jambi, Indonesia

ABSTRACT: Child-friendly school program is one of the educational policies in Indonesia. The implementation of child-friendly schools needs an evaluation as part of the repair process of the policy. This study aims to develop, validate, and measure the reliability of an evaluation questionnaire for child-friendly schools in Indonesia. The questionnaire was developed based on child-friendly school standards. The questionnaire was piloted on 290 elementary school high-grade students. The content validity index using Aiken's V formula was 0.87. Construct validity with exploratory factor analysis formed four factors (1) CFS policy; (2) Implementation of the learning process; (3) Child participation; and (4) Parent/guardian participation. The alpha Cronbach coefficient of 0.70 was sufficient to indicate the reliability of the questionnaire. This evaluation questionnaire is considered good enough to be used as an evaluation instrument for child-friendly schools in Indonesia

Keywords: Child-friendly school, validity and reliability, evaluation questionnaires

1 INTRODUCTION

The 1945 Constitution of the Republic of Indonesia stated that "every child has the right to survival. Grow and develop, and have the right to protection from violence and discrimination." This provision is operationally regulated in Article 54 of the Basic Law on Child Protection, which states that "Children in and within the school environment must be protected from acts of violence committed by teachers, school administrators, or their friends in the school concerned, or other educational institutions.

A report entitled *Global Prevalence of Past-Year Violence Against Children: A Systematic Review and Minimum Estimate* concluded that in 2014 the highest violence against children occurred in Asia, which was 64% or around 715.5 million children (aged 2–17) victims per year. Based on data from the Program for International Students Assessment (PISA) in 2018, the case of child abuse in Indonesia was ranked 5th in the world out of 78 countries. In 2020, The Indonesian Child Protection Commission (KPAI) recorded 119 cases of bullying against children. This number has increased from previous years, which ranged from 30–60 cases per year.

The child-friendly school program is one of the programs created by the government to ensure the safety and security of children in the school environment. This program is carried out to create a child-friendly school environment. Child-friendly schools can be interpreted as educational institutions that can facilitate and empower children's potential so that children can grow and develop, participate in, and be protected from acts of violence and discrimination. Schools must also create adequate programs and create a conducive and educative environment (Asrorun *et al.* 2016). In line with this, child-friendly schools can also be interpreted as "a safe, clean and healthy and shady school inclusive and comfortable for the physical, cognitive, psychosocial development of girls and boys, including children who need special education and/or special service education" (Supiandi *et al.* 2012).

*Corresponding Author: tatiwulandari@uinjambi.ac.id

 DOI: 10.1201/9781003322054-17

The implementation of child-friendly schools requires evaluation to obtain recommendations for improvement. A good evaluation instrument needs to be created to get accurate results. This research was conducted to develop, validate, and measure the reliability of the questionnaire instrument for evaluating child-friendly schools in Indonesia.

2 METHODS

This study uses the research and development method to develop, validate, and measure the reliability of a child-friendly school evaluation instrument. The instrument developed was a questionnaire based on the standards of child-friendly schools in Indonesia. It is the guideline for the implementation of CFS compiled by the Ministry of Education and Culture of the Republic of Indonesia. The fifth- and sixth-grade elementary school students were used as a population. It is because researchers consider that they are more familiar with the school situation. The instrument was tested on 290 fifth- and sixth-grade elementary school students. The validity is obtained through several stages, namely content validity and construct validity. The reliability of the instrument in this study uses the *Alpha Cronbach formula.*

3 RESULTS

3.1 *Content validity*

The content validity of the instrument in this study uses Aiken's V formula by involving three experts as validators. The Aiken's V formula calculates the content validity coefficient based on the results of the assessment of three expert panels. Aiken's V formula (Azwar 2012) can be seen below:

$$V = \sum s/[n(c-1)]$$

s = r – lo lo = the lowest score of validity (1) c = the highest validity rating score (5)

r = number given by rater

n = number of raters/rater

The coefficient of validity of the 43 questions on the questionnaire sheet calculated by Aiken's V formula is 0.87; this value indicates that the questionnaire sheet is valid.

Table 1. Items and content validity coefficient.

Has your friend ever (not in a joke or a sport) hit, slapped, kicked, threw, scratched, pinched, bit, grabbed your hair, pulled your ear, or forced something on you?	0.92	Has the teacher ever favored boys over girls or vice versa in the classroom?	0.75
Have your teachers or principals ever punish you by hitting, slapping, kicking, throwing, scratching, pinching, biting, pulling your hair, pulling your ears, or forcing something on you?	0.92	Has the teacher ever ignored you while studying in class?	0.83
Has a teacher or the principal ever insulted or hurt your feelings?	0.83	Has the teacher ever failed to give you an assessment of the work you are doing?	0.92
Has the teacher ever given a punishment in the form of adding schoolwork?	0.83	Has the teacher ever scored a discrepancy between you and your friend?	1.00
Have you ever made a mistake and was then given a punishment that you didn't think was appropriate?	1.00	Have your teachers or principals prevented you from participating in extra-curricular activities at school?	0.92
Has the teacher ever advised you to be diligent in attending school?	0.83	Have your teachers or principals forced you to participate in extra-curricular activities that you didn't like?	0.92

(continued)

Table 1. Continued

Has the teacher ever asked a student the reason for not attending school?	0.92	Have you or your friends attended meetings related to school rules with the principal, teachers, and other school staff?	0.92
Have you ever done a fundraiser for one of your underprivileged friends?	1.00	Has your name been listed on the board/ committee of an activity at school?	0.92
Have you or your friends smoked in the school environment?	0.83	Have you ever made a suggestion about school rules?	0.92
Have the principal, teacher, security guard, school guard, or cleaning staff ever smoked in the school environment?	1.00	Have your suggestions been heard by the principal, teacher, or other school staff?	0.83
Have you or your friends ever used illegal drugs in the school environment?	0.75	Have you ever made a complaint about the problem you were experiencing to the principal, teacher, or other school staff?	0.83
Has the principal, teacher, security guard, school guard, or cleaning staff ever used illegal drugs in the school environment?	0.67	Have you ever helped your friend to report a problem to the principal, teacher, or other school staff?	0.92
Have you ever been forbidden by the principal, teacher, security guard, school guard, or cleaning staff to worship at school?	0.75	Have your parent/guardian listened and responded when you vent at home?	0.92
Has the teacher ever forbidden you to worship during class hours?	0.75	Have your parents/guardians ever accompanied you to study at home?	0.92
Have you ever been unable to worship at school because no prayer facilities available?	0.92	Have your parents/guardians forbade you to take private lessons for extracurricular activities or other school activities?	0.75
Has your friend ever bothered you while you were praying?	1.00	Have your parents/guardians ever accompanied or escorted you to private lessons or took part in extra-curricular activities at school?	0.83
Has the master ever forbade you to eat or drink when you asked permission to do so?	0.83	Have your parents/guardians ever watched you play gadget/mobile at home?	0.83
Have your teachers or friends ever distributed food or drinks when you didn't bring them to school?	0.92	Have your parents/guardians ever forbade you to open some applications or sites on your gadget /mobile at home?	1.00
Have you ever prayed together with the teacher before and after studying in class?	0.92	Have your parents/guardians ever asked who you are friends with on social media (Facebook, Instagram, WhatsApp, Tik-Tok, etc.)?	0.75
Have you ever held a flag ceremony every Monday at school?	0.75	Have your parents/guardians ever asked how you were doing at school?	0.83
Has the teacher ever come late to class?	1.00	Have your parents/guardians contacted your teacher to ask how you are doing?	0.92
Has the teacher ever scolded you while studying?	0.83		

3.2 *Construct validity*

Construct validity was carried out to show that the test results were able to reveal a theoretical construct to be measured. The factor analysis used to ensure the construct validity of the instrument is exploratory factor analysis with the help of SPSS software.

3.3 *Initial check*

The initial check uses the Kaiser-Meyer-Olkin (KMO) measure of sampling value indicator, which must be > 0.5 with Bartlett's test significance that must be < 0.05.

Table 2. KMO and Bartlett's test.

KMO and Bartlett's Test		
Kaiser-Meyer-Olkin Measure of Sampling Adequacy		.680
Bartlett's Test of Sphericity	Approx. Chi-Square	2.557E3
	Df	820
	Sig.	.000

This value indicates a sufficient value because the KMO value is more than 0.5, which means it is quite feasible for factor analysis to be carried out. The next step to increase the KMO value is to eliminate items with a correlation of less than 0.5 (< 0.5) on the *anti-diagonal diagonal image correlation.* By eliminating two items that do not meet the criteria, namely item number 11 and item number 20, the KMO value of 0.680 is obtained with a significance level of 0.000.

3.4 *Factor extraction*

The next step is factor extraction of the existing data. Factor extraction helps to determine the number of factors that represent the construct being measured (Pett et al. 2003). From several extraction methods, this study uses principal components analysis (PCA). This is done because SPSS tools use PCA as the *default.* There are several ways of determining the number of extracted factors. One of the ways is eigenvalues > 1.

3.5 *Factor rotation*

It allows the items with high *factor loading* to be grouped into certain components or factors. Some references suggest a *loading factor* > 0.30 (McCauley et al. 1994), while others suggest a higher *loading factor* value of > 0.50 (Satyadi & Kartowagiran 2014; Wijanto 2007). Responding to differences like this, in this analysis, some of these possibilities as the best solution were tried.

The way to facilitate the completion of items with a high factor load grouped into certain factors is determined by items with a load > 0.30. The result of factor rotation is the formation of 5 components using 41 items (reduced by 2 items from the previous 43 items). The next problem that arises is that the solution of these factors is difficult to interpret because the distribution of items that should be grouped in one factor spreads to several factors or components. Costello and Osborne (2005) state that if some items *load* on the wrong factor, it indicates the formation of an inappropriate *factor structure.* After trying several possibilities with unsatisfactory results, it was finally chosen to extract four factors by limiting the factor load to each item of $\geq$ 0.30.

3.6 *Naming of factors*

Pett, Lackey, and Sullivan (2003) assert that the factor analysis process needs to involve the naming of the factors that give identity to these factors. The items that are grouped in the first factor are items that ask about CFS policies that can be known by students as respondents so that this factor is then given the name factor or component CFS policy. The items that are grouped in the second factor contain questions related to the process of implementing the learning process that is felt by students, so this factor is named implementation of the learning process. The items grouped in the third factor contain questions related to education and child rights-trained educators in CFS, so this third factor is named the child participation factor. The last factor is the fourth factor, it contains items with questions related to the participation of parents/guardians in the implementation of CFS related to students, so this factor is named the parent/guardian participation factor.

In more detail, the results of the factor analysis are presented together with the selected items with a factor loading above 0.30 in Table 3 below:

Table 3. Factor analysis results.

Name of Factor	Items
CFS Policy	2, 3, 5, 6, 7, 8, 9, 12, 13
Implementation of the Learning Process	14, 15, 16, 18, 19, 21, 22, 23, 24
Child Participation	25, 26, 27, 28, 31, 32, 34, 35
Parent/Guardian Participation	36, 38, 39, 40, 41, 42, 43

3.7 *Instrument reliability*

Reliability is the extent to which a measurement result can be trusted so that it can provide constant results (Arikunto 2006; Azwar 2012).

The results of the reliability estimation obtained from the questionnaire instrument with 33 items can be seen in Table 4 below:

Table 4. Reliability statistics.

Reliability Statistics	
Cronbach's Alpha	N of Items
.708	33

With an alpha coefficient of 0.708, it can be said that this questionnaire instrument can be used to evaluate the implementation of CFS policies in elementary schools in Sleman Regency.

4 CONCLUSION

Another consideration in construct validity to form the four factors is the suitability of the initial concept based on the recommendations of Nunnally and Bernstein (1994) described by Pett, Lackey, and Sullivan (2003). According to them, there is no easy way to determine which factors to extract. Whatever factor solution is reached, according to them, it should not depend on statistical criteria alone, but should also be theoretically appropriate. The final criterion for determining the number of factors is whether or not it is easy to interpret the initial extraction procedure or after the factors have been rotated to reach a clearer factor solution.

The child-friendly school program in Indonesia requires evaluation to obtain some recommendations for improvement in its implementation. This program evaluation requires valid and reliable instruments to obtain measurement results. The results of this study contain the criteria needed for a valid and reliable instrument to be used as an evaluation instrument for child-friendly schools in Indonesia.

REFERENCES

Arikunto, S. & Safruddin, C. 2004. *Evaluasi PROGRAM Pendidikan*. Jakarta: Bumi Aksara

Arikunto dan Jabar 2007. *Evaluasi Program Pendidikan, Pedoman Teoritis Praktis Bagi Praktisi PendidKan*. Jakarta: Bumi Aksara.

Asruron, N. S., & Luthfi, H. 2016. *Panduan Sekolah & Madrasah Ramah Anak*. Jakarta: Erlangga.

Azwar, S. 2012. *Tes Prestasi: Fungsi dan Pengembangan Pengukuran Prestasi Belajar*. Yogyakarta: Pustaka Pelajar.

Costello, A. B., & Osborne, J.W. 2005. Best practices in explanatory factor analysis: four recommendations for getting the most from your analysis. *Practical Assessment, Research & Evaluation*.

McCauley, C. D., Ruderman, M. N., Ohlott, P. J., & Morrow, J. E. 1994. Assessing the developmental components of managerial jobs. *Journal of Applied Psychology*.

Pett, M. A., Lackey, N. R., & Sullivan, J. J., 2003. *Making Sense of Factor Analysis*. USA: Sage Publication

Republik Indonesia. 1945. *Undang Undang Dasar Republik Indonesia*

Retnawati, H. 2016. *Analisis Kualitatif Instrumen Penelitian*. Yogyakarta: Parama Publishing

Satyadi, H. Dan Kertowagiran, B. 2014. Pengembangan instrumen penilaian kinerja guru sekolah dasar berbasis tugas pokok dan fungsi (The assessment instrument development of elementary school teachers' performance based on task and functions). *Jurnal Penelitian dan Evaluasi Pendidikan*. Tahun 18, No. 2.

Setyawan, D. 2015. KPAI: Pelaku Kekerasan Terhadap Anak Tiap Tahun Meningkat.

Sugiyono. 2012. *Metode Penelitian Kuantitatif, Kualitatif dan R&D*. Bandung: Alfabeta

Supriadi. 2012. Petunjuk Teknis Penerapan Sekolah Ramah Anak Kementerian Pemberdayaan perempuan dan Perlindungan Anak Republik Indonesia.

Wijanto, S. H., 2007. *Structural Equation Modeling dengan Lisrel 8.8*. Yogyakarta: Graha Ilmu.

Religion, Education, Science and Technology towards a More Inclusive and Sustainable Future – Rahiem (Ed.)

Challenges of mastering information and communication technology literacy competence for teachers in the age of digital learning

Reksiana*, A. Zamhari, M. Huda, D. Rosyada & A. Nata
UIN Syarif Hidayatullah, Jakarta, Indonesia

ABSTRACT: In today's digital era, there are still many teachers who are not literate in information technology and do not yet have information and communication technology (ICT) literacy competence. So teachers are required to carry out learning innovations. This study aims to identify, analyze, and interpret all challenges to ICT literacy competencies in the digital learning era faced by teachers. This study uses a qualitative method with a systematic literature review (SLR). The stages in the SLR method are reviewing, identifying, and analyzing journal articles with certain themes on the Google Scholar platform systematically, and in each process following the steps or procedures that have been set. The results of the research show several challenges for teachers in mastering ICT literacy competencies in learning: (1) Teachers are not proficient in using IT-based learning tools. (2) Information and communication technology literacy is a new competency that is a necessity in teaching and learning activities in the digital era. (3) Teachers need to improve ICT literacy competence by training and competency development through digital literacy training programs. The synthesis in this research found the biggest challenge for teachers – having to master digital literacy competency procedures, namely: accessing, selecting, understanding, analyzing, verifying, evaluating, distributing, producing, participating, and collaborating in utilizing the information obtained in learning, so that teachers are able to innovate learning.

Keywords: teacher, teacher competency, digital literacy, information and communication technology, digital era

1 INTRODUCTION

Currently, in the all-digital era, it cannot be denied that in the world of education, the challenges faced are becoming increasingly complex because education is required to adapt to the challenges of an era of technological advancement and technological integration. Therefore, digital literacy learning needs to be implemented to build digital literacy competencies for teachers and students to form Human Resources (HR) who have information and communication technology literacy competencies to advance education. (Kominfo 2020:https://gln.kemdikbud.go.id). According to Asari *et al.* (2019) and Dhimas *et al.* (2021), in the era of digital learning, teachers need to be literate in digital literacy competencies because teachers and students still do not understand how to effectively use digital information media. In addition, teachers in Indonesia still have problems making technology-based learning media because of internal and external factors. According to Priyanto (2020), with the changes in the current era of Industry 4.0, there is a need for modernization of

*Corresponding Author: reksiana.19@mhs.uinjkt.ac.id

DOI: 10.1201/9781003322054-18

education by revitalizing various things, and one of them is the teacher aspect. Teachers are required to master digital technology competencies. According to Maulana (2020), teachers are required to be able to adapt to technology that is currently growing rapidly. In addition, teachers must be able to innovate and be creative in developing learning tools in the classroom. (Kominfo 2020:https://gln.kemdikbud.go.id)

Regarding information and communication technology literacy in education, some researchers (Agus Sulistyo & Ismarti 2022; Anggeraini *et al.* 2019; Ashari 2019; Djaja *et al.* nd;) concluded that teachers were still lacking in mastering ICT competencies in learning and were ill-prepared for the changes brought about by the digitalization of education.

The research conducted by Djaja *et al.* nd, about online learning does not discuss in detail teachers' challenges in mastering ICT literacy in the digital era. However, the research by Agus Sulistyo and Ismarti (2022) was more focused on the challenges of Islamic education in the 4.0 era. Apart from that, the research also examines the urgency of media literacy. However, it does not discuss the stages of digital literacy in learning in detail, which should also be included.

Ashari (2019) discussed more about student challenges and changes in attitudes they experienced in entering the 4.0 era and did not focus on discussing teacher challenges in the digital learning era. However, the research also included what researchers will look for and discuss regarding challenges faced by teachers in mastering ICT literacy and the procedures for the stages of ICT literacy competency in learning. Accordingly, this study supports the findings of Adam-Turner (2017). This study shows how digital literacy is a solution to challenges for teachers in the digital era. This research also supports the theory given by Schield (2013), Ridsdale *et al.* (2015), Grillenberger and Romeike (2018), Spengler (2015), and Listiaji and Subhan (2021) in response to challenges in the 4.0 era regarding ICT literacy. In their findings, they stated that the current educational environment offers access opportunities, such as information, tools, and learning resources for teachers to educate and teach using digital literacy to face the challenges of the 4.0 era. With some of the reasons that were found, the systematic literature review (SOR) with the theme, *The Challenges of Mastering Information and Communication Technology Literacy Competencies for Teachers in the Digital Learning Era*, is interesting and relevant for study.

2 METHODS

This research uses the library research method with SLR. The data collection technique in this research involves five stages: (1) finding relevant literature data (search process) (Randolph, 2009); (2) evaluating data, theory, information, and research results; (3) identifying themes and gaps between theory and conditions in the field, if any, and analyzing them according to the research questions (Randolph 2009); (4) constructing the concept and structure of the results of the theme study; and (5) compiling the findings from the literature review.

The research procedure is as follows:

(1) First Stage – Creating research question:
  RQ 1: *What are some teacher challenges in mastering competence for teachers in the era of digital learning?*
  RQ 2: *How do teachers improve their mastery in the era of digital learning?*

(2) Second Stage – Search process: At this stage, the researcher tries to determine a clear topic and seeks to collect and search for data using a search engine, such as www.google.com with the site address www.google.scholar.com. as the main research data to be able to answer research questions.

(3) Third Stage – Appropriate and inappropriate (inclusion and exclusion) criteria: (i) Data used is the 2018–2021 timeframe; (ii) Data obtained using keywords according to the title; (3) The data is obtained from www.google.scholar.com.

(4) Fourth Stage – Quality assessment (QA): The QA is done from a list of problem formulations. This stage is used to evaluate the data that has been found based on the question of quality assessment criteria.
(5) Fifth Stage – Data analysis: Evaluating data, theory, information, and research results.

3 RESULTS

The results of the SLR regarding how teachers challenge the competence of information and communication technology literacy in the digital learning era are found. There are 28 articles that can be used as material for analysis in this study. Furthermore, using a content analysis study, it was shown that as many as 7 research articles stated that not all teachers in schools had information and communication technology literacy skills; 9 articles stated that literacy on information technology is now a basic literacy that has become a necessity in teaching and learning activities; and 12 articles stated that teachers need to improve their technological literacy competencies with training and development in the ICT field.

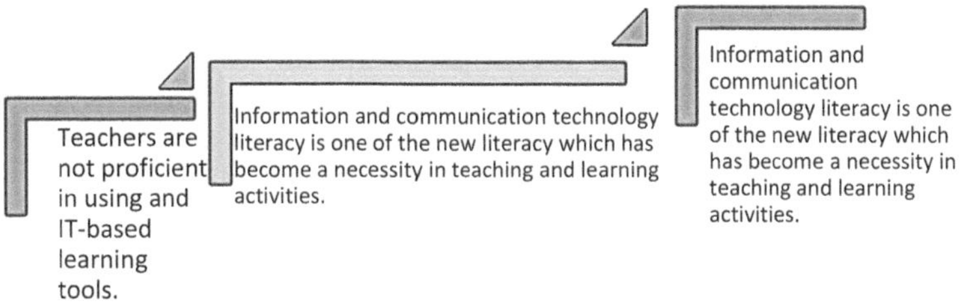

Figure 1. Research findings.

3.1 *Teachers are not proficient in using IT-based learning tools*

According to Asari *et al.* (2019) there are still many teachers who have not mastered digital literacy because teachers and students do not understand how to effectively utilize digital information media in learning. Djaja *et al.* nd; Dhimas Arsyad Arrajiv, Miftahul Arifah, & Tyas Wahyuningsih (2021), in their findings, show that individual skills in information technology literacy need to be improved in the world of education. Studies by Maulana (2020) and Priyanto (2020) state that teachers have not mastered literacy and technology competencies. Retnaningsih (2019) and Amirudin (2019) said in their findings that the big challenge for teachers today is IT mastery, professionalism, learning creativity, time mismatch with learning load, and teacher monotony.

3.2 *Information and communication technology literacy is a new competency that has now become a necessity in teaching and learning activities*

Helaludin (2019), Susilo (2019), and Astini (2019) state that the current need of teachers in the digital era is mastering ICT literacy. According to Maulana (2020) teachers not only need to master the ability to operate digital devices and software, but also complex skills, such as production skills, photovisual skills, hypertextuality skills, etc. Bahri *et al.* (2021), in their study, showed that it is imperative for teachers to optimize literacy development in the classroom and its integration into subjects. The findings of Agus Sulistyo and Ismarti (2022) state that the biggest challenge for a teacher is to be literate in technological literacy. (Riadi 2019). Ashari (2019) mentioned that in dealing with changes in student behavior in the 4.0 era, teachers must be literate in digital literacy.

3.3 *Teachers need to improve ICT competence by training and participating in workshop activities*

Teachers need to improve their digital literacy skills by upgrading themselves, one of which is developing themselves by participating in training and workshops. This is to create a social order with critical and creative mindsets and views, collaboration, communication, innovation, and problem solving. Ariyati (2020) and Roman (2019) state that the competency needed by teachers and students is ICT literacy and is improved by means of teachers having to take part in digital literacy training. According to Anggeraini *et al.* (2019) and Rachmawati *et al.* (2021), teacher ICT skills can be improved with training on ICT. In their findings, Zulfa Hazizah and Henry Aditia Rigianti (2021) and Adam-Turner (2017) mentioned that teachers need to increase their own capacity in the field of ICT literacy by participating in training and workshops.

4 DISCUSSION

The lack of teacher expertise in using IT-based learning tools is a problem not only in Indonesia. This also happens in countries such as Malaysia (Kaur & Singh 2014). Teachers in Malaysia are still experiencing difficulties mastering ICT and require intensive training in the use of ICT. Apart from Malaysia, teacher problems in mastering ICT also occur in countries such as Spain, in cities such as Carrascal, Cantilan, Madrid, Carmen, and Lanuz. Teachers in several of these cities experience difficulties using ICT in language subjects. (Correos 2014). Correos recommends that teachers be given intensive training on ICT.

In mastering ICT competencies comprehensively, teachers are not only required to be able to present, filter, process, and evaluate data and information, but are also required to be more creative, active, collective, and build knowledge through digital media (Janíková & Kowaliková 2018; Savage 2006; Schield 2013). Ferrari (2013) mentions five stages in mastering digital literacy competencies: (1) Identify, find, retrieve, store, organize, and analyze digital information and assess its relevance and purpose; (2) Communication through digital or online tools; (3) Content creation, which includes creating and editing new content; (4) Security, which includes personal protection, data protection, digital identity, and security measures; (5) Problem-solving, which includes identifying needs, making the right decisions according to goals or needs, etc.

Skantz-åberg *et al.* (2022) mention seven aspects related to teachers' professional digital competence: (1) technological competence; (2) content knowledge; (3) attitude toward the use of technology; (4) pedagogical competence; (5) cultural awareness; (6) critical approach; and (7) professional engagement, with technological and pedagogical competence being the most prominent. According to Asari *et al.* (2019), there are different stages of digital literacy competencies for teachers: (1) accessing; (2) selecting; (3) understanding; (4) analyzing; (5) verifying; (6) evaluating; (7) distributing; (8) producing; (9) participating; and (10) collaborating. According to Gündüzalp (2021), these digital literacy competencies can be divided into basic and advanced levels. The two levels are summarized in ten stages: (1) accessing; (2) selecting; (3) understanding; (4) analyzing; (5) verifying; (6) evaluating; (7) distributing; (8) producing; (9) participating; and (10) collaborating.

In facing the challenges of information and communication technology literacy competencies for teachers in the digital learning era, teachers must improve competence by participating in training related to data and information literacy (Spengler 2015). In line with this view, Adam-Turner (2017) mentioned that there are a number of things that must be done by teachers in Malaysia, such as teachers must use the computer often and practice and keep abreast of information. This needs to be done so that students can use any technology in teaching. With the application of ICT literacy in learning, student learning outcomes can improve.

5 CONCLUSION

The findings from the SLR research show that the biggest challenge for teachers today is that they do not yet have ICT competence or digital literacy and mastery of procedures. In addition, it is no less important that teachers attend ICT competency training and development. Teachers and students need to be given education regarding procedures for implementing digital literacy in everyday life. From these challenges, the main causal factors were found, namely the lack of facilities and infrastructure used in learning and the indifferent attitude of teachers toward the development of learning systems in the digital era. However, the analysis in this study found that the ICT literacy competency procedures that must be mastered by teachers are accessing, selecting, understanding, analyzing, verifying, evaluating, distributing, and producing.

REFERENCES

Adam-Turner, N., 2017. *Digital Literacy Adoption with Academic Technology: Namely Digital Information Literacy to Enhance Student Learning Outcomes*. ProQuest Dissertations and Theses, 292.

Agus Sulistyo and Ismarti, 2022. Urgency and strategy for strengthening media and digital literacy in learning Islamic religion. *At Turots: Journal of Islamic Education*, 3(2), 51–61.

Anggeraini, Y., Faridi, A., Mujiyanto, J., and Bharati, DAL, 2019. *Digital Literacy: Impacts And Challenges In Language Learning*. UNNES Postgraduate National Seminar, 388.

Astini, NKS, 2019. The importance of information and communication technology literacy for elementary school teachers to prepare the millennial generation. *Proceedings of the Dharma Acarya National Seminar*, 1 (2018), 113–120.

Asyari, F., 2019. Challenges of PAI teachers entering the west heritage, *Muslim Journal*, 4 (2).

Bahri, NS, Rakib, M., Said, MI, and Hasan, M., 2021. The Influence of digital literacy and entrepreneurial behavior on small business performance (Study on culinary business in jeneponto regency), 1 (2).

C. Correos, CTC, 2014. Teachers' ICT literacy and utilization in english language teaching, 'ICT & innovations in education. *'ICT & Innovations in Education' International Journal International Electronic Journal Cite*, 2 (1).

Dhimas Arsyad Arrajiv, Miftahul Arifah B. B, Tyas Wahyuningsih, K. & LER, 2021. The level of student learning independence in view from the implementation of digital literacy of sukoharjo 2 public high school students. *School Cultural Literacy Bulletin*, UMS, 55–64.

Djaja, S., Economics, S., and Unej, F., nd *Expectations and Challenges of Online Learning Teachers*. 2016, 10(2).

Ferrari, A., 2013. *DIGCOMP: A Framework for Developing and Understanding Digital Competence in Europe*. Spain: Luxembourg: Publications Office of the European Union.

Grillenberger, A. and Romeike, R., 2018. Developing a theoretically founded data literacy competency model. *ACM International Conference Proceeding Series*.

Helaludin, 2019. Increasing technology literacy capabilities in efforts to develop educational innovation in higher education. *Pendais*, I (1), 44–55.

Maulana, MA, 2020. Challenges of islamic religious education teachers in facing developments in the industrial Age 4.0. *EDURELIGIA: Journal of Islamic Religious Education*, 4(1), 88–100.

Priyanto, A., 2020. Islamic education in the industrial revolution era 4.0. *J-PAI: Journal of Islamic Religious Education*, 6(2), 80–89.

Retnaningsih, D., 2019. Teacher challenges and strategies in the era of the industrial revolution 4.0 in improving the quality of education. *Proceedings of the National Seminar: Education Policy and Development in the Industrial Revolution Era 4.0*. (September), 23–30.

Riadi, A., 2019. Challenges of Islamic religious education teachers in the 4.0 era Akhmad. *Journal: Azkiya*, 2(1), 1–10.

Ridsdale, C., Bliemel, M., Kelley, DE, and Matwin, SS, 2015. *Strategies and Best Practices for Data Literacy Education Knowledge Synthesis Report Strategies and Best Practices for Data Literacy Education Knowledge Synthesis Report View project Mentored Undergraduate Research and Identity Development View project (November)*.

Schield, M., 2013. Multi-domain analysis by FEM-BEM coupling and BEM-DD Part I: Formulation and implementation. *Applied Mechanics and Materials*, 353–354 (6), 3263–3268.

Skantz-åberg, E., Lantz-andersson, A., Lundin, M., Williams, P., Skantz-åberg, E., Lantz-andersson, A., Lundin, M., and Williams, P., 2022 Teachers 'professional digital competence: an overview of conceptualisations in the literature. *Cogent Education*, 9(1).

Spengler, S., 2015. *Educators' Perceptions of a 21st Century Digital Literacy Framework*. Walden University, 200.

Susilo, H., 2019. *Effects of Digital Literacy and Affective Learning in Islamic Religious Education Influences of Digital Literacy and Islamic Information Literacy on Affective Learning Outcomes of Islamic Religious Education Students of SMA N 1 Kendal Influences of Digital Literacy and Information Literacy*.

Religion, Education, Science and Technology towards a More Inclusive and Sustainable Future – Rahiem (Ed.)
© 2024 the Author(s), ISBN: 978-1-032-56461-6
Open Access: www.taylorfrancis.com, CC BY-NC-ND 4.0 license

The effect of self-awareness and critical thinking on students' ability to do scientific activities

A. Syukri* & Sukarno
UIN Sulthan Thaha Saifuddin, Jambi

ABSTRACT: This study aims to determine the effect of self-awareness and critical thinking skills on the ability of physics education students in carrying out scientific activities. The approach used in this research is qualitative, with the data analysis technique used being a product-moment correlation. The F test was used to find out how much confidence there was in the influence between variables. Based on the data and analysis carried out, it was found that self-awareness had a positive effect on the ability of physics education students to carry out scientific activities, with a correlation of 0.845 in the "high" category. As for the variable of the influence of critical thinking skills on the ability to carry out scientific activities of physics education students, the correlation was 0.818 in the "high" category. In addition, the F test for the two variables, $0.00 < 0.05$, means that the two variables simultaneously have a positive effect on the ability of physics education students to carry out scientific activities simultaneously. Therefore, lecturers and teachers are advised that in the learning process, it is better to use models, methods, or strategies that can stimulate the development of self-awareness and critical thinking skills of students so that their ability to carry out scientific activities can be improved.

Keywords: self-awareness, critical thinking, scientific activities, physics education students

1 INTRODUCTION

Self-awareness (SA) is one of the important factors in human success and satisfaction in living daily life. This is in line with the opinion of Dariyo (2017) that SA provides positive support for life satisfaction. According to Akbar et al. (2018) there is a positive relationship between a person's level of religiosity and SA. In this study, it was explained that the higher the level of SA, the more religious they would be. Another study also stated that SA also plays a role in increasing a person's awareness of protecting himself from a certain disease (Sukarno et al. 2021). Cakici (2018) and Kreibich et al. (2022) state that SA affects critical thinking skills (CTSs) and a person's ability to solve problems.

Referring to the description above, it can be understood that SA has a positive effect on the success of an individual in almost all areas of life. Therefore, the ability of SA must be developed by teachers and lecturers in the learning they impart inside and outside the classroom. For this reason, qualified teachers and lecturers, who can develop the full potential of students, are needed (Olsen 2021). SA can be increased by learning folk songs (Fadillah et al. 2021), applying peer counselor techniques (Yuliasari 2020), problem-oriented learning (Eriyani et al. 2021), and the use of Gestalt counseling (Sari et al. 2019).

In addition, another factor that is very influential in the success of an individual's life is CTS. For example, in learning activities, thinking skills are very influential on student achievement. Nurfitriyanti (2020) and Amto (2019) state that CTS has a positive effect on students' learning outcomes. Sharfina (2019) found that CTS affects a person's ability to make decisions and be influential in social life (Nugraha et al. 2020). Therefore, it can be understood that CTS has a positive impact on human life. The efforts to improve CTS can be made by involving technology

*Corresponding Author: ahmadsyukriss@uinjambi.ac.id

 DOI: 10.1201/9781003322054-19

in learning (Septiyani et al. 2020), using teaching based on interconnection integration (Suparni 2020), and using problem-based learning (Adiwiguna et al. 2019; Afriansyah et al. 2020).

Another factor that plays an important role in the success of humans in the future is carrying out scientific activities (Sci-A). Throughout their life, humans have never been separated from various Sci-A, ranging from observing appropriate activities, seeking information, or solving a certain problem, all of which are Sci-A. Achadah (2020) and Rella (2020) state that Sci-A is integrated into systematic actions and behavior. Milasari et al. (2021) also state that Sci-A are integrated into systematic actions and behavior, which are later known as the scientific method and include observation, problem formulation, fact-finding, and data analysis.

For physics education, students and even the general public cannot escape Sci-A. Therefore, efforts to examine what factors affect Sci-A of students must continue. Unfortunately, empirical research related to this is still very limited. Several studies related to the ability of Sci-A were carried out, among others, by Aulia et al. (2020) on scientific reasoning. He stated that scientific reasoning is intended as the ability to think systematically and logically to solve problems using the scientific method, including the process of evaluating facts, making predictions and hypotheses, determining and controlling variables, designing and conducting experiments, collecting data, analyzing data, and drawing conclusions. In addition, research on the factors that affect a person's ability to carry out Sci-A is still relatively small.

Partially, several studies were found related to Sci-A, for example, the ability to identify and solve problems. Felani *et al.* (2018) and Suprapto (2018) found that students' problem-solving abilities were within sufficient criteria and could improve their ability to conduct experiments. The study also did not explain what factors affect the ability to identify and solve problems and the ability to conduct experiments. This study seeks to determine the influence of SA and CTS on the ability of physics education students in carrying out Sci-A. The research will also see how strong the two variables are in influencing the ability of physics education students to carry out Sci-A. This research will have implications for teachers and the learning process, namely that there are internal factors of students that play a role in the success or failure of Sci-A. Each variable used in the study is limited to certain indicators: The SA indicator used in the study is the respondent's ability to determine: strengths and weaknesses, encouragement, values, and their impact on others that can guide individuals in making the right decisions. The CTS indicators used in this study are: (1) looking for a clear statement of each question; (2) finding reasons; (3) trying to know the information well; (4) using a credible source and mentioning it; (5) trying to stay relevant to the main idea. Sci-A is limited to five aspects of ability: observation, classification, measurement, data collection, and making simple reports.

2 METHODS

This study uses a quantitative approach with data acquisition techniques carried out by tests. The study involved 18 students from the physics education department at UIN Sulthan Thaha Saifuddin Jambi. The SA test instrument used in this study was a questionnaire with an ordinal scale of 25 questions. Based on the scores obtained, they were grouped into high (score 76–100), medium (score 50–75), and low (score 25–49) category. As for CTS variables, the test instrument used is in the form of five essay questions with a kinematics physics concept background. The instrument of students' abilities in carrying out Sci-A is in the form of an assessment rubric. Furthermore, based on the scores obtained in SA and CTS, they were grouped into high (score 11–15), medium (score 5–10), and low (score 0–5) category. Furthermore, the F test was used to assess the level of influence between variables. The F test was used to determine the significance of the effect of the two independent variables on the dependent variable. The population, as well as the sample in the study, are students of physics education of semester 4.

3 RESULTS AND DISCUSSION

Based on the results of the tests/measurements carried out using the instruments that have been prepared, then quantification is carried out on each research variable. This is done to perform a classification or category of ability in each variable.

Table 1. Self-awareness and critical thinking skills in scientific activities.

Aspect	Score	Number	Percentage (%)	Category
SA	76–100	9	50	High
	51–75	8	44,4	Moderate
	25–50	1	5,6	Low
	Total	18	100	
CTS	11–15	11	61,1	High
	6–10	5	27,7	Moderate
	0–5	2	11,2	Low
	Total	18	100	
Sci-A	11–15	8	44,4	High
	6–10	8	44,4	Moderate
	0–5	2	11,2	Low
	Total	18	100	

Based on Table 1, it can be explained that students' SA, CTS, and Sci-A are generally good. Thus, it can be said that the ability of physics students in terms of SA, CTS, and the ability to carry out Sci-A is relatively good. Furthermore, to determine the correlation between variables, a correlation test was conducted using Pearson Correlation:

Table 2. Correlation between variables of self-awareness, critical thinking skills, and Sci-A.

		SA	CTS	Sci-A
SA	Pearson Correlation	1	.845**	.870**
	Sig. (2-tailed)		.000	.000
	N	18	18	18
CTS	Pearson Correlation	.845**	1	.818**
	Sig. (2-tailed)	.000		.000
	N	18	18	18
Sci-A	Pearson Correlation	.870**	.818**	1
	Sig. (2-tailed)	.000	.000	
	N	18	18	18

**Correlation is significant at the 0.01 level (2-tailed).

Based on the value of sig. (2-tailed) between the variables, SA and Sci-A is $0.000<0.05$. Thus, it can be understood that there is a significant correlation between the variables SA and Sci-A. Furthermore, for the CTS variable, the value of sig. is $0.000<0.5$. This means that there is also a significant correlation between CTS and Sci-A. Referring to the R_{value}, it is known that the calculated R_{value} between variables is 0.870>from the R_{table}, which is 0.468. This shows that there is a significant correlation between SA and Sci-A. The calculated R_{value} on the CTS against the Sci-A is 0.818>from the R_{table}, which is 0.468. Thus, it can be said that there is a significant correlation between CTS and Sci-A.

Table 3. Model summary.

Model	R	R Square	Adjusted R Square	Standard error in estimation
1	.870[a]	.841	.838	1.191

[a]Predictors: (constant), SA, CTS.

Table 4. Model ANOVA[b].

Model	Sum of Squares	Df	Mean Square	F	Sig.	
1	Regression	888.297	2	444.149	313.111	.000[a]
	Residual	55.322	16	1.419		
	Total	943.619	18			

[a]Predictors: (constant), SA, CTS
[b]Dependent Variable: scientific activity

Based on the summary model table above, the calculated R_{value} is 0.870>from R_{table}, which is 0.468. This shows that there is a significant contribution between the variables SA and CTS and the variable Sci-A. Thus, it can be said that 87% of the Sci-A of physics students is influenced by the ability of SA and CTS, and the remaining 3% is influenced by other factors.

From the ANOVA table above, it can be known that the sig. 0.000, which means less than (<) 0.05, it can be said that SA and CTS simultaneously affect the Y variable. This is also reinforced by the calculated F_{value} 313.111>F_{table} 3.22. This means that the two variables, SA and CTS, simultaneously affect the Sci-A variable. Referring to the data and the results of the correlation analysis above, the three variables, SA, CTS, and Sci-A are shown to be interrelated and influence each other simultaneously. It can be said that the ability of physics students in carrying out Sci-A is strongly influenced by their SA and CTS. According to the F test, SA and CTS can affect students' ability to carry out Sci-A by more than 80%. This finding answers the question of whether SA affects students' ability to carry out Sci-A. Given that, so far, there has been no research related to it. So far, research related to SA has been associated with a person's psychology or personality, for example, aggressive, disciplined, and cheating behavior (Helawati et al. 2022; Shodiqin 2020) and employee performance (Even *et al.* 2022; Helawati et al. 2022; Safariland *et al.* 2021; Shodiqin 2020) have reviewed 31 research articles related to SA, none of which has been associated with Sci-A. If there is research that links SA with Sci-A, it is still partial, for example, in terms of communication skills (Nahzatun et al. 2021) and goal-based problem solving (Kreibich et al. 2022). Therefore, this research has opened new avenues for further research to look at the role of SA in carrying out Sci-A.

This study also answers the influence of CTS on students' abilities in carrying out Sci-A. Many studies related to CTS have been carried out previously, for example, the role of critical thinking in religious tolerance (Nugraha et al. 2020), the effect of CTS on learning outcomes (Nurfitriyanti et al. 2020), decision-making abilities (Nurfitriyanti et al. 2020), and communication skills (Noor & Ranti 2018). However, there has been no research that has clearly stated the effect of CTS on the ability to carry out Sci-A. Therefore, this research has contributed to providing a new direction of research, namely linking CTS with the ability to carry out Sci-A, Considering SA and CTS have been shown to influence students' ability to carry out Sci-A (Table 4). This also proves that the three variables, SA, CTS, and the ability to carry out Sci-A are interrelated with each other. An increase or decrease in one variable will be followed by another variable. Thus, teachers/lecturers must be able to implement and develop various learning models that have the potential to improve these three variables.

4 CONCLUSION

Based on the data and analysis carried out, it was found that SA had a positive effect on the ability of physics education students in carrying out Sci-A with a correlation of 0.845 in the high category. As for the variable of the influence of CTS on the ability to carry out Sci-A of physics education students, the correlation was 0.818 in the high category. In addition, the F test for the two variables is 0.00<0.05, which means that the two variables simultaneously have a positive effect on the ability of physics education students to carry out Sci-A simultaneously. Based on ANOVA (it is known that the sig. 0.000, which means less than 0.05), it can be said that SA and CTS simultaneously affect Sci-A. This is also reinforced by the calculated F_{value} 313.111 > F_{table} 3.22. This means that the two variables, SA and CTS, simultaneously affect Sci-A variables.

REFERENCES

Achadah, & Fadil, M. (2020). Filsafat Ilmu: Pertautan aktivitas ilmiah, metode ilmiah dan pengetahuan sistematis. *Jurnal Pendidikan Islam*, *4*(1 Juni), 131–141.

Adiwiguna, Dantes, & Gunamantha, (2019). Pengaruh model PBL berorientasi STEM terhadap kemampuan berpikir kritis dan literasi sains siswa. *Jurnal Pendidikan Dasar Indonesia*.

Afriansyah, E. *et al.* (2020). Mendesain soal berbasis masalah untuk kemampuan berpikir kritis matematis calon guru. *Mosharafa: Jurnal* . . . , 9, 239–250.

Akbar, M., Amalia, R., & Fitriah, I. (2018). Hubungan religiusitas dengan self awareness mahasiswa program studi bimbingan konseling UAI. *Jurnal Al-Azhar Indonesia Seri Humaniora*.

Amto, A., Ertikanto, C., & Nyeneng. (2019). Pengaruh keterampilan berpikir kritis melalui pembelajaran berbasis aneka sumber belajar terhadap hasil belajar fisika siswa. *Jurnal Pendidikan Fisika*.

Aulia, G., *et al.* (2020). Profil tingkat penalaran ilmiah siswa SMA pada materi ekosistem. *Biodik*, *6*(2), 176–186.

Cakici, D. (2018). Metacognitive awareness and critical thinking abilities of pre-service EFL teachers. *Journal of Education and Learning*, *7*(5), 116.

Carden, J., Jones, R., & Passmore, J. (2022). Defining self-awareness in the context of adult development: A systematic literature review. *Journal of Management Education*, *46*(1), 140–177.

Dariyo, A. (2017). Peran self-awareness dan ego support terhadap kepuasan hidup remaja tionghoa. *Psikodimensia*.

Eriyani, T., *et al.* (2021). Edukasi 3M dalam meningkatkan self-awareness terhadap penyebaran Covid-19 di SMKN 4 Garut. *Kumawula: Jurnal Pengabdian Kepada Masyarakat*, *4*(1), 98.

Fadillah, S. *et al.* (2021). Peningkatan SA anak usia 5-6 tahun melalui pembelajaran lagu daerah riau. *Pernik*, *4*(1), 100–104.

Felani, Ramdhani, & Hendriana, (2018). Kemampuan mengidentifikasi dan merumuskan masalah bangun datar serta minat belajar. *JPMI (Jurnal Pembelajaran Matematika Inovatif)*, *1*(3), 229.

Helawati, *et al.* (2022). Pengaruh self awareness terhadap perilaku menyontek. *Al Husna*, *3*(1), 31–46.

Kreibich, *et al.* (2022). The Role of self-awareness and problem-solving orientation for the instrumentality of goal-related means. *Journal of Individual Differences*, *43*(2), 57–69.

Milasari, *et al.* (2021). Filsafat ilmu dan pengembangan metode ilmiah. *Filsafat Indonesia*, *4*(3), 217.

Nahzatun, *et al.* (2021). Relationship between self confidence and self awareness with effective communication in students. *Indonesian Psychological Research*, *03*(July), 109–120.

Noor, F., & Ranti, M. G. (2018). Kemampuan berpikir kritis dan komunikasi matematis siswa SMP. *Math Didactic: Jurnal Pendidikan Matematika*.

Nugraha, I., Maslihah, S., & Misbach, I. H. (2020). Keterampilan berpikir kritis dan perannya terhadap toleransi beragama murid SMA. *Mediapsi*, *6*(2), 119–131.

Nurfitriyanti, *et al.* (2020). Pengaruh kemampuan berpikir kritis, adversity quotient dan locus of control terhadap prestasi belajar matematika. *JKPM (Jurnal Kajian Pendidikan Matematika)*.

Olsen, B. (2021). Teacher quality around the world: what's currently happening and how can the present inform the future? *European Journal of Teacher Education*, *44*(3), 293–294.

Rella Turella. (2020). *Modul Belajar Mandiri: Metode Ilmiah, Materi dan Perubahannya*. 19–40.

Safariani, D. *et al.* (2021). The Role of self-awareness on employee performance. *International Journal of Sciences and Technologies*, *29*(2), 663–666.

Sari, *et al.* (2019). The use of gestalt counseling to improve students' SA. *Alibkin* *7*(1), 1–15.

Septiyani, D. Y., Haji, S., & Widada, W. (2020). Meningkatkan kemampuan berpikir kritis calon guru matematika . . . di universitas bengkulu. *Jurnal Pendidikan Matematika Raflesia*.

Shodiqin, Rina. (2020). *Self Awareness dan Perilaku Agresi Pada Masyarakat Saat Pendisiplin Protokol Kesehatan Oleh Petugas. 001*.

Sukarno, S. *et al.* (2021). Investigating capabilities of science and information literacy. *3397| International Journal of Pharmaceutical Research, 12*.

Suparni, S. (2020). Upaya meningkatkan kemampuan berpikir kritis mahasiswa. *Jurnal Derivat: Jurnal Matematika Dan Pendidikan Matematika*.

Suprapto. (2018). Peningkatan kemampuan melakukan percobaan dan prestasi belajar ipa melalui *"Pentas Diskotek."* 1–10.

Yuliasari, H. (2020). Pelatihan konselor sebaya untuk meningkatkan self awareness terhadap perilaku beresiko remaja. *Jurnal Psikologi Insight*, *4*(1), 63–72.

Religion, Education, Science and Technology towards a More Inclusive and Sustainable Future – Rahiem (Ed.)

Distance learning in the perspective of Sadd al-Żarāi'during the Covid-19 pandemic

A. Arifin*, R. Yazid, I. Sujoko & H. Hasan
UIN Syarif Hidayatullah, Jakarta, Indonesia

S. Muttaqin
STAI Alhikmah, Jakarta, Indonesia

ABSTRACT: This study aimed to look at distance learning policies from *Sadd al-Żarāi'* point of view. This policy was taken by the government not long after Covid-19 hit the country. Problems also arise, many are in favor of the policy, but not a few are against it. The reasons for these opponents are varied but generally relate to the level of effectiveness of the learning model which is not optimal. This research is qualitative by using the ushul fiqh approach, and the existing data is presented in a descriptive-analytical manner. The results of this study reveal that the distance learning policy set by the government is not contrary to Islamic law. According to *Sadd al-Żarāi'*, this policy needs to be taken and implemented by the whole community, especially students. This is nothing but to prevent greater confusion if other policies are implemented, such as face-to-face learning. The *mafsadah* in question is the threat of safety that leads to death.

Keywords: *Sadd al-Żarāi'*, distance learning, coronavirus

1 INTRODUCTION

In this time of the pandemic, distance learning is needed. This type of learning system can teach students to be wise in using and accessing the internet, interacting with available content, including interacting with instructors and teachers in different rooms, gaining knowledge and insight – which is broader than face-to-face learning- during learning, gaining individual understanding and enhance a different learning experience than usual (Bruggen 2005). It seems that, of course, the teacher or instructor provides teaching and examples that are tailored to students. Indeed, distance learning will take place well if the teacher has the competence to do it. In this case, the teacher plays an important role in the distance learning process (Oktaviani et al. 2021).

The problem that arises is that this distance learning model is by no means an effective option based on research. The majority of researchers agree that face-to-face learning that emphasizes direct communication is the most effective model in the learning process (Tang & Chaw 2013). A study in Canada stated that more than 50% of students who had participated in the study experienced a decline in achievement when not doing face-to-face learning (Alam & Jackson 2013). This shows that the face-to-face learning model is still effective because of the lack of obstacles and the avoidance of limitations. This then becomes a separate problem in society, especially among the Muslim community. Some of them even complain about the implementation of this learning model, as summarized by CNBC Indonesia as follows: "Stop distance/online learning for basic education, especially in elementary schools @DPR_RI @DPDRI !!! Psychologically, children of this age still need direct education from their teachers..." (Hasibuan 2020).

*Corresponding Author: azis.arifin20@mhs.uinjkt.ac.id

Parents of students feel that online learning, even though it is carried out during a pandemic, is by no means effective. Various problems were raised such as their lack of understanding of how to study distance, their inability to buy internet data quota, and so on (Hasibuan 2020). On the other hand, according to many classical to contemporary Islamic scholars, the protection of the soul is considered one of the points of *maqāṣid al-sharī'ah*. Among those who stated this were al-Ghazali (Kamali 2008), al-Qarafi (d. 1285 H), Ibn al-Subki (d. 1370 H), Ibn Ali al-Shaukani (d. 1834 H) (Kamali 2008), and Ibn Taimiyah (d. 1328 H) (Taimiyyah 2004). Protection of the soul according to some scholars' views must be prioritized and prioritized over the protection of religion, like being able to admit to being a kāfir in a state of urgency, including eating carrion in an emergency and so on (Al-Munjid 2019). This means that the safety of the soul is at the highest level in the ḍarūriyyah hierarchy. Thus, comparing the safety of the soul with the smooth education of children should be questioned again.

This is what prompted us to do this research. We are intrigued to see the distance learning policy[1] from the point of view of Sadd al-Żarāi' as one of *adillah al-aḥkām* which is *mukhtalaf fīh* (a legal proposition that is not agreed upon by some scholars). We see the potential for *mafsadah* in this policy, but on the other hand, the resulting harm seems to be even greater if the policy is not implemented.

2 LITERATURE REVIEW

There has been a lot of research on distance learning, especially recently. However, the existing studies only focus on students' perceptions of distance learning during the pandemic (Mseleku 2020; Tacoh 2020) and the implementation of the distance learning policy (Adedoyin & Soykan 2020; Ferraro et al. 2020; Oktaviani et al. 2021). As for the *Sadd al-Żarāi'*, it seems that there is no research on a similar topic from that perspective. We only find research with a *Sadd al-Żarāi'* perspective on other topics, such as seeing the potential benefits of postponing weddings during a pandemic in the *Sadd al-Żarāi'* review (Amin 2020). In some of the studies mentioned, none of them seem to have attempted to view distance learning as a policy that needs to be re-examined in other perspectives, such as Islamic law. Therefore, we view that this problem is still a problem that must be solved, especially from an academic point of view.

3 METHODS

Based on the analysis perspective, the research method used is qualitative. The use of qualitative methods is based on the author's goal of understanding, describing, obtaining findings and generating hypotheses (Merriam & Tisdell 2016). The literature review and content analysis are also the method used as a reference in this research. The data sources of this research are joint decree of 4 ministers regarding learning guidelines during the COVID-19 pandemic, books, internet news portals, and related journals. The data collection technique used in this research is to use documentation techniques. This technique is done by searching and reading data from data provider centers, such as libraries, online pages or other places that provide data (Merriam & Tisdell 2016). It was analyzed using the theory of *maqāṣid al-sharī'ah* to see the benefits and *mafsadah* aspects of this distance learning policy. This research is qualitative in nature by using the *Uṣūl Fiqh* approach, while the existing data is presented in an analytical descriptive manner. Based on this explanation, the goal to be achieved in this study is to look at distance learning policies from *Sadd al-Żarāi'* point of view. Because this research is

[1]This regulation is based on a joint decision between the Minister of Religion, Minister of Health, Minister of Education, Research and Technology, and Minister of Home Affairs, regarding learning guidelines during the COVID-19 pandemic which was signed on December 21, 2021.

qualitative, the author uses the Miles and Huberman model in managing data analysis, data condensation, data presentation, and drawing and verifying conclusions (Miles *et al.* 2014).

4 FINDINGS

4.1 *Deepen theory and measuring mafsadāt and maṣlaḥāt of distance learning*

The definition of distance learning continues to change from time to time. This is firmly motivated by various factors, especially technology. Some experts have defined it well, but we argue that the definition proposed by Teaster and Blieszner is more flexible and appropriate. According to them, distance learning is a learning process carried out by educators and students separately, in space, maybe even time by using certain supporting tools (Teaster & Blieszner 1999).

As it is known that the Coronavirus (Covid-19) is a virus that can cause damage to the function of organs in the body to lead to death. A study conducted by researchers at New York University Grossman School of Medicine said that viruses that attack vital organs are responsible for causing fatalities (Sumartiningtyas 2021). Based on this information, it can be concluded that this virus can attack and aggravate the organs it attacks. This means that this virus is not directly the cause of the death of a patient, but is a comorbid (become a comorbid disease). Keeping the soul is something that must be done by humans, especially Muslims as their endeavor to carry out other obligations to maintain survival. Protecting the soul is part of the *maqāṣid al-sharī'ah* points agreed upon by all scholars. In other terms referred to as *hifż al-nafs*, it belongs to the *maslaḥah al-ḍarūriyyah* category (Al-Jābirī 1996). Based on this explanation, it can be mapped that after being weighed and classified there are two types of damage with different capacities from each other, namely distance learning with risks as mentioned and face-to-face learning with safety threats. In this case, we try to weigh it with the rule which reads: "*When faced with two mafsadah, then avoid the larger mafsadah by taking the lighter mafsadah.*"

This rule can be applied as in the case of a person who has a severe injury to his knee. So if you prostrate, then the wound will get worse. The rule can also be applied to the distance learning phenomenon as examined in this paper, by considering the *mafsadah* caused by the implementation of the distance learning policy with face-to-face learning. The mafsadah generated by the first policy only touches on minor aspects without threatening the safety of life in the slightest. Meanwhile, if the second policy is implemented, health and life can be at stake. Thus, the *mafsadah* on the implementation of the second policy has a greater weight than the *mafsadah* on the implementation of the first policy.

4.2 *Sadd al-Żarāi' study viewing distance learning policy*

Some of the information in the Joint Decree explains how important the implementation of distance learning is for students during this Covid pandemic. Among them are the following:

> *"Bahwa kesehatan dan keselamatan semua warga satuan pendidikan merupakan prioritas utama yang wajib dipertimbangkan dalam menetapkan kebijakan pembelajaran di masa pandemi COVID-19."* (Kemdikbud 2021)

(That the health and safety of all unit residents education is a top priority considered in setting policy learning during the COVID-19 pandemic).

This point is the opening of this decision letter. This is proof of the government's commitment to prioritize public health and safety in the midst of this deadly virus attack, especially in the education unit environment.

> *"Pendidik yang tidak diperbolehkan atau ditunda menerima vaksin COVID-19 karena memiliki komorbid tidak terkontrol atau kondisi medis tertentu berdasarkan keterangan dokter, pelaksanaan tugas pembelajaran/bimbingan pendidik dilakukan melalui pembelajaran jarak jauh."* (Kemdikbud 2021)

(Educators who are not allowed or postponed received the COVID-19 vaccine due to comorbidities uncontrolled or certain medical conditions based on a doctor's statement, carrying out duties learning/guidance of educators is carried out through distance learning).

How important is the health and safety of education unit residents, educators who have health problems so that they are hampered from receiving vaccines are required to conduct distance learning to ensure these interests.

> *"Menghentikan sementara penyelenggaraan pembelajaran tatap muka terbatas pada tingkat satuan pendidikan dan dialihkan menjadi pembelajaran jarak jauh selama 14 (empat belas) hari, apabila: a) terjadi klaster penularan COVID-19 di satuan pendidikan."* (Kemdikbud 2021)

(Temporarily suspend maintenance face-to-face learning is limited to the level of education unit and transferred to distance learning for 14 (fourteen) days, if: a) a cluster of COVID-19 transmission occurs in the unit education).

Even if face-to-face education is initially held, then this is found to create clusters of the spread of COVID in the educational environment, then the learning method must be changed to distance learning for at least the next 14 days. This is intended to anticipate the possibility of a wider spread of the virus.

What has been stated in the Joint Decree is actually in accordance with the principle of Sadd al-Żarāi'. Sadd al-Żarāi' as an analytical knife of law diggers is claimed to have a close relationship with the benefit aspect. Abu Zahrah defines al-Żarāi' as something that mediates the realization of what is forbidden (Al-Zuhailī 1986). Imam Malik and Imam Ahmad believe that al-Żarāi' is the foundation of the fiqh proposal (Al-Zuhailī 1986). Based on this, distance learning in the perspective of Sadd al-Żarāi' means preventing the occurrence of harm in the form of endangering the health and safety of the soul based on benefit. Distance learning is an effort to break the chain of the spread of the Coronavirus due to mobility restrictions. Distance learning during a pandemic is important to do as a form of protection for oneself, as well as protection for the souls of others.

This is following the terms and conditions as stipulated concerning indicators of benefit and mafsadah in the process of making Islamic law. Light *mafsadah* or in this case can also be referred to as *maṣlaḥah* such as difficulties in monitoring the tasks assigned to children; burdensome study time; inadequate availability of facilities (devices); poor network connection; high cost of internet quota; the low intensity of teacher-student interaction; loss of effectiveness in understanding the material; the achievement index is decreasing; and the psychological condition of students is disturbed, is if compared to a bigger *mafsadah* such as the loss of life, are being an important reason for distance learning policies to be issued and implemented by the entire community, especially Muslims without doubting its legal validity from an Islamic perspective. Distance learning does not contradictory with the qaṭ'i argument, in fact, this policy is in line with the qaṭ'i proposition which requires Muslims to study and also maintain the safety of their souls. In addition, this policy is also not based on mere speculation that life will be threatened. This has been proven not only in Indonesia but in all countries in the world. In addition, distance learning policies are the only solution to support students in obtaining an education. Therefore, public safety can be well guaranteed because it avoids face-to-face learning which risks the occurrence of crowds and the massive spread and transmission of the Covid-19 virus. Thus, al-Żarāi' in the matter of distance learning is included in the category of *al-Żarāi'* which is dominant in bringing harm.

5 CONCLUSION

Distance learning is an important option that deserves to be implemented to support education during this pandemic. Now, Muslims need not hesitate to comply with and carry out these policies as well as possible. This is because distance learning policies do not conflict with Islamic law. *Sadd al-Żarāi'* becomes a sharp analytical knife with *maqāṣid al-sharī'ah* as a reinforcement for the emergence of this policy. This is because distance learning is an effort to prevent the public from spreading infection with the Coronavirus which can endanger lives. Distance learning is not only a shield for protection for oneself but also the souls of others. The *mafsadah* arising from distance learning is smaller than face-to-face learning. For this reason, as long as the pandemic is still ongoing, distance learning is the right option to continue to apply for the sake of continuing education and public safety.

This research is limited to *Sadd al-Żarāi's* review of distance learning policy. Other views of Islamic law can be highlighted to see the validity of this policy.

REFERENCES

Adedoyin, O.B. and Soykan, E. (2020) 'Covid-19 pandemic and online learning: the challenges and opportunities', *Interactive Learning Environments* [Preprint]. doi:10.1080/10494820.2020.1813180.

Al-Jābirī, M. 'Ābid (1996) *Al-Dīn wa al-Daulah wa Taṭbīq al-Sharī'ah.* Beirut: Markaz Dirāsāt al-Waḥdah al-'Arabiyyah.

Al-Munjid, M. Ṣāliḥ (2019) *al-Ḍarūriyyāt al-Khams wa al-Khilāf fī Taqdīm Ḥifẓ al-Dīn 'alā Ḥifẓ al-Nafs, Islamqa.* Available at: https://islamqa.info/amp/ar/answers/307202 (Accessed: 10 October 2008).

Al-Zuhailī, W. (1986) *Uṣūl al-Fiqh al-Islāmī.* Damaskus: Dār al-Fikr.

Alam, S. and Jackson, L. (2013) 'A case study: Are traditional face to face lectures still relevant when engineering course?', *IJEP*, 3(4), pp. 9–15. doi:http://dx.doi.org/10.3991/ijep.v3is4.3161.

Alam, S.O. (2020) *Studi Ungkap Virus Corona Covid-19 Bisa Jadi Penyebab Langsung Kematian, detikhealth.* Available at: https://health.detik.com/berita-detikhealth/d-5097388/studi-ungkap-virus-corona-covid-19-bisa-jadi-penyebab-langsung-kematian (Accessed: 21 October 2008).

Amin, M.N.K. Al (2020) 'Menakar nilai kemanfaatan dari penangguhan walimat al- 'Ursy di masa darurat COVID-19 melaluiSadd Adz-Dzari'ah', *Ulumuddin: Jurnal Ilmu-ilmu Keislaman*, 10(1).

Bruggen, J. Van (2005) 'Theory and practice of online learning', *British Journal of Education Technology*, 36 (1), pp. 111–120. doi:10.1111/j.14678535200500445.

Ferraro, F.V. *et al.* (2020) 'Distance learning in the COVID-19 era: Perceptions in southern italy', *Education Sciences*, 10. doi:10.3390/educsci10120355.

Hasibuan, L. (2020) *Saat 'Emak-emak' Pada Protes Belajar Online Ribet!, CNBC Indonesia.* Available at: http://www.cnbcindonesia.com/tech/2020072016311537174071/saat-emak-emak-pada-protes-belajar-online-ribet (Accessed: 10 October 2008).

Kamali, M.H. (2008) *Shari'ah Law An Introduction.* Oxford: Oneworld Publication.

Kemdikbud (2021) Keputusan Bersama Menteri Pendidikan dan Kebudayaan, Menteri Agama, Menteri Kesehatan dan Menteri Dalam Negeri tentang Panduan Pembelajaran di Masa Pandemi Coronavirus Disease *2019 (Covid-19), Kemdikbud.* Available at: https://www.kemdikbud.go.id/main/files/download/2b7a3531e4b5551&ved=2ahUKEwjT_ (Accessed: 9 October 2008).

Merriam, Sharan B., and Elizabeth J. Tisdell. *Qualitative Research a Guide to Design and Implementation.* 4th ed. San Fransisco: Jossey-Bass, 2016.

Miles, Matthew B., A. Michael Huberman, and Johnny Saldaña. *Qualitative Data Analysis a Methods Sourcebook.* 3rd ed. California: SAGE Publication, 2014.

Mseleku, Z. (2020) 'A literature review of e learning and e-teaching in the era of covid-19 pandemic', *International Journal of Innovative Science and Research Technology*, 5(10).

Muḥammad, M. bin (2007) *Al-Mumti' fī Qawā'id al-Fiqhiyyah.* Riyad: al-Mamlakah al-'Arabiyyah al-Sa'ūdiyyah.

Oktaviani, N. *et al.* (2021) 'Implementasi pembelaran daring di masa pandemi covid-19 berdasarkan perspektif guru sekolah dasar', *Jurnal Review Pendidikan Dasar: Jurnal Kajian Pendidikan dan Hasil Penelitian*, 7(2), pp. 86–93.

Tacoh, Y.T.B. (2020) 'Perpektif mahasiswa terhadap pendekatan pedagogi spiritual dalam pembelajaran daring', *Perspektif Ilmu Pendidikan*, 34(2), pp. 67–80. doi:doi.org/10.21009/PIP.342.1.

Taimiyyah, A. bin (2004) *Majmū' Fatawā.* Madinah: Mujamma' al-Malik Fahd li Ṭāba'ah li Muṣḥaf al-Sharīf.

Tang, C.M. and Chaw, L.Y. (2013) 'Readiness for blended learning: Understanding attitude of university students', *International Journal of Cyber Society and Education*, 6(2), pp. 79–100. doi:doi:10.7903/ijcse.1086.

Teaster, P. and Blieszner, R. (1999) 'Promises and pitfalls of the interactive television approach to teaching adult development and aging', *Educational Gerontology*, 25(8), pp. 741–754.

Religion, Education, Science and Technology towards a More Inclusive and Sustainable Future – Rahiem (Ed.)
© 2024 the Author(s), ISBN: 978-1-032-56461-6
Open Access: www.taylorfrancis.com, CC BY-NC-ND 4.0 license

Educational model in Tzu Chi school in terms of multicultural education dimensions

Nukhbatunisa*, M.R. Lubis, A. Zamhari, A. Khoiri & W. Triana
UIN Syarif Hidayatullah, Jakarta, Indonesia

I.Y. Palejwala
International Islamic University of Indonesia

ABSTRACT: Various models of multicultural education in each school and their approaches are interesting to be studied. This study analyzes the education model in a Tzu Chi school, which includes a variety of cultures, races, and religions, and integrates with the dimensions of multicultural education (James Banks 2006). This study uses a descriptive qualitative method, with interviews with religious teachers at the school. The results of this study indicate that the five dimensions that have been declared are fulfilled: (1) Dimensions of material integration by applying the mandatory material, namely learning humanist culture; (2) The knowledge dimension is fulfilled with a complete understanding of the child and is reflected in the attitude of respect, respect, and tolerance between races, cultures, and religions; (3) The dimension of reducing prejudice is fulfilled, seeing the teacher directing students in organizing various religious events with cooperation; (4) The dimensions of education are the same with the form of providing the same places of worship for all religions and not leaning toward one religion; (5) The dimension of school culture empowerment in the form of group learning culture implemented in Tzu chi schools. The researcher concludes that although the education model is humanist and respectful of each other, it does not reduce each other's religious beliefs because the dimensions of harmony are exclusive.

Keywords: dimensions multicultural education, Tzu chi school, humanist culture

1 INTRODUCTION

As immigrant populations are growing rapidly in a growing number of countries, educational policymakers and researchers are focusing more on the importance of multicultural competence demonstrated by students. As a result, multicultural education in a broad sense has emerged as a critical policy issue in an increasing number of countries worldwide (Yun Kyung Cha 2017). The Latin word "educate" is where we get the word "education," which means to "teach," "raise," or "bring forth the latent forces of the kid." Durkheim defined schooling as "the action conducted by the more experienced ages upon the individuals who are not yet prepared for public activity." Its goal is to instill in the child the moral, academic, and physical qualities that are required of them by both the general public and the environment for which they are specially designed (Mohammad et al. 2021) The idea of "tolerance" is linked with the idea of "multiculturalism," which is treated as a collection of thoughts and deeds of various social entities (statutory bodies, for example) aiming at equal development of diverse cultures, bridging different population groups in many social spheres, equal employment opportunities, and provision of educational options (Grishaeva 2012).

*Corresponding Author: nukhbatunisa01@gmail.com

DOI: 10.1201/9781003322054-21

The objectives of multicultural-based education as disclosed by Skeel (1995), which can be used as a benchmark for multicultural-based schools, are as follows:

(1) Schools' responsibility to function for the existence of students from various backgrounds; (2) to help students belonging to different cultural, racial, ethnic, and religious groups in building positive treatment; (3) provide student resilience by teaching them decision-making and social skills social; (4) helping students in building intercultural interdependence and giving a positive image about group differences (Junaidi 2018).

In Indonesia, the term multicultural is generally modern for the community. Multicultural instruction that has been created in Indonesia could be a frame of decentralization and territorial independence arrangements. Multicultural instruction is required to be an instruction that will be able to change the learning framework, not fair a custom in reexamining learning materials (Affandi 2019). Analyzing multicultural instruction in Indonesia can be seen from the five multicultural measurement benchmarks (James Banks 2004). Specifically: substance integration, information development preparation, preference decrease, a value instructional method, and an enabling school culture and social structure. The hypothesis can be utilized as a benchmark for multicultural-based schools. For more subtle elements, the creator will join a figure cited from the book (James Bank 2014). The analyst chose the protest of investigating to analyze multicultural instruction in Indonesia with different multicultural measurements advanced (James Banks 2004), specifically a school with different religions in it which is found in Jakarta Indonesia specifically the Indonesian Buddhist Tzu Chi School.

The beginning of Tzu Chi's work was in 1991 when Master Cheng Yen wanted to help flood victims in Bangladesh, outside the Tzu Chi center in Taiwan. Since then its global mission has helped 11 million victims of wars, floods, earthquakes, and other disasters (Mark O'Nail 2010). The missions emerge from the human spirit, namely: love (charity mission), great compassion (health mission), great joy (humanist cultural mission), and great equanimity mission (educational mission). While the word Tzu-Chi is: Tzu-Phi-Si-Se. Tzu means: Giving love sincerely without regrets. Phei means resilience in experiencing difficulties and reproaches. Si means mutual respect, respect, and love, and the word Se means: When we give, don't demand anything in return; this concept was used by Master Cheng Yen to be persistent in building hospitals, schools, and other organizations (Yen 2017). From this concept emerged the Tzu Chi school, one of which is now in Indonesia.

The mission of Tzu Chi education is education to form a whole person, not just teaching knowledge and skills, but also manners and human values. This pattern is developed to guide students as the forerunner of the nation in the future that is reliable. At Tzu Chi educational institutions, teachers and students study together. Apart from studying knowledge, they learn to develop love and benevolence in everyday life. The Tzu Chi School of Love has levels of education from playgroup KB/TK, SD, SMP, and SMA. Seeing that Tzu Chi's social and educational activities do not place boundaries between religion and ethnicity, as well as the curriculum and learning methods taught there, are respectful, and acknowledge other cultures and religions, the researcher made the Tzu Chi school the object of research in the application of multicultural education that may be used as an example of lessons or vice versa for criticism for the school.

Previous research conducted by (Muary & Ismail 2017) revealed that the Tzu Chi Organization was also built based on humanity regardless of ethnic, religious, or racial background so that that the organization is open to anyone who wants to join a humanitarian mission. Finally, Tzu Chi built a framework of a multireligious, racial, and ethnic social organization and built good relations, not only with the Buddhist sects but also with other religions. Tzu Chi is a religious organization that builds a more modern, rational organizational system with a mission of universal values. Likewise, research conducted by Nugraha, Hanim, and Siswono (2020) on the application of humanism in learning in the first grade of the Tzu Chi Private Elementary School taught how students must humanize humans, especially in the school and family environment. And many activities are held in it such as religious activities and the habit of being respectful of each other. In this way, the creators are exceptionally inquisitive about making Tzu Chi School a question of research with multicultural values in it which is able afterward to be analyzed with 5 measurements agreeing to James Banks (2004) whether the Tzu Chi School fulfills all viewpoints of its measurements or where the deficiencies of the school which at the point gotten to be input and assessment material for Tzu Chi School.

2 METHODS

This study uses a descriptive qualitative research design according to Bogdan dan Taylor (1975) A qualitative study, according to the definition, is a research procedure that generates descriptive data in the form of written or spoken words or sentences from people and observable behavior (Suwardi Endraswara 2006) This study uses a descriptive qualitative method because it describes the multicultural education model applied at the Tzu Chi School and the attitudes or behaviors that describe the harmony between ethnic cultures and religions in it.

Data for this study is gathered through participant observation, in-depth interviews, and documentation. The validity of qualitative research data includes data objectivity, the validity of internal-external data, and reliability. Ensuring the credibility of internal data includes extending the participation of researchers in the field, increasing the persistence of observations, triangulation, analysis of negative cases, peer examination through discussion, availability of references, and member checks (Iskandar 2009). In collecting data the writer interviewed the religious teachers from the school. The information investigation strategy agreed to (Iskandar 2019) employ 3 stages: (a). Information lessening, this handle is the collection of inquiries about information by an analyst to urge a part of information. (b). Information show, what is implied by information show is compiling information and entering it in several frameworks or list categories and displaying it with narrative text.(c). Drawing conclusions and after that being confirmed, this handle is the final prepare after information lessening and data display, in concluding whereas analysts have the opportunity to induce transitory input and examine them with colleagues, as it were after the comes about of the inquiry about is demonstrated genuine, analysts can conclude expressive frame as inquire about the report.

3 RESULTS

In this paper, the author integrates the theory (Banks 1993) that explains that education multiculturalism has five interconnected dimensions that can assist teachers in implementing programs to respond to student differences (student). (a) Content integration dimension (content integration): This dimension is used by teachers to provide information with "key points" of learning by reflecting on different materials. Teachers, in particular, integrate content from learning materials into the curriculum from a variety of perspectives. (b) Knowledge construction dimension (knowledge construction): Some dimensions in which teachers assist students in understanding various points of view and formulate conclusions that are influenced by the discipline of knowledge they possess. (c) Prejudice reduction dimension; The teacher works hard to help students develop positive attitudes toward different groups.

If the attitudes and behavior of activities at Tzu Chi School are measured by the dimensions that have been revealed by Banks (2006), it can be seen that according to Tzu Chi Religion Teacher Hadawiyah Lubis (2021); First, the Tzu Chi School has integrated multicultural material with a humanist approach by holding compulsory subjects, namely Tzu Chi humanist culture learning, or ethics lessons, in which it teaches how to respect parents, tolerate each other, love each other, love the young, etc. According to Suherman (2021) in realizing character education and implementing humanist culture with the teacher's strategy of working with the coordinator of humanist culture education, this collaboration aims to provide guidance and direction regarding the habits of students in everyday life including home, the coordinator of humanist culture education coordinates the assessment of student behavior through the observations of related teachers, it encourages teachers to be more actively involved in routine, spontaneous and exemplary activities. Both in the classroom and outside the classroom, teachers, and coordinators of humanist culture also conduct evaluations, thus this dimension has been fulfilled by the Tzu Chi School. The second dimension applied can be measured through their accuracy in understanding the meaning of differences so that mutual respect between races, ethnicities, cultures, and religions is applied among fellow students.

The third, dimension is reducing prejudice, and efforts to help students to do positive things about group differences. This can be seen during holiday activities carried out by each religion. The implementation of the religious day event in Tzu Chi School holds a major event of choice for five religions in one year, for example, Muslim students choose the Prophet's birthday event which is held together, of course, Muslim students cannot do it themselves so for non-Muslims they are required to help organize the birthday, such as participating in designing the Kaaba and vice versa; during the time of Christmas for Christians, apart from Christian students, others are obliged to help, for example, participate in wrapping gifts, which will be placed under the Christmas tree; Chinese New Year is celebrated for Confucian beliefs, other than Confucians they can prepare red lanterns and others. If in Islam the law of saying Merry Christmas or Lunar New Year is still under debate, to avoid this, religions do not say it but are shown by the behavior of helping, acknowledging, and appreciating between religions, the attitude of helping each other that exudes the value of tolerance and respect between religions. Fourth, the dimension of the same education. This dimension also includes education, which is intended to shape the school environment into various types of groups, such as ethnic groups, women, and students with special needs, in order to provide an educational experience with equal rights and equal learning opportunities (Ibrahim 2013) Tzu chi schools provide the same facilities for all races, ethnicities, and religions. One of them, the places of worship provided, from mosques for Muslims, temples for Buddhists, and churches for Christians are all provided and welcome to worship according to each individual's religion on Friday. Fifth, the dimensions of empowering school culture and social structure. Teachers who get effective professional development in cooperative learning gain insight into how to implement cooperative activities in the classroom and make links to the model's theoretical foundations (Patricia Roy 1998) this dimension has been fulfilled because the learning model carried out in Tzu chi is cooperative (Cooperative Learning). In tzu chi schools, cooperative learning is a teaching and learning strategy that emphasizes shared attitudes or behavior in working or helping others in an organized structure of cooperation in groups of two or more people (Nugraha et al. 2020).

4 DISCUSSIONS

In addition to being a benchmark for the dimensions of multicultural education above, there are three aspects that must be understood by children, including aspects of attitude, namely cultural awareness and sensitivity, cultural tolerance, and skills to avoid conflict. Aspects of knowledge in the form of knowledge of languages, intercultural traditions, and values in various religions. Aspects of learning that seek to correct distortions, stereotypes, and misunderstandings about ethnic groups in instructional media textbooks, respecting different values (Ujang Syarip 2018) Tzu chi school uses an exclusive dimension of harmony, namely absolute belief in the teachings of the religion it adheres to and does not open oneself to seek other truths (Ridwan Lubis 2020) This can be seen by assessing students and teachers who still hold fast to tolerance but still believe in the religion they adhere to, even though organizing the big day events together, will not interfere with each other's beliefs and still respect and appreciate between religions and religions. ethnicity.

As for the implementation of humanist culture, in addition to holding a coordinator for humanist education, the Tzu chi school held a Tzu Chi Humanist Culture seminar. This seminar is held once a year before entering new teaching, intended for all teachers so that teachers also understand the meaning of religious tolerance and do not look at each other's ethnicity, culture and culture and instill an attitude of sympathy, and empathy for others. The training is directly from Taiwan which instills the concept of "All are equal and one family" as well as the word bullying (for Muslims such as hadith or the words of scholars) which reads "In this life, there are two things that we cannot delay: respecting parents and doing good" they interpret doing good by mutual tolerance between races and religions. The curriculum used refers to the National Education System. However, there are additional subjects that become the selling point of Tzu Chi culture, namely learning the humanist culture of Tzu Chi, or ethics lessons, which teaches how to respect parents, tolerate each other, love each other, and love the young and others (Hadawiyah 2021).

5 CONCLUSION

In its application, tzu chi schools have met the standards of multicultural education with a humanist approach. description of attitudes and values in the school. This research involves limited sources, uses one research method, and is not carried out for a long period, but this research is the initial entry point for further research, researchers suggest using a larger sample and using combined qualitative and quantitative research.

REFERENCES

Afandi (2018), Mewujudkan pendidikan multikultural di Indonesia. *Sebuah Kajian Pen- didikan Multikultural Di Berbagai Negara.* Tanjungpura University, Article October, h.7.

Banks, James A, (2014) *An Introduction to Multicultural Education.* United States: Library of Congress Catalog-in-Publication Data, h.35

Banks, James A, Cherry A. M (2010), *Multicultural Education Issues and Perspective*, United States of America: WILEY, h. 20.

Buku Profil TC Indonesia 2013 plus cover' (2013) Tzu Chi Yayasan Buddha Indonesia, pp.25.

Grishaeva, E.B. (2012) Multiculturalism as a central concept of multiethnic and polycultural society studies, *Journal of Siberian Federal University.* Humanities & Social Sciences, pp.97.

Ibrahim, R. (2013) Pendidikan multikultural: Pengertian, prinsip, dan relevansinya dengan Tujuan Pendidikan Islam, *Addin*, Vol. 7, No. 1, Februari 2013, pp. 143–144.

Iskandar (2009) *Metedeologi Penelitian kualitatif.* Jakarta: Gaung persada (GP) Press, pp. 151–153.

Junaidi (2018), Model pendidikan multikultural, al-insyiroh Vol 2, Nomor 2, 2018, pp.67.

Lubis, M. Ridwan (2020) *Merawat kerukunan: Pengalaman Indonesia.* Cetakan Pe. Edited by Z.M. Murodi. Tangerang Selatan: UIN Jakarta Press, 2017, pp.16.

Mark O'Nail (2010) *Tzu Chi Serving with Compassion.* John Wiley & Sons (Asia), pp. 10.

Mohammad, S., Faruqe Jubaer, O. and Hoque, L. (2021) The concept of education: A western rationalist approach, *International Journal on Integrated Education*, pp. 139.

Muary, R. and Ismail, R. (2017) Gerakan Sosial Budha Tzu Chi Pasca Reformasi di Kota Medan Post-reformation Budha Tzu Chi Social Movement in Medan, pp.257

Nugraha, F., Hanim, W. and Siswono, E. (2020) 'Penerapan Humanisme dalam Pembelajaran di Sekolah Dasar', *Indonesian Journal of Educational Counseling*, 4(2), pp. 117–124. doi:10.30653/001.202042.138.

Patricia Roy (1998) Professional development for coorperative learning *"Staff Development that makes a difference".* Albany: State University of New York Press, pp. 87.

Skeel Dorothy J. (1995) *Elementary Social Studies: Challenges For Tomorrow's World.* Harcourt Brace College Publisher.

Suherman (2021) *Monograf Implementasi Kebijakan Pendidikan Karakter.* Sumatera barat: CV. Insan Cendekia Mandiri, pp. 265.

Suwardi Endraswara (2006) *Metode, Teori, Teknik, Penelitian Kebudayaan, Ideologi, Epistimologi dan Aplikasi.* Tangerang Selatan: PT. Agromedia Pustaka, pp. 85.

Ujang Syarip (2018) *Menumbuhkan Pendidikan multikultural pada peserta didik melalui pembelajaran di kelas.* Sukabumi: Budi Mulia CV, pp. 147.

Yen, M.C. (2017) Ide Dan Gagasan Filsafat Humanis Master Chen Yen Hotmatua Paralihan', 14(1), *STAI Sumatera Barat.*

Yun Kyung Cha (2017) *Multicultural Education Policy in the Global Institutional Context.* Singapore: Library of Congress Control Number, pp. 13.

Religion, Education, Science and Technology towards a More Inclusive and Sustainable Future – Rahiem (Ed.)

Innovative diffusion and social penetration in education: WhatsApp as a medium of communication for thesis guidance between lecturers and students

E. Hadiyana*, Fahrurrozi & M. Fakhri
UIN Mataram, Indonesia

ABSTRACT: This study aims to describe the use of WhatsApp as a medium of communication between lecturers and students for thesis guidance. Analyzing how students connect with supervisors using WhatsApp is relevant, including consultation for thesis guidance. The subjects of this research are 45 students who have been undergoing thesis guidance since 2022. This study uses the theory of diffusion of innovation and the theory of social penetration to examine the data. According to the findings of this study, the longer students receive thesis guidance from lecturers via WhatsApp, the closer their relationship will be. On the other hand, some students fail to build communication and close relationships because they rarely get a response from the lecturer. This is due to the tight schedule of lecturers outside of teaching hours. In addition, the adoption of WhatsApp as an innovation for thesis guidance has not been utilized by all lecturers and students. The tendency to reject this innovation is caused by the convenience and habits of students. Several lecturers and students tend to be more comfortable with face-to-face thesis guidance.

Keywords: WhatsApp, communication media, thesis, lecturers and students

1 INTRODUCTION

Digitalization has been shaping modern society's imagination more and more since the 1990s. At the time, digital media had become one of humankind's obsessions. A seemingly infinite cycle that currently permeates the daily lives of billions of people involves trading emails, messages, texts, and WhatsApp messages, as well as liking and commenting on social media pages (Gabriel & Paola 2018).

The pace of technological advancement has accelerated in many ways. Industries, including those related to education, have developed because of technological advances (Morsidi et al. 2021). The use of technology for communication and information collection makes it significant in education. Take distant learning over the internet as an example. The utilization of contemporary technical instruments is preferred by learners. The employment of learning media greatly influences student responses, where these cutting-edge technical tools contribute to the effectiveness of learning. The utilization of digital media will make the learning process more efficient, interactive, and creative, as well as boost students' interest in and motivation for learning (Johnston et al. 2015).

WhatsApp is a highly well-liked program that can be found on all smartphone platforms today. In 2009, WhatsApp first made its mobile messaging service available to the general public. In just a few years, WhatsApp has soared in popularity across the globe. One in seven people in the world uses WhatsApp, according to a report by the company at the beginning of 2016. In contrast, WhatsApp's user base expanded to 1.3 billion in 2018, and each user

*Corresponding Author: Endanghatake18@gmail.com

DOI: 10.1201/9781003322054-22

sent and received an average of 65 billion messages daily. In the meantime, 88.7% of the population of Indonesia uses WhatsApp, which ranks third in the globe and is the most popular social media network in that nation, according to a Hootsuite (We Are Social) survey (Riyanto 2022). WhatsApp is a social media platform that provides instant messaging services that are also equipped with features such as sending text, images, videos, and voice messages (Rosenberg & Asterhan 2018).

Through the WhatsApp platform, instant messaging has become a primary avenue of communication and a means to maintain relationships (Blabst & Diefenbach 2017). WhatsApp's appeal has been linked to its ability to mimic the best face-to-face communication and the sense of intimacy it fosters because group members' messages flow in unison (Matassi et al. 2019). With the WhatsApp application, it will be easy for students to undertake the thesis guidance process. The guiding technique does not simply employ verbal communication, which is only through the narration of words by the professor, so that students have more conversations since they are also involved in exchanging information with lecturers and even fellow guidance buddies. In addition, students can also ask the professor in writing to address their concerns without having to meet the guidance lecturer directly (Ahmad et al. 2020).

As part of the thesis writing process, it is the job of a lecturer to ensure that students are able to generate good theses and are ready to be examined (Phillips & Pugh 2005). To date, there has been no formal guideline from UIN Mataram regarding the maximum number of times students can do thesis guidance. In addition, many authors find that many students do not realize the significant role of a professor; students only know the lecturers' duties when instructing and delivering help to students. But lecturers also have other obligations, such as personal development, community service, research, and academic support. On the other hand, lecturers also have duties for their families (Gregory 2011).

Previously, research related to the WhatsApp application had been published quite a lot in both national and international publications, but research related to the use of WhatsApp as a communication medium for thesis guidance for lecturers and students had not been found. The difference between the researcher's research and other research lies in the purpose of the research. Researchers emphasize the adaptation process of lecturers and students interacting during the thesis counseling process on WhatsApp.

The trend of utilizing WhatsApp Messenger for students and lecturers is inseparable from the findings of the author's observations on the growth rate and development of communication technology in students of the Islamic Communication and Broadcasting study program in 2022 at UIN Mataram. The goal of this study is to describe the usage of WhatsApp as a means of communication between professors and students of thesis counseling for the Islamic Communication and Broadcasting study program.

1.1 *Diffusion of innovation theory*

The notion of diffusion of innovation relates to how to transmit invention in a social system through specified channels over a set period. According to Rogers, members of the social system must go through four phases in making social decisions. At the first level, participants of the social system must comprehend the notion of innovation and its roles. The second stage is persuasion. At this stage, they establish an attitude toward innovation (optimistic or destructive) (beneficial or detrimental). The third step is decision-making, that is, deciding whether or not to embrace the innovation. And the last step is confirmation, that is, the implementation of innovations (Mahat et al. 2022).

The rate of innovation adoption is affected by several factors, such as communication routes, features of the social system, and the role of communicators. According to Rogers (1962) and Schiffman and Kanuk (2010), there are four concepts in innovation theory that might impact the adoption rate. (1) Relative advantage: How does a novel innovation differ from a prior one in terms of quality? The benchmark is how a person experiences the immediate effects of innovation. The sooner an invention is embraced, the greater the apparent profit. (2) Compatibility, the degree to which adopters' demands and societal values are considered while evaluating an invention. Innovation won't be embraced by society if it is deemed improper. (3) Complexity: The degree to which an innovation is deemed comparatively challenging to grasp and use. Difficulties in understanding and using innovations will be a hurdle in speeding the adoption of innovations. (4) Trialability: Where an invention is tried on a limited scale, through trials, adopters might find out the advantages and downsides of the innovation before it is wholly adopted (Syahadiyanti & Subriadi 2018).

1.2 *Social penetration theory*

Altman and Taylor were the ones who first proposed the social penetration idea. According to Altman and Taylor, communication is essential for fostering and preserving interpersonal connections. As a result, the theory of social penetration clarifies the significance of self-disclosure, communication, and intimacy in forming interpersonal relationships. SPT analyzes the development of interpersonal interactions in sociological literature. The "onion theory" refers to the various levels and types of information shared between the parties. The ability of each person to convey their attitudes toward one another, individual traits, ideas, and feelings will determine the nature of their relationship (Mangus et al. 2020). According to the penetration theory, self-disclosure can range from superficial to intimate. Broad and deep social penetration are two alternative conceptualizations of the same term. Depth refers to the level of closeness experienced during the social penetration process, which determines how comfortable a person is with disclosing particular areas of their private lives (Wulandari 2021). For instance, someone might mention their hometown as a surface-level detail but solely discuss their thoughts regarding parenting (Pennington 2021).

2 METHODS

To address current issues, this study adopts a qualitative descriptive method. This approach was chosen because, using the theory of social penetration, the researchers attempted to describe the phenomenon of communication between professors and students in the thesis guidance of the Islamic Communication and Broadcasting study program at Mataram State Islamic University. Students and lecturers of the Islamic Communication and Broadcasting Study Program, UIN Mataram, became the research subjects. From the Islamic Communication and Broadcasting study program, 45 students participated. In this research, purposive sampling is used (Etikan et al. 2016). In this study, observation, interviewing, and documentation approaches were employed to obtain data. Researchers utilize the observation method by traveling immediately to the research location and noting significant items linked to the research data required (Denscombe 2017).

3 RESULTS AND DISCUSSION

3.1 *Why professors and students should use WhatsApp as a guidance tool*

According to the findings of the researcher's study, which involved conducting interviews with 10 WhatsApp users, it was discovered that WhatsApp was used both as a communication tool and as a source of information. According to Rogers (2003), several factors influence how quickly a product or concept gets adopted, one of which is that the idea offers benefits and is in line with the adopter's values, beliefs, and requirements (Vargo et al. 2020). Because it is thought to be capable of supplying information, lecturers and students have adopted the WhatsApp program. This is partly due to WhatsApp's appealing and functional design, which meets the needs of modern society. It is also used for entertainment purposes, such as sharing photos or pictures without time or space restrictions, and as a substitute between lecturers and students to carry out thesis guidance. The WhatsApp application can also make interaction easier because users can add additional friends to their WhatsApp groups, making it more straightforward for them to get to know one another. Developing an interpersonal communication process will be crucial for users since it will make it simpler and more convenient for them to communicate freely. WhatsApp creates social connections between targeted individuals, but only if they are connected to the service. The social penetration theory predicts that users of the WhatsApp program will go through a process of recognizing and adapting to one another. Students from UIN Mataram KPI experienced this. It is necessary for them to get to know one another because the guidance group includes not only KPI students but several students from other study programs. How the process relates to others has been outlined in the social penetration theory. Numerous progressive methods take place. Lecturers and students both experience this. Discussion is the time for adaptation. However, students continue to pay attention to language and morals when conversing. The existence of this adaptation process fosters a sense of belonging among students.

3.2 *Lecturers and students use WhatsApp as a medium for exchanging information*

Furthermore, based on observation findings in the guidance group, the supervisor usually delivers information about the thesis guidance timetable, thesis improvement, social critique, outpouring of heartfelt humor, and motivational lines. This is in keeping with what is indicated by Altman and Taylor (1975). WhatsApp communication as a medium is starting to be observed by many people and will undoubtedly cause many perceptions regarding the usage of WhatsApp media as a communication medium, especially among students. The theory of social penetration examines the different interpersonal behaviors that occur in evolving interpersonal relationships (Taylor & Altman 1975), precisely the amount of information sharing. This step applies to this study, where lecturers and students share information relevant to the thesis guidance schedule and the difficulties faced during the guidance procedure. Preliminary observations of the study suggested that students were often frustrated if their lecturers could not be found due to the preoccupation of lecturers with different heavy academic matters. The lack of time-intensity and boring guidance approaches will only make students unenthusiastic about guidance. On campus, researchers typically find many students waiting for lecturers without confidence in the lecturer's presence. Students typically, before coming to campus, have called the lecturer first, but sometimes their telephones are not picked up and their SMS is not answered. On the other side, several students also canceled their meetings with lecturers. The lecturers were ready to guide, but the students did not attend. This makes the communication between lecturers and students poor. This dilemma then makes old students complete their thesis. To achieve this, students must maintain, improve, and develop harmonic communication with their lecturers. Such as using WhatsApp as a medium for thesis consultation because lecturers are tough to find.

In addition, through the results of an interview with Vera Yuniar, he explained that the contact he made with the supervisor was not short, but the communication lasted a long time and continued until the thesis was ready to be verified. This thesis writing guidance, which usually lasts for months, makes interaction and communication closer between students and supervisors. In this process, WhatsApp makes students feel close to the lecturer so that when they meet face-to-face it is not awkward; even the students do not hesitate to say hello first. This is in line with what was said by Taylor and Altman (1975): close interpersonal relationships become intimate when there is a thorough flow of information and the length of time spent talking (Moon et al. 2019).

Students also commonly experience challenges when writing a thesis. According to the informant's confession, they are confused about determining the proper procedure, do not understand how to examine data, and even have problems determining the central theory. Finally, students typically approach the lecturers to ask for solutions to their concerns. Therefore, the availability of this problem-solving activity strengthens the interaction between lecturers and students. Students are then not bashful about discussing the challenges they experienced during the study process. Self-disclosure is described as giving personal information about oneself to others (Collins & Miller 1994), which contributes to the development of relationships in the idea of social penetration (Skjuve et al. 2021).

However, on the other side, the formation of relationships in the thesis guidance process using WhatsApp does not have much influence because some students also do not get any reaction from their supervisors. In reality, their chats are rarely read, or even merely read without response. This arises because instructors have many other duties, such as community service, attending academic activities, and even being busy with family. Finally, college students are unable to build more intimate connections. In the end, WhatsApp failed to become a communication medium between lecturers and students.

On the other hand, WhatsApp is not so efficiently employed as a medium for thesis guidance. Several students during the interview admitted to experiencing this; they prefer face-to-face guidance. Not only students but also lecturers. There are some senior lecturers who want pupils to be mentored personally because most senior lecturers do not use WhatsApp very much. WhatsApp is utilized only as a channel of communication to ask about the location of lecturers. As a result, as stated by Takada and Jain (1991); Tellefsen and Takada (1999), social groups made up of people from the same culture have varied dispositions to adopt or reject innovations (Desmarchelier & Fang 2016). Differences linked to the adoption of the WhatsApp application are attributable to various variables of habit and comfort of professors and students. There are academics who prefer in-person mentoring because their spare time is relatively limited. On the other hand, instructors who have functional responsibilities on campus and do not have much free time often use WhatsApp as a medium of guidance.

In the background of the preparation of the thesis, students typically face tension. The form of stress that students commonly face is academic stress. During the thesis preparation process, several barriers are experienced by students, both external and internal. One of them is problems with the lecturers (Nabila & Sayekti 2021). To lessen the risks posed, stress management is essential through physical relaxation, happy emotions, positive thoughts, positive behaviors, positive relationships, and spirituality (Nyumirah & Novianti 2021). Building a positive relationship through WhatsApp with a lecturer is one form of stress management carried out by students in this study. That is, by expressing the issues they confront during the thesis writing process. In this stage, pupils take out self-disclosure. This is very significant in interpersonal connections since it is a way of disclosing personal information that is not yet known (Faidlatul Habibah et al. 2021).

4 CONCLUSION

The findings of this study provide information that WhatsApp is considered user-friendly and broadly facilitates students' communication with lecturers during thesis guidance. In addition, WhatsApp is a medium of communication for lecturers and students. The intimacy of relationships is built during the guidance process. On the other hand, there are some students who fail to build communication skills because the lecturers are very busy. WhatsApp is widely used as an innovation in learning because it is considered efficient and time-saving. But there are some senior lecturers who do not understand technology and more often receive direct thesis guidance. The focus of this research is still divided, so the results of the research are not maximal enough. The hope is that the use of WhatsApp as a learning medium will continue to be innovated at several universities. The use of cheap data and its easy features will make WhatsApp continue to be used in the future. Moreover, students feel the new innovation to be helpful.

REFERENCES

Ahmad, S., Zulfikar, T. and Hardiana, F. (2020) 'The use of social media whatsapp among english education students for solving thesis writing problems', *Humanities & Social Sciences Reviews*, 8(3), pp. 447–455. doi: 10.18510/hssr.2020.8348.

Astarcioglu, M. A. *et al.* (2015) 'Time-to-reperfusion in STEMI undergoing interhospital transfer using smartphone and WhatsApp messenger', *American Journal of Emergency Medicine*, 33(10), pp. 1382–1384. doi: 10.1016/j.ajem.2015.07.029.
Blabst, N. and Diefenbach, S. (2017) 'WhatsApp and wellbeing: A study on WhatsApp usage, communication quality and stress', *Proceedings of the 31st International BCS Human Computer Interaction Conference (HCI 2017) (HCI)*. doi: 10.14236/ewic/hci2017.85.
Denscombe, M. (2017) *The Good Research Guide for small-scale Social Research Projects*. 6th ed. London: Open University Press.
Desmarchelier, B. and Fang, E. S. (2016) 'National culture and innovation diffusion. Exploratory insights from agent-based modeling', *Technological Forecasting and Social Change*, 105, pp. 121–128. doi: 10.1016/j.techfore.2016.01.018.
Etikan, I., Musa, S. A. and Alkassim, R. S. (2016) 'Comparison of convenience sampling and purposive Sampling', *American Journal of Theoretical and Applied Statistics*, 5(1), pp. 1–4. doi: 10.11648/j.ajtas.20160501.11.
Faidlatul Habibah, A., Shabira, F. and Irwansyah, I. (2021) 'Pengaplikasian teori penetrasi sosial pada aplikasi online dating', *Jurnal Teknologi Dan Sistem Informasi Bisnis*, 3(1), pp. 44–53. doi: 10.47233/jteksis.v3i1.183.
Gabriel, B. and Paola, M. (2018) *Myths and Counter Hegemonic Narratives in Digital Media History*. New York: Routledge.
Gregory, S. T. (2011) 'Black faculty women in the academy: History, status, and future', *The Journal of Negro Education*, 70(3), pp. 124–138.
Johnston, M. J. *et al.* (2015) 'Smartphones let surgeons know WhatsApp: An analysis of communication in emergency surgical teams', *American Journal of Surgery*, 209(1), pp. 45–51. doi: 10.1016/j.amjsurg.2014.08.030.
Mahat, N. S. *et al.* (2022) 'E-Procurement adoption in the malaysian construction sector: Integrating diffusion of innovations and theory of planned behaviour framework', *Jurnal Kejuruteraan*, 34(03), pp. 347–352. doi: https://doi.org/10.17576/jkukm-2022-34(3)-01.
Mangus, S. M. *et al.* (2020) 'Examining the effects of mutual information sharing and relationship empathy: A social penetration theory perspective', *Journal of Business Research*, 109(December 2019), pp. 375–384. doi: 10.1016/j.jbusres.2019.12.019.
Matassi, M., Boczkowski, P. J. and Mitchelstein, E. (2019) 'Domesticating WhatsApp: Family, friends, work, and study in everyday communication', *New Media and Society*, 21(10), pp. 2183–2200. doi: 10.1177/1461444819841890.
Moon, H. *et al.* (2019) 'Peer-to-peer interactions: Perspectives of Airbnb guests and hosts', *International Journal of Hospitality Management*, 77(August), pp. 405–414. doi: 10.1016/j.ijhm.2018.08.004.
Morsidi, S. *et al.* (2021) 'WhatsApp and its potential to develop communication skills among university students', *International Journal of Interactive Mobile Technologies*, 15(23), pp. 57–71. doi: 10.3991/ijim.v15i23.27243.
Nabila, N. and Sayekti, A. (2021) 'Manajemen stres pada mahasiswa dalam penyusunan skripsi di institut pertanian bogor', *Jurnal Manajemen dan Organisasi*, 12(2), pp. 156–165. doi: 10.29244/jmo.v12i2.36941.
Nyumirah, S. and Novianti, E. (2021) 'Manajemen stres di masa pandemi Covid-19', *Buletin Kesehatan: Publikasi Ilmiah Bidang kesehatan*, 5(2), pp. 91–100. doi: 10.36971/keperawatan.v5i2.91.
Pennington, N. (2021) 'Extending social penetration theory to Facebook', *The Journal of Social Media in Society*, 10(2), pp. 325–343. Available at: https://www.thejsms.org/index.php/JSMS/article/view/973/549.
Phillips, E. and Pugh, D. S. (2005) *How to get a PhD: A handbook for students and their supervisors*. Maidenhead: Open University Press.
Riyanto, A. D. (2022) *Hootsuite (We are Social): Indonesian Digital Report 2022*. Available at: https://andi.link/hootsuite-we-are-social-indonesian-digital-report-2022/.
Rosenberg and Asterhan (2018) 'WhatsApp, Teacher?', Student Perspectives on Teacher-Student WhatsApp Interactions in Secondary Schools. *Journal of Information Technology Education: Research*, 17(06), pp. 205–226. doi: 10.28945/4081.
Skjuve, M. *et al.* (2021) *My Chatbot Companion – a Study of Human-Chatbot Relationships*.
Syahadiyanti, L. and Subriadi, A. P. (2018) 'Diffusion of Innovation Theory Utilization Online Financial Transaction: Literature Review', *Journal of Economics and Financial Issues*, 8(3), pp. 219–226. Available at: https://www.econjournals.com/index.php/ijefi/article/view/6438/pdf.
Vargo, S. L., Akaka, M. A. and Wieland, H. (2020) 'Rethinking the process of diffusion in innovation: A service-ecosystems and institutional perspective', *Journal of Business Research*, 116 (December 2018), pp. 526–534. doi: 10.1016/j.jbusres.2020.01.038.
Wulandari, E. (2021) 'Utilization of the tiktok video application as a means of showing existence and self-disclosure of teenagers on social media', *International Journal of Social Science And Human Research*, 4(9). doi: 10.47191/ijsshr/v4-i9-48.

Religion, Education, Science and Technology towards a More Inclusive and Sustainable Future – Rahiem (Ed.)

Integration and interconnection between Al-afidah and neurosciences in the human learning process

Syukri*, E.A. Subagio & Zulyadain
UIN Mataram, Indonesia

ABSTRACT: This paper revealed that Qur'ān Surah 16:78 about the process of human learning seems dichotomous with neuroscience, even though both have something in common with each other. The purpose of this study was to find new scientific varieties resulting from the hybridization of the interpretation of al-Quran and neuroscience. This research is an interdisciplinary study with a qualitative approach in the form of library research and human neurosurgery. The data taken from the existing literature, the relevant verses of the Qur'ān, and scientific evidence of neuroscience were analyzed using the content analysis approach. The results of the research revealed that the term *al-afidah* corresponded to human neurosurgery because the learning process from the cochlea (*as-sama'*), the retina, and the dermis nerves (*al-abṣār*), and all of the neurons transfers the signal to the frontal lobe (*al-afidah)* to be processed in the cerebral cortex to determine the extent whether it is a logical science or not.

Keywords: *al-afidah*, neuroscience, cerebrum, and human learning

1 INTRODUCTION

In the early 21st century, Western and Muslim scientists were intensively researching the human brain. Johnson (2014) pointed out that neuroscience is relatively new to education and educational psychology. Western scientists recently paid heed to research on the integration and interconnection of the human brain and neuroscience with educational science and its application in the classroom (Edelenbosch et al. 2015; Masson & Brault Foisy 2014). On the contrary, Muslim scientists are struggling to identify the word *'aqal* (brain) in the Qur'ān and seeking to research integration and interconnection with neuroscience (Suyadi 2019; Suyadi & Jailani 2021).

To date, the statements of the Qur'ān and neuroscience about the human learning process seem dichotomous despite their intertwined relationships. Researchers believe that the term *al-afidah* in al-Qur'ān Surah an-Nahl 78 has been integrated and interconnected with neuroscience in the human learning process. Gredler (2009) opined that the neocortex consists of four lobes, each of which is involved in processing a different type of information from other parts of the brain. Visual information from the thalamus is processed further in the occipital lobe, from which auditory information is then processed in the temporal lobes, while the parietal lobes process information about tactile sensation. Then, each lobe serves as a communication link between the prefrontal cortex, which is involved in goal setting and planning. This shows that the terms *al-afidah* and prefrontal cortex both lead to the final

*Corresponding Author: syukyun19@gmail.com

DOI: 10.1201/9781003322054-23

decision that the science being processed is true and convincing science. Therefore, there is an urgent need to study the integration and interconnection between al-Qur'ān and neuroscience in the human brain.

2 METHODS

The research approach is hybrid research, namely qualitative research in the form of library research (Creswell 2014) and human neurosurgery research in the form of performing surgery on the human brain. Two sources of data were used in this study, verses from the Qur'ān related to *al-afidah*, educational books and journals in the field of neuroscience, and results of observations of human neurosurgery. Three kinds of techniques were used to collect data: (a) the *Maudhu'iy* method, which is a way of analyzing the themes of the verses of the Qur'ān related to the human learning process (al-Farmawi & al-Hayy 1977); (b) neurosurgery method, which was done by observing the working relationship of the cerebrum with the cerebellum and brainstem; and (c) the interrelationships between the temporal lobes, occipital lobes, parietal lobes, and frontal lobes with cerebral and prefrontal cortex networks. The three methods set out to look for the integration and interconnectedness of the verses of the Qur'ān with the overly complex structure of human neural networks.

3 FINDINGS

There are two findings from this study. First, according to Surah an-Nahl/16 verse 78, hearing function (*as-sama'*) is the first step in the process of human learning and is a source of information or knowledge. The information is then verified by the four organs – eyes, nose, tongue, and skin (*al-abṣār*), and the outcome of this verification is transmitted to the brain for thorough processing so that it can make the final decision that the information or knowledge verified by the senses can be regarded as a convincing scientific fact (*al-afidah*) (Syukri 2021). In general, the word *afidah* in the al-Quran is always preceded by actual things. The word *afidah* is mentioned in the al-Quran 15 times in Surah 17:36, 28:10, 53:11, 25:32, 11:120, 23:78, 67:23, 32:9, 46:26, 6:110, 6:113, 14:37, 14: 43, and 104:7. The functions of *as-sama', al-abṣār*, and *al-afidah* become functions that are mutually integrated and interconnected with each other to obtain a convincing human understanding.

Second, according to the findings in human neurosurgery, the human learning process starts with the function of the cochlea in the inner ear, which then sends the signal to the temporal lobe. To see the object, as evidenced by the retina, it sends the signal to the occipital lobe. Based on the results of interviews with two students from the Islamic University of Mataram said that teaching with the help of trees to explain the meaning of history, they still remember the meaning even though they studied this in 2018 (Interviewed with Ais Saputra, October 2, 2022). On the other hand, a patient named Nur at a mental hospital in Surabaya is unable to remember how much money he spent the day before (Observation, September 20, 2022). This shows that the brain can remember well when something is seen and touched.

Furthermore, human learning has a strong understanding of the obligation to touch by the dermis the object or objects as a source of strengthening knowledge, which is then sent a signal to the parietal lobe, and the neocortex and prefrontal cortex are the core of the highest human function in utilizing serves mechanically between nerve cells (Gredler 2009, p. 79). Finally, the following figure shows the integration and interconnection of *al-afidah* with neuroscience.

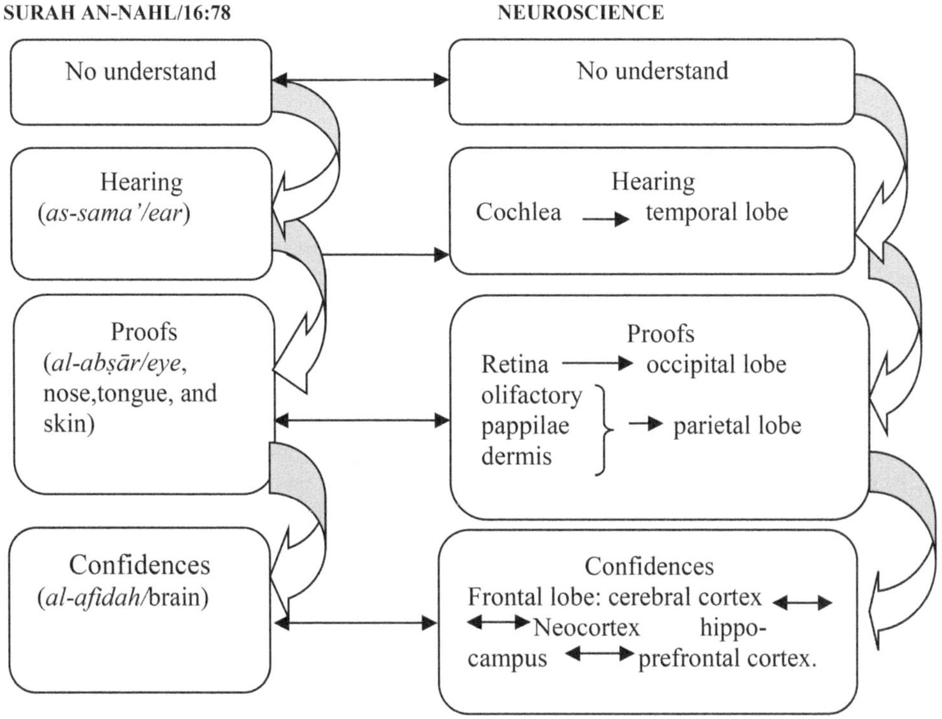

Figure 1. Human learning process.

The figure above shows the integration and interconnection of human learning between *al-afidah* and neuroscience, which are interconnected, because both *al-afidah* and neuroscience, especially the frontal lobe, are to be supported by the five senses as a source of information from the cochlea (*as-sama'*) and factual knowledge, which predominantly comes from the retina and the dermis (*al-abṣār*), then the cochlea sends signals to the temporal lobe, the retina sends signals to the occipital lobe, and the dermis sends signals to the parietal lobe. The three lobes all send their respective signals to the frontal lobe to be processed through the interconnection between the cerebral cortex and neocortex with the hippocampus and decided by the prefrontal cortex to be a logical and convincing scientific truth (*al-afidah*).

4 DISCUSSIONS

The commentators (*mufasir*) agreed that Surah an-Nah/16: 78 explains the process of human learning, starting from not knowing to hearing function (*as-sama'*) through the ear to get information (al-Marāghī 1981), but by the word *al-abṣār*, the researcher means evidence, not only sight, as is generally the case with commentators, but also includes smell, taste, and touch. This is reinforced by al-Ṭabāṭabā'ī that the word *al-abṣār* can mean smell, taste, and touch. He said that the senses that support the function of hearing and evidence are sight, touch, taste, and smell (al-Ṭabāṭabā'ī 1991) This means that the word *al-abṣār* can be interpreted as seeing, smelling, tasting, and touching. While the word *al-afidah* generally interprets the heart (Ibn Kathīr 1991), researchers interpreted it as confidence. The researcher's opinion was strengthened by al-Rāzī's statement that *al-afidah* is the acquisition of true knowledge and convincing science (al-Qurtūbiy 1990; al-Rāzī 1990; al-Ṭabāṭabā'ī

1991; Hayyan 1993); al-Ṭabāṭabā'ī (1991) interpreted *fuad* as thinking and deciding confidently without any hesitation (1991, 93). We have interviewed Prof. Thib Raya (a lecturer in Tafsir), who said that the meaning of *fuad* is the heart, but if you have another opinion and it is supported by a commentator (mufasir), there is no problem in believing in that (Interview with Thib Raya, October 4, 2022). Thus, the word *al-afidah* is more suitable to mean confidence, and this word implies the process of processing stages of information – from the *as-sama'* and *al-abṣār* functions then to *al-afidah* as the final function to determine whether the knowledge obtained is true and convincing or not. In this case, the word *al-afidah* conforms with the meaning in neuroscience. However, some Muslim scientists disagreed with the researchers that neuroscience is more accurately equated with the word *nāsiyah*, which means the crown or brain (Suyadi 2019). They believe that the word *nāsiyah* is suitable to mean the crown or brain, as stated in Surah al-'Alaq/96:16. Regarding the position of *fuad*, we are of the opinion that *fuad* or *al-afidah* is in the brain. This idea is supported by Qurtūbiy, who says that *fuad* has its place in the brain (*ad-dimāg*) (al-Qurtūbiy 1990).

However, the results of research in neuroscience have demonstrated that the function of hearing information is established when the newborn is in the womb. Rose's research (2005, p. 80) demonstrated that very slow waves of electrical activity can be seen at the brain's surface of babies in the womb about the third prenatal month and that continuous electrical activity seems to occur in the brain stem about the fourth month. This activity indicates that the neurons are signaling to each other at an early stage in the development of the fetus. After birth, the human brain develops rapidly because synapses in the visual auditory cortices attain a maximum density that is 150% of adult levels between age 4 and 12 months (Huttenlocher 1994). Furthermore, the human brain undergoes a lengthy period of development that continues into the early twenties. Growth in the brain refers to the lengthening and branching of the neuronal axon. This process occurs in different regions, including the subcortical structures such as the thalamus, hippocampus, amygdala, and cerebellum. The hippocampus fulfills a critical role in learning and memory. In contrast, the neocortical lobes are primarily engaged in processing and integrating different types of information. This statement is reinforced by Ormrod's research results (2016), which concluded that the frontal lobes are active when we pay attention to and think about new information and events, and all of the lobes of the cortex may be active to a greater or lesser extent in interpreting new input in light of previously acquired knowledge.

5 CONCLUSIONS

Vaughn et al. (2020), proved that the parietal lobe/precuneus and visual cortex are attributed to working memory, executive function, and visual attention. According to Sloan and Norrgran (2016), the hippocampus organizes short-term memories for long-term storage. The short-term memories are moved to the neocortex, where they are arranged, connected, and consolidated into long-term memory. The prefrontal cortex also plays a role in the storage of memory, adding considerable value to the memory. Finally, the prefrontal cortex adds content such as human appropriateness, judgment, usefulness, and connectedness. The cerebral cortex, neocortex, and prefrontal cortex in neuroscience *al-afidah* have the same function in the human learning process. to judge and to convicted a logical science and strong memory. Therefore Tommerdahl (2010) suggested the need to build a bridge between neuroscience and practical learning in the classroom. The result of Degen's research (Degen 2011) proved that the learning process of using students' lean experiences in a thought-provoking environment can lead to effective teacher teaching. The results of Caine et al. (2009) research proved that to develop learning strategies to stimulate thinking, the way of learning is adapted to the design of the brain to learn naturally.

Finally, this research is not without limitations. Further research needs to collect a large amount of data through observation and narratives.

REFERENCES

al-Farmawi,A. and al-Hayy, 1977. al-Bidayah fi al-Tafsir al-Mawdu'i: Dirasat Manhajiyat Mawdu'iyah. *Maktabah Jumhuriyah, Mesir.*

al-Marāghī, A.M., 1981. *Tafsīr al-Marāghī, Juz 14.* Beirut.

al-Qurtūbiy, A.A.M. bin A. al-A., 1990. *Tafsir al-Qurtubi al-Jāmi' li Ahkami al-Qur'ān, Juz 14.* Beirut: Dār al-Sya'abi.

al-Rāzī, F. al-D., 1990. *Tafsīr al-Kabīr aw Mafātiḥ al-Ghayb, Juz 20.* Beirut.

al-Ṭabāṭabā'ī, S.M. Ḥusayn, 1991. *Al-Mīzān fī Tafsīr al-Qur'ān, Juz 13.* Beirut.

Caine, R.N., Caine, G., McClintic, C.L., and Klimek, K.J., 2009. *12 Brain/Mind Learning Principles in Action: Developing Executive Functions of the Human Brain.*

Creswell, J.W., 2014. *Research Design: Qualitative, Quantitative, and Mixed Methods Approaches.*

Degen, R.J., 2011. *Brain-Based Learning: The Neurological Findings About the Human Brain that Every Teacher Should Know to be Effective.*

Edelenbosch, R., Kupper, F., Krabbendam, L., and Broerse, J.E., 2015. *Brain-based Learning and Educational Neuroscience: Boundary Work.*

Gredler, M.E., 2009. *Learning and Instruction: Theory into Practice.*

Hayyan, M. bin Y. al-S., 1993. *Tafsir al-Bahri al-Muhit, Juz 5.* Beirut: Dār al-Kutub al-Ilmiyah.

Huttenlocher, P.R., 1994. *Synaptogenesis in Human Cerebral Cortex.*

Ibn Kathīr, A.-I.A. al-F.I., 1991. Tafsīr al-Qur'ān al-Aẓīm.

Johnson, A., 2014. *Educational Psychology: Theories of Learning and Human Development.*

Masson, S. and Brault Foisy, L.-M., 2014. *Fundamental Concepts Bridging Education and the Brain.*

Sloan, D. and Norrgran, C., 2016. *A Neuroscience Perspective on Learning.*

Suyadi, S., 2019. Hybridization of islamic education and neuroscience: Transdisciplinary studies of 'Aql in the Quran and the brain in neuroscience. *Dinamika Ilmu.*

Suyadi, S. and Jailani, M., 2021. The concept of 'Aql and brain in the Quran and neuroscience: A concept analysis of Nāṣiyah in Salman's Tafsir.

Syukri, S., 2021. *Teori Belajar al-Bayan Berbasis al-Qur'an.*

Tommerdahl, J., 2010. *A Model for Bridging the Gap Between Neuroscience and Education.* Oxford Review of Education, 97–109.

Vaughn, A.R., Brown, R.D., and Johnson, M.L., 2020. Understanding conceptual change and science learning through educational neuroscience. *Mind, Brain, and Education.*

Religion, Education, Science and Technology towards a More Inclusive and Sustainable Future – Rahiem (Ed.)

Shifting education and technology through the lens of constructivism

A. Saeful*
STAI Binamadani, Tangerang, Indonesia

M. Huda, S.U. Masruroh & D. Khairani
UIN Syarif Hidayatullah, Jakarta, Indonesia

S. Muttaqin
STAI Alhikmah, Jakarta, Indonesia

ABSTRACT: This paper examines education and technology within the framework of constructivism. The theory of constructivism does not view knowledge as a fact or information passively received by learners. It is more of a formulation of actively acquired knowledge. This paper uses a qualitative descriptive approach, in which the existing data is explained and analyzed in depth. This research uses various types of literature that are relevant to the problem being discussed, both from books and journal articles. The paper reveals that constructivism learning theory can be aligned with technology-based learning, where in the process educators act as facilitators. Constructivism learning theory is a theory that can be developed on an ongoing basis, especially in a contemporary context that is no stranger to technology. This theory can be a solution to the development of technology-based learning.

Keywords: technology, education, constructivism

1 INTRODUCTION

Education is an important means of acquiring knowledge (Samsul Hadi 2017). Today, education has shifted to a new era often referred to as the digital age. In this era, technology is an important thing that must be embraced and mastered (its use) by all parties involved in education. In other words, the use of technology is very important in today's world of education (Hasyim 2018).

On the other hand, technology, to some extent, has become one of the indicators for measuring the quality and development of education. Thus, assuming an educational institution that does not have technological progress is backward may not be entirely wrong (Srinivasan & Khrisna 2017) because the current context considers not only technology as a supporting tool in educational institutions, but also as the main need of such institutions.

The development of technology has changed human behavior toward the world today. They have adapted and adjusted from technological illiteracy toward technological literacy. This perspective on the change in human behavior is in line with the philosophical view of progressivism, which believes that human beings have the ability to overcome problems in themselves (Barnadib 1997), including those related to changing times and the use of

*Corresponding Author: achmadsaeful@stai-binamadani.ac.id

 DOI: 10.1201/9781003322054-24

technology. Thus, no matter how advanced the age of human beings on this earth, they will be able to adapt to changes. On the other hand, progressivism places great emphasis on a forward-looking view, that is, considering past events as a record of the past that its existence is useful to be a support for the present life.

Returning to the educational context and passing it through with the philosophical theory of progressivism, it can be said that the form of learning changing from conventional (behavioristic) to modern/contemporary (constructivism) mode is not entirely contradictory. Thus, the change in the current form of learning that puts technology first is also important from the philosophical perspective of progressivism, especially since this kind of change can make students more dynamic and creative in learning. This assumption is in line with Gutek's (1974) view that progressive learning can foster initiative for learners and build creative reasoning on it. Therefore, the use of technology in education helps students grow and build these initiatives and creative reasons.

2 METHODS

This research uses a qualitative descriptive approach, in which data are obtained, described, and analyzed in depth (Zuriyah 2006). This is a type of library research that collects data from relevant sources, both from books and journal articles. The data collection method in this study was carried out by selecting and grouping data related to the research title. After this method of data collection is carried out, the author does two things: (i) Identify problems and develop them in the form of fundamental questions related to the problems to be studied; (ii) Analyze a wide variety of data that has been collected, read, and observe using inductive techniques. The use of inductive techniques is used as an initial foothold in conducting this research. Data analysis techniques use qualitative content analysis techniques. In this analysis, all the analyzed data is in the form of text. In this case, it is in the form of texts related to the title of the study. Qualitative content analysis is used to find, identify, and analyze texts or documents to understand the meaning, significance, and relevance of the text or each document under study, to give birth to a clear understanding (Afifudin & Beni Ahmad Saebani 2009). The study includes the questions: What is the role of technology in education? How is the actualization of technology in the learning model of constructivism done?

3 RESULTS AND DISCUSSION

Technology can be understood as a tool used by humans to produce something (Syukur, 2008). Technology can also be perceived as an auxiliary tool for humans to make their work easier or more efficient (Arifin & Setiyawan 2012). Technology is often associated with information because one of the functions of technology is to search for information. In the educational environment, technology is widely used to find information, such as looking for references to learn and accessing actual information about certain educational issues (Lestari 2018).

3.1 *Education, knowledge, and technology*

Education is inseparable from knowledge; just like the two sides of a coin define the coin itself. This is mainly because the main purpose of education, in addition to creating human resources/students with noble character, is to have knowledge (Hendrowibowo 1994). In today's modern era, knowledge travels very fast. New things and all kinds of information and knowledge can be easily accessed by anyone. This condition is something that cannot be separated from the existence of technology (Dwiningrum 2012). The existence of technology provides easy access for everyone to various sciences that are developing today.

With the rapid development of knowledge due to technology, educational institutions must adjust to the development of existing technology (Jamun 2018). This attitude is in line with the theory developed by the philosophy of progressivism that requires educational institutions to open themselves to all existing developments (Muhmidayeli 2011), including technological developments. Educational institutions should not restrict students from using technology and be apathetic toward it. Apathy or ignorance toward technology, in some cases, is still a problem in education, especially in Indonesia.

The development of technology, especially in the field of education, has created a condition under which every educational stakeholder must adapt and adjust to it, as well as take advantage of it. This means that every educational institution is required to have reliable human resources in the field of technology, both at the level of educators, students, and educational personnel. Technology in education can also change the learning situation, which was initially monotonous and focused on the teacher, to become more dynamic and student-oriented. Through technology, teachers can provide creative learning, such as giving questions through the Kahoot application, providing material using virtual applications (Zoom, Google Meet, and Google Classroom), and providing learning and assignments through podcast media (Jamun 2018).

On the other hand, technology can also be a means for educators and students to access knowledge because technology can facilitate this process (Andriani 2015). This type of practice is possible in learning; for example, educators can ask students to write resumes using certain data (books and articles) available online. Technology allows educators and students to find a variety of additional resources to enrich their knowledge and support their learning process.

The existence of technology in education requires open-minded human resources who care about its development. They should not be apathetic about it and develop a positive interest in it. Awareness of the important use of technology can provide new experiences for educators and students in learning (Jamun 2018).

3.2 *Education and technology from the lens of constructivism*

The use of technology in education is very much in line with the theory of constructivism. This theory requires everyone (learners) to actively participate in building and developing knowledge for themselves (Schunk 2012). Constructivism requires building knowledge independently. In this context, learners must actively participate in learning, not just mechanically collect facts, and take responsibility for their own learning outcomes. In addition, students also make reasoning for what they have learned by looking for meaning, comparing it with what they already know, and constructing all the knowledge they already have.

However, the position of the educator is only to help provide learning facilities and situations so that the construction process of students in acquiring knowledge can be carried out and become more effective. Knowledge cannot be transferred from educators to learners, as long as learners are not active in constructing their own knowledge through reasoning activities (thinking deeply) (Suparlan 2019).

In reasoning activities, the entire process of knowledge produced by learners is unlikely to come entirely from educators. If the educator tries to convey concepts and ideas to the learners, then the delivery must be interpreted and constructed by the learners themselves through their various experiences. Sometimes the knowledge taught by educators is not captured properly by students. Knowledge must be interpreted and constructed by the learners themselves. This shows the accuracy of the constructivism view, which states that knowledge cannot simply be transferred to learners, but must rather be interpreted and constructed.

The constructivism approach greatly allows for better learning spaces for student engagement. Thus, students can be better able to actualize themselves in learning, not be afraid to express opinions, and get used to conveying new knowledge obtained from the results of thinking and reasoning activities. In different expressions, the constructivism approach used in learning will allow the learners to elaborate on and explore each piece of knowledge they gain. (Bruning et al. 2004) The implication is that learners can avoid any form of knowledge that is doctrinal in nature. The ability to avoid doctrinalization is due in constructivism, as the learning center is no longer controlled by the educator, but the center is in the learner.

In learning constructivism, students are always trained to be individuals who can think innovatively. This innovative thinking ability is carried out by educators by giving project-based tasks, such as asking students to make environmental-based papers or artworks or videos that contain motivational or positive messages. These kinds of tasks will stimulate the learners' ideas (thinking power) to be creative and innovative in completing the tasks.

In addition, another achievement that the learning constructivism approach wants to produce is the independence of students (students) from educators (teachers). The attitude of dependence in learning is an attitude that will make students uncreative and will have an impact on their inability to use their minds fully (De Kock *et al.* 2004). One of the main keys to learning is to build creativity, without which learning will be monotonous and boring. It is this kind of learning condition that constructivism rejects. Because the estuary of constructionism itself is independence in learning.

According to Jean Piaget, one of the figures who is considered the founder of constructivism (Sridevi 2008), students have the ability to adapt to their environment. Thus, the learning model with a constructivism approach is suitable for building creativity in each student. This is where educators need to realize that the learner is not an empty vessel that must be filled with water continuously, but rather deserves to be seen as a human being having a mind, and that with that thought he is able to develop his creative reasoning. So, in constructivism learning, the main task of the teacher is to act as a facilitator whose role is to assist students in carrying out learning, while the active process of finding knowledge is handed over to students. Thus, from the perspective of constructivism, learners are directed to be more active in acquiring knowledge under the supervision of teachers or educators. The role of the educator is to ensure the validity of knowledge if the student acquires false or misleading knowledge. In this case, the educator can supervise and direct him on the right track, especially when students use technology unwisely. (Rangkuti 2014; Wing & Mui 2002).

The use of technology in constructivism learning, on the one hand, can benefit learners because they can adapt to the development of the times that today cannot be separated from the use of technology. On the other hand, when its use is not directed, it can make learners stuck on behaviors of a negative nature. The presence of technology gives birth to negative influences on students, such as being used to accessing negative news or sites or spreading fake news (hoaxes), and it is more widely used as a means of playing than learning. Thus, although constructivism provides learning independence to students through the use of technology as a means of learning, the presence of educators is still necessary. Their existence (educators) can be a shield to counteract the negative behavior of learners toward the use of technology.

In the current context, constructivism learning models are more relevant and applicable, particularly within the framework of the use of technology in education. Through the use of technology in educational institutions, students become more liberated to actualize their abilities (without being limited by classes) in acquiring knowledge. In other words, the use of technology will lead students to become more independent in carrying out the learning process to gain knowledge. Independence in learning is a core concept of constructivism learning theory.

4 CONCLUSION

Technology develops over time and affects every aspect of human life, including education. Therefore, educational institutions must adapt and adjust in accordance with technological developments. Technology should be used to support the learning process because technology can help students and teachers carry out the learning process more easily. More importantly, the use of technology provides more opportunities for students to explore the world through online sources. This is an important idea that can be achieved using technology. Theoretically, the learning process that provides more freedom for students to build and acquire knowledge is in line with the theory of constructivism learning. In the learning of constructivism, knowledge is acquired, discovered, and built by learners through active questions. Thus, the use of technology in education, from the perspective of constructivism, is very supportive of creating a better collaborative learning environment. The study of technology-based constructivism learning theory is a very interesting study. However, further studies in this field are important to carry out. The more studies this semcam is carried out, the more it strengthens the theory presented in this paper.

REFERENCES

Abbeduto, Leonard. 2004. *Taking Sides: Clashing Views on Controversial Issues in Educational Psychology*. Third Edition, McGraw Hill/Dushkin.

Andriani, Tuti. 2015. Information and communication technology-based learning systems. *Socio-Culture: Communication Media of the Social and Cultural Sciences*. 12 (1): 120–132.

Arifin, Z. and Adhi Setiyawan. 2012. *Development of Active Learning with ICT*. Yogyakarta: Skripta Media Creative.

Barnadib, Imam. 2012. *Philosophy of Education: Systems and Methods*. Yogyakarta: Andi.

Bruning, R., Schraw, G., Norby, M., & Ronning, R. 2004. *Cognitive Psychology and Instruction*. Upper Saddle River. NJ: Prentice Hall.

Corbin, J. and Strauss, A. 2019. Grounded theory research: Procedures, canons, and evaluative criteria. *Qualitative Sociology* 13 (1): 21.

Dale, H Schunk. 2012. *Theories of Educational Perspective Learning*. Yogyakarta: Student Library.

De Kock, A., Sleegers, P., and Voeten, M.J.M. 2005. New learning and choices of secondary school teachers when arranging learning environments. *Teaching and teacher education* 21, 799–816.

Gutek, G.L. 2015. *Philosophical Alternatives in Education*. Chicago: University of Chicago.

Hadi, Samsul. 2017. The right to get an education: An epistemological review and Axiology of Educational Philosophy. *Palapa: Journal of Islamic Studies and Educational Sciences*,5 (2): 68–80.

Hasyim, Harwati. 2018. Application of technology in the digital era education. *International Journal of Research in Counseling in Education* 1 (2): 1–3.

Hendrowibowo, L. 2013. Scientific studies on the science of education. *Horizons of Education*. Jakarta: Kencana.

Jamun, Yohannes Maryono. 2018. The impact of technology on education', *Journal of Education and Culture* 1 (1): 50.

Lestari, Sudastri. 2018. The role of technology in education in the era of globalization. *Journal of Islamic Religious Education: Edureligia* 2 (2): 95.

Muhmidayeli. 2012. *Philosophy of Education*. Bandung: Refika Aditama.

Nguyen, N., Williams, J. and Nguyen, T. 2012. The use of ICT in teaching tertiary physics: Technology and pedagogy. *Asia-Pacific Forum on Science Learning and Teaching* 13 (2): 1–19.

Rangkuti, N.A. 2014. Constructivism and mathematics learning, *Journal of Darul 'Ilmi* 2 (2): 25–35.

Ratnaya, I Gede. 2011. Negative impact of the development of informatics and communication technology and how to anticipate it. *Undiksha* 8 (1): 17–28.

S.I.A, D. 2012. *Basic Social and Cultural Sciences*. Yogyakarta: UNY Press.

Sridevi, K.V. 2008. *Constructivism in Science Education*. New Delhi: Discovery Publishing House Pvt Ltd.

Srinivasan, J., and Khrisna, S. 2017. Teaching and Learning in the Digital Era. *International Journal of Science, Humanities, Management and Technology*. 3 (3): 22–35

Suparlan. 2019. The Theory of Constructivism in Learning. *Islamika: An Islamic Journal* 1 (2): 75–88.

Syukur, N. F. 2012. *Educational Technology*. Semarang: RaSail Media Group.

Religion, Education, Science and Technology towards a More Inclusive and Sustainable Future – Rahiem (Ed.)

A scoping review: Nonsuicidal self-injury during emerging adulthood for a more empathetic understanding

Pihasniwati*, H.L. Muslimah & R.R. Diana
UIN Sunan Kalijaga Yogyakarta, Indonesia

A. Mujib, D. Saepudin & U. Kultsum
UIN Syarif Hidayatullah Jakarta, Indonesia

ABSTRACT: DSM-V Part III's inclusion of nonsuicidal self-injury (NSSI) as a new mental disorder necessitates more research into the detailed mapping study of emerging adults (EA) as a new developmental stage identified as a high-risk group for NSSI. This scoping review aims to explore literature studies and inform the evidence of nonpsychiatric NSSI in the EA period (aged 18 to 29) for a more empathetic understanding in developing assessment, intervention, and communication between parties. The articles about DSH/SH/SI/NSSI on EA were published between at least the last ten years, reporting on empirical data/peer-reviewed/literature reviews from trusted databases. After selection by inclusion–exclusion criteria and the review process by the PRISMA-P method, nine were included in this study. The NSSI related to reasons, motivations, and its factors, along with implications for empathetic support, is presented.

Keywords: NSSI, emerging adulthood, related factors of NSSI, motivations/functions of NSSI

1 INTRODUCTION

Self-harm was believed to be only a symptom of a larger psychiatric disorder, as stated in the DSM IV-TR (Diagnostic and Statistical Manual of Mental Disorders IV-Text Revision), where self-harm was related to other psychiatric diagnoses such as borderline personality disorder, depression, or schizophrenia. Currently, as the prevalence of self-harm increases with the complexity of its phenomenon in various parts of the world, which shows that self-harm is not always associated with a diagnosis of a particular psychiatric disorder and is not always associated with the intention to die, a new criterion has emerged. This new criterion is considered sufficient to describe the functional, motivational, and emotional aspects that are typical, so it was proposed to the Childhood Disorders and Mood Disorders working group to be included as a disorder in the fifth edition of the DSM (Zetterqvist 2015). Nonsuicidal self-injury disorder (NSSID) is listed in DSM-V Part-III as a new mental disorder that requires further study (American Psychiatric Association 2013). NSSI is done intentionally, but not without committing suicide. NSSI is the intentional, direct, self-inflicted destruction of body tissue without suicidal intent and for social disapproval (Cipriano et al. 2017). In addition to causing direct physical damage, in the medium and long term, it also predicts more severe emotional problems, such as increased anger, guilt, shame, and isolation, academic difficulties for students, and accidental death, and is a strong predictor for suicidal behavior (Kim L. 2006).

*Corresponding Author: pihasniwati@uin-suka.ac.id

DOI: 10.1201/9781003322054-25

One of the most vulnerable age groups involved in NSSI is early young adults (emerging adulthood/EA). This age group emerged along with the development of industrial/urban communities, which experienced a unique period of development where they had reached maturity in terms of physical appearance and sexuality, with very diverse backgrounds, educational paths, and occupations that occurred in the age range of 18 to 29 years (Arnett et al. 2014). Arnett describes EA as a transition from late adolescence to early adulthood, with the task of developing self-autonomy, exploring identity, and initiating and building relationships. The three characteristics that mark this phase are accepting personal responsibility, making independent decisions, and being financially independent. As a transition period from adolescence to adulthood, this age is full of pressures and demands for roles that are completely different from adolescence. Rapid changes in many important aspects, including personality, social, and academic aspects, occur in this phase and often have the potential to cause stress and are prone to stress (Arnett 2020). Considering the problems mentioned, the questions to be explored are as follows: (1) How is the NSSI phenomenon in the EA period based on the evidence? (2) What is the reason and motivation for being engaged in NSSI? (3) What are the risk factors and protective factors for NSSI in emerging adulthood?

2 RESEARCH METHODS

The scoping review approach in this study was guided by the PEOS Framework and PRISMA flowchart. Data for the scoping review was gathered based on the scoping review methodology. The stages were carried out according to Arksey and O'Malley (2005): review focus, identification of relevant studies, description of the process, identification of literature with PRISMA flowcharts, data extraction, and mapping or scoping. It was carried out using the PEOS (Problem, Exposure, Outcome, and Study Design) framework, as mentioned in the table below.

The use of PEOS helps in identifying the key concepts in the focus of the review, developing appropriate search terms to describe the problem, and determining inclusion and exclusion criteria (Bettany-Saltikov 2010). The sample is limited to journal articles – literature reviewed, peer-reviewed, and empirical study of NSSI (and the related terms) on EA included in psychology and related disciplines or the relevant ones (social sciences and medicine and public health), with sub-disciplines (such as psychiatry, public health, and epidemiology) (distribution based on cases/number), published in January 2011 to May 2022. The literature was collected through scientific databases from electronic search engines such as APA PsycInfo–American Psychological Association, PUB.MED–NCBI National Library of Medicine, and publisher (SpringerLink) databases (such as Springer Nature (BMC), Sagepub, Taylor & Francis Group), as well as journals from Indonesia indexed in SINTA (Science and Technology Index) and Portal Garuda, and two from some other databases related to this theme. In this study, Table 1 contains the data extraction or charting from the scoping review.

Table 1. PEOS (Problem, exposure, outcome, and study design) framework for scoping review.

Population and Problems	Exposure	Outcome or Themes*	Study Design
NSSI on emerging adulthood (EA) (as the transition from adolescent to young adult) Aged range 17/18 to 29	a. Experiences b. Maladaptive Behaviors c. Correlates of NSSI d. Psychological Distress e. Risk Factors f. Reason/motivation/ functions	a. Reason, function, and or motivation b. Risk and Protection Factors	a. Literature review b. Peer-review c. Empirical study

Table 2. Inclusion and exclusion criteria of the published articles for the selection process.

Inclusion criteria	Exclusion criteria
(1) Articles published in the last 10 years (2012–2022) in a trusted database (2) Psychology journals, or relevant ones in this discipline, (namely Social Sciences and Medicine and Public Health), with sub-disciplines (namely Psychiatry, Public Health, and Epidemiology); Articles in English or Indonesian (3) Type of article – Reporting on empirical data/ peer-reviewed/literature review. Empirical study articles based on participant research subjects from self-harm (and related terms) are preferred (4) People with DSH/SH/SI/NSSI in the stage of emerging adulthood aged 18–29*	(1) DSH/SH/SI for suicidal intent. (2) Other irrelevant (3) NSSI that is not part of a larger psychiatric disorder: NSSI with clinical major illness at risk who could not participate in the study actively/ excluding the self-harm intending to suicide, also excluding the major clinical symptoms that have a high risk.

3 RESEARCH FINDINGS AND DISCUSSION

3.1 *Factors associated with NSSI and the reasons/motivations for EA engaged in NSSI*

As a new mental disorder that requires further study and the recognition of the high susceptibility of EA to NSSI involvement, it is important to build a deeper understanding of the factors associated with NSSI, the reasons and motivations for NSSI involvement, and the function of NSSI in EA. The stress experienced by young adults, which if failed to be managed properly, can lead to pathological solutions, one of which is self-harm, as they feel difficulty with emotional regulation – a pathological approach that NSSI needs for emotional regulation and distress (Favazza 2012). Here we present the factors related to NSSI, including persistent factors (age and gender), internal factors of psychological aspects, and external factors including things that are out of the control of the subject, to explore the causes of NSSI in EA. This data was obtained from 10 articles that included the eligibility criteria. The present scoping review was conducted to understand emerging adulthood (because at this age people enter the life-crisis period and the prevalence of NSSI is high around EA) and to know the factors related to NSSI in EA for a more empathetic understanding to develop more helpful pre-assessment and pre-intervention approaches.

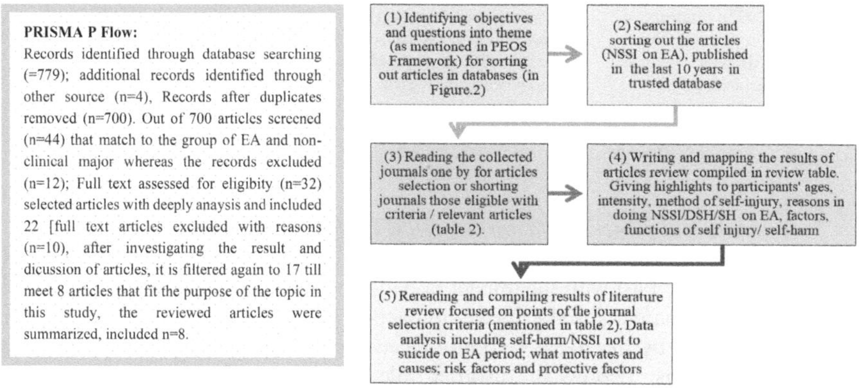

Figure 1. Data extraction of scoping review in this study.

Table 3. Factors and functions related to NSSI/DSH on emerging adulthood based on evidence.

Data source and participants	Related factors to self-harm/self-injury	Reason/motivation/function of NSSI
Mahtani et al. (2018): 384 Australian EA with a history of NSSI (81.3% female, 18.7% male. Of them 235 Caucasians, 97 Asians, and others)	Current shame-related functioning and perceived history of parental invalidation during childhood and adolescence period that indirectly leads to psychological distress, shame proneness, and internalized shame coping.	Individuals' internalized shame coping (by self-attack, social withdrawal); Intrapersonal: self-punishment. InterP.-other: revenge, InterP-self: sensation seeking, self-care, toughness, autonomy
[3] Fadhila & Syafiq (2020): 2 female students aged 21 years. Early age for self-injury: 19 years old by cutting the skin (age at interview, 21), and doing some methods of NSSI	Traumatic life events (including sexual harassment) in the past turned to guilt; Breaking-up with boyfriend brings up negative feelings: emotionally hurt, losing/missing someone, no longer have a place to talk, stressed, pessimistic about future; feeling unworthy-unloved; following friends with the same tendency in doing self-injury.	Response to problems to release negative emotions such as displacement; release stress by cutting hands; memory triggered trauma leads to self-injury; Feeling relieved, calmer, and relaxed after self-injury but feelings such as fear of getting negative stigma from others after self-injury
[4] Daukantaitė et al. (2021): The 10 years follow up after 8th grade approximately 13/14 years old + 10 = 23/24 years old, the incidence of NSSI decreases from about 40% in adolescence to 18.7%	Stable recurrent NSSI exhibited significantly higher levels of stress, anxiety, NSSI, and emotional regulation difficulties at 10 years (during adolescence to young adulthood).	Some of the young adults with NSSI since their adolescents who perform NSSI (both rarely/repeatedly unstable, or recurrently stable) had various mental health problems in 10 years. (longitudinal study)
[5] Singhal et al. (2021): drawn from 1.574 EA aged 18–25 years, 58.32% female from 19 educational institutes, city of Bengaluru, India. 353 [22.40%] who engaged in at least one NSSI behavior in the past 12 months	NSSI behavior is related to sub-group profiles (multimethod, exclusively severe, female minor, male minor and experimental NSSI).; NSSI behavior is influenced by gender, severity, frequency, diversification, age of onset, and NSSI function.	Multimethod and Exclusively Severe NSSI showed significantly higher psychological distress and difficulty with emotion regulations.
[6] Elvina & Bintari (2021): 311 Indonesian EA aged 18–29, of them 40.2% (N = 125) had engaged in NSSI and majority still engaging in NSSI since past year.	Religious coping affecting NSSI, Negative religious coping is significantly correlated with NSSI severity, whereas positive religious coping (PRC) is not a significant predictor of any functions of NSSI	Overcoming negative emotions, NSSI as coping with stress
[7] Widyawati & Kurniawan (2021): 77,7% (N = 199) females and 12,3% (N = 28) males aged 18 to 25 years old have self-harm	Social media with self-harm contents exposure; gender: women are more likely to be affected	the more exposed to social media exposure related to self-harm behavior, the higher the individual's desire to do self-harm.
[8] Arinda & Mansoer (2021): 12 young adult women engaged in NSSI, N = 4 (aged 21–22) participated fully in study [college students and business women]	Feeling emptiness; Problems arising from family and social pressure including treason, unappreciated work; negative response from others (problems with parents and friends: prone to be blamed, neglected, ignorance)	NSSI as a coping strategy for dealing with feelings of emptiness and negative emotions; as expression of negative feelings lead to problem with parents (e.g. mother, in this study)
[9] Kruzan *et al.* (2022): Students at an undergraduate university in the United States Midwest, 448 experienced past-year NSSI	Poor self-concept clarity; Poor self-perception is associated with increased NSSI through self-blame coping when body-regards are low/average.	Blaming and punishing self; low body appreciation; poor self-concept

3.1.1 *Risk factors and protective factors of NSSI on emerging adulthood*

The internal factors include personal habits, problems experienced, feelings within self (traumatic experiences in the past with the impact of bad experiences in childhood); depression and anxiety disorders; self-hate and self-criticism impacting self-esteem and self-hatred (self-hatred can also be influenced by the social response received); feelings of emptiness, impulsiveness, and unstable emotions; depressive disorder, alcohol use, and forensic history on the younger age/adolescents; negative religious coping may influence NSSI involvement in EA.

The individuals who engage in NSSI tend to have less stable emotions (barriers in regulating emotions internally), impulsiveness, and psychological distress. The external factors reported are related to the surrounding social environment/family/friendship, such as friends with the same tendency; problems arising from family and social pressure, who feel underestimated and experienced compliance threats (e.g., by parents); and social media exposure about self-harm.

Our result is consistent with prior research from Turner *et al.* (2014), who analyzed the reasons for discontinuing the association with NSSI. It was found that resilience-related reasons for discontinuing NSSI were associated with greater hope, social support, adaptive coping, and prospectively abstaining from NSSI three months later. While vulnerability-related reasons were associated with greater psychopathology and dysfunctional coping and predicted more chronic and severe NSSI.

Research conducted on community-based participants, both individually and in groups, found that women are more susceptible to NSSI (Widyawati & Kurniawan 2021; Wilkinson 2022). Women have a greater tendency to bet influenced by social media exposure and engage in self-harm behavior (Widyawati & Kurniawan 2021). Partial mediation of this gender/age effect by psychological distress provides further support for the reduction of psychological distress as an important component of treatment for NSSI (Wilkinson 2022).

In women, NSSI appears to increase from early adolescence, peaking in mid-adolescence, and then gradually decreasing. In men, however, NSSI is stable across all ages, where age patterns differ markedly between genders (Wilkinson et al. 2022). In a cross-study, the quality of parent and peer relationships was related to the likelihood of engaging in NSSI among Asians, but the set of variables was not significant in the history of NSSI among Caucasians (Turner *et al.* 2015). Although the meaning, context, and rationale of NSSI may differ in developing and developed countries, NSSI is almost as common in developing countries as in developed countries (Mannekote Thippaiah *et al.* 2021).

Young adults interpret their NSSI experiences as difficulties in regulating negative emotions as a result of inappropriate distribution of distress, and they find self-injury as a way to get relieved (Arinda & Mansoer 2021; Saputra *et al.* 2019). Self-harm could be a powerful form of communication that could not be spoken by those who engaged in it. In their studies, Motz (2009) and Munson (2012) included various clinical cases handled by doctors who treat individuals with self-injurious behavior. Socially, we may think of self-harm as a suicide attempt, but there is something to be conveyed beyond this – a response to problems to release negative emotions as displacement (Fadhila & Syafiq 2020).

3.2 *Implications for the more empathetic support for NSSI on EA based on evidence*

Based on the result of the scoping review focusing on participants' age period at EA, related factors to self-harm/self-injury, and the reason/motivation/function of NSSI, we provide our analysis for a more empathetic understanding, as mentioned in Table. Our analysis lists the implications of empathy from each study. According to the findings of Mahtani et al. (2018), the therapeutic intervention could be more attentive to the adaptive shame regulation strategies. Sensitivity assessment and intervention to traumatic experiences and peer coping effects through NSSI. Assessment of sensitivity to traumatic experiences, feeling loneliness, and victims of sexual abuse/harassment; giving more love and affection support from the closest persons/family (Fadhila & Syafiq 2020). Furthermore, sensitivity to having NSSI history early/younger age, because it can be a risk factor for negative mental health impact till 10 years later, NSSI resulted in negative outcomes in later years (Daukantaitė et al. 2021). and it impacts on dangers (Daukantaitė et al. 2021). For the assessment, a study by Singhal et al. (2021) could have implications for comprehensive assessment frameworks, early identification of more vulnerable groups, and planning of targeted interventions. As well as the assessment sensitivity of negative religious coping on EA. Practitioners should be more aware of negative religious coping with EA that is harmful to NSSI; so positive religious coping may be needed (Elvina & Bintari 2021). The assessment sensitivity in the scope of family, work, and social environment; leveraging social support in NSSI interventions; and communication among family members could be protective (Arinda & Mansoer 2021). Prevention and inhibition of social media exposure related to self-harm content; teaching self-skills to become more wiser in responding to content received on social media with a healthy mentality (Widyawati & Kurniawan 2021). Development of interventions related to mindfulness, emotion regulation, and reducing impulsivity. Implication for empathetic support in treatments involving the body and self-perceptions (Kruzan *et al.* 2022) related to self-concept (e.g., body regards) to reduce NSSI behavior. A positive regard for the body can be a protective factor when individuals perceive they feel worthless.

The quality of the parent-child relationship during the EA period is a history of the early experiences of parent-child bonding. Attachment theory suggests that positive emotional attachment to older people during early childhood facilitates the development of social relationships with others and in response to stressful events (Mikulincer & Shaver 2009). As reviewed, problems in the family are things that play a role in NSSI. The most specific family problems in influencing participants for NSSI are related to the attitudes and behaviors of family members. Ignoration, excessive supervision and control, and blame from parents make them feel uncomfortable and trigger NSSI behavior (Arinda & Mansoer 2021). Family tends to support high levels of negative emotions, which leads to an increase in intentional self-harm/DSH (Arinda & Mansoer 2021; Victor et al. 2019). So involving family understanding in the process of preventing or reducing self-injury behavior helps by paying attention to the other side as well, such as the influence of close friends or romantic relationships (Arinda & Mansoer 2021; Fadhila & Syafiq 2020), just as social media content exposure may lead to self-harm (Widyawati & Kurniawan 2021), and the individual personality on emotional regulation who engage on DSH (Daukantaitė et al. 2021; Favazza 2012). A study with 10 years of follow-up found that NSSI behavior at least once or more in adolescence becomes an indicator of poor mental health vulnerability in young adults (Daukantaitė et al. 2021).

Although cultural differences may have a major influence on the prevalence and nature of NSSI, little is known about NSSI in non-Western countries, nor have the instruments been modified to measure the extent, causes, and methods of performing DSH/SH/SI/NSSI in different cultures and countries (Brown & Witt 2019). This study may contribute to the growing literature that emphasizes the importance of understanding NSSI in EA and the motivations and factors related to empathetic support.

4 CONCLUSION

Family and social pressures and feelings of emptiness are the risk factors, but the most prominent are related to traumatic experiences, unstable/impulsive emotions, anxiety, depression, and self-criticism. It is important to note that traumatic experiences can be a risk factor in later years. The function and motivation of NSSI are related to the challenges and needs of psychological development, but developmental tasks, challenges, and issues are important for the relevance of the developmental period before EA. Reasons or motivations in NSSI has a history that can be interconnected in the range of life experiences from childhood and adolescence to early adulthood. Women need early prevention because of the tendency for NSSI at a younger age. The intrapersonal function is more prominent than the interpersonal function. The most prominent risk factors are related to traumatic experiences, unstable/impulsive emotions, anxiety, depression, and self-criticism. EA risk factors are inseparable from experience before the EA period.

In this study, we recommend that the ability to accept, hope, social support, and adaptive coping can be significant protective factors. Greater psychopathology, dysfunctional coping, and social media exposure of self-harm should be inhibited. Reducing emotional stress and psychological distress due to self and environmental demands plays the most prominent function. This function has a strong context with developmental tasks and needs, including in the EA period such as issues of love, future, social support, career, and family. The development of interventions related to mindfulness, emotion regulation, and reducing impulsivity is also recommended. The results are expected to be useful for further research, such as developing assessments, formulating prognoses, and developing interventions and communication between interested parties, such as the closest persons, families, doctors, nurses, and psychologists or professionals who work with mental-health problems. Reducing emotional stress and psychological distress due to self and environmental demands plays the most prominent function. This function has a strong context with developmental tasks and needs, including in the EA period, such as romance, future, social support, career and family issues. The development of interventions related to mindfulness, emotion regulation, and reducing impulsivity is recommended.

REFERENCES

American Psychiatric Association. 2013. *Diagnostic and Statistical Manual of Mental Disorders*, 5th edn. Washington, DC: American Psychiatric Association

Arinda, O. D., & Mansoer, W. W. D. 2021. NSSI (Nonsuicidal Self-Injury) of emerging adults in Jakarta: An interpretative phenomenological approach. *Jurnal Psikologi Ulayat*, *8*(1), 123–147.

Arksey, H., & O'Malley, L. (2005). Scoping studies: Towards a methodological framework. *International Journal of Social Research Methodology*, 8, 19–32.

Arnett, J. J., Žukauskiene, R., & Sugimura, K. 2014. The new life stage of emerging adulthood at ages 18–29 years: Implications for mental health. *The Lancet Psychiatry*, *1*(7), 569–576.

Brown, R.C., & Witt, A. 2019. Social factors associated with Non-Suicidal Self-Injury (NSSI). *Child and Adolescent Psychiatry and Mental Health*, 13,23. https://doi.org/10.1186/s13034-019-0284-1.

Cipriano, A., Cella, S., & Cotrufo, P. 2017. Nonsuicidal self-injury: A systematic review. *Frontiers in Psychology*, 8, 1946. https://doi.org/10.3389/fpsyg.2017.01946

Daukantaitė, D., Lundh, L. G., Wångby-Lundh, M., Claréus, B., Bjärehed, J., Zhou, Y., & Liljedahl, S. I. 2021. What happens to young adults who have engaged in self-injurious behavior as adolescents? A 10-year follow-up. *European Child & Adolescent Psychiatry*, *30*(3), 475–492.

Elvina, N. & Bintari, D. R. 2021. An exploration of indonesian emerging adults Non-Suicidal Self-Injury (NSSI) functions and religious coping. *Jurnal Psikologi Malaysia*, *35*(2)

Fadhila, N. & Syafiq M., (2020). Pengalaman psikologis self-injury pada perempuan dewasa awal. *Character: Jurnal Penelitian Psikologi*, *07*(03). (2020).

Faried, L., Noviekayati, I., & Saragih, S. 2019. Efektivitas pemberian ekspresif writing therapy terhadap kecenderungan self-injury ditinjau dari tipe kepribadian introvert. *Psikovidya*, *22*(2), 118–131.

Favazza, A. R. (2012). Nonsuicidal self-injury: How categorization guides treatment. *Current Psychiatry*, *11* (3), 21–25.

Kim L., G. (2006). Risk factors for deliberate self-harm among female college students: The role and interaction of childhood maltreatment, emotional inexpressivity, and affect intensity/reactivity. *American Journal of Orthopsychiatry*, *76(2)*, 238–250. https://doi.org/10.1037/0002-9432.76.2.23

Mahtani, S., Melvin, G. A., & Hasking, P. 2018. Shame proneness, shame coping, and functions of nonsuicidal self-injury (NSSI) among emerging adults: A developmental analysis. *Emerging Adulthood*, 6(3), 159–171. https://doi.org/10.1177/2167696817711350

Mikulincer, M. & Shaver, P.R. 2009. The attachment and behavioral systems perspective on social support. *Journal of Social and Personal Relationships*, *26*(1), 7–19.

Singhal, N., Bhola, P., Reddi, V., Bhaskarapillai, B., & Joseph, S. 2021. Non-Suicidal Self-Injury (NSSI) among emerging adults: Sub-group profiles and their clinical relevance. *Psychiatry Research*, 300, 113877.

Victor, S. E., Hipwell, A. E., Stepp, S. D., & Scott, L. N. (2019). Parent and peer relationships as longitudinal predictors of adolescent non-suicidal self-injury onset. *Child and Adolescent Psychiatry and Mental Health*, 13, 1. https://doi.org/10.1186/s13034-018-0261-0

Widyawati, R. A. & Kurniawan, A. 2021. Pengaruh paparan media sosial terhadap perilaku self-harm pada pengguna media sosial emerging adulthood. *Buletin Penelitian Psikologi dan Kesehatan Mental (BRPKM)*, *1*(1), 120–128. http://dx.doi.org/10.20473/brpkm.v1i1.24600

Wilkinson, P. O., Qiu, T., Jesmont, C., Neufeld, S., Kaur, S. P., Jones, P. B., & Goodyer, I. M. (2022). Age and gender effects on non-suicidal self-injury, and their interplay with psychological distress. *Journal of Affective Disorders*, 306, 240–245. https://doi.org/10.1016/j.jad.2022.03.021

Zetterqvist, M. 2015. The DSM-5 diagnosis of nonsuicidal self-injury disorder: A review of the empirical literature. *Child and Adolescent Psychiatry and Mental Health*, *9*(1), 1–14

Religion, Education, Science and Technology towards a More Inclusive and Sustainable Future – Rahiem (Ed.)
© 2024 the Author(s), ISBN: 978-1-032-56461-6
Open Access: www.taylorfrancis.com, CC BY-NC-ND 4.0 license

"The school does not belong to me:" Involving parents in an Islamic boarding school in Indonesia

D.E. Ginanto*, Alfian, K. Anwar & Noprival
UIN Sulthan Thaha Saifuddin, Jambi, Indonesia

K. Putra
Prince Sattam Bin Abdulaziz University, Saudi Arabia

K. Yulianti & T. Mulyadin
Binus University, Indonesia

ABSTRACT: Despite the widespread practice and research on parental involvement (PI) in developed countries, the literature suggests that Indonesia needs to do more to encourage PI in education, both in boarding schools and in traditional settings. The purpose of this study is to uncover the implementation of PI using Epstein's (2009) framework at an Indonesian boarding school. We interviewed teachers, principals, and parents. The results show that even when parents live apart from their children, the parents can still be involved in the upbringing of their children in the boarding school. Both parents at home and teachers at school were able to carry out five of Epstein's six types of PI, particularly in involvement in academic and non-academic activities. This study implies that Epstein's (2009) PI models, e.g. Parenting, learning at home, communicating, volunteering, decision-making and collaborating with the community, could be used as alternative guidance for involving parents and surrounding communities in boarding school settings.

Keywords: Boarding school parent, parental involvement, parental involvement framework, students' success

1 INTRODUCTION

Most boarding schools have made a lot of initiatives to get parents involved in their children's education (Sgro 2006). Boarding schools are distinctive academic institutions that offer various educational advantages that ordinary public schools are unable to provide. This includes education systems that are available full day, allowing students and teachers to communicate regularly. Students have more additional curricular pursuits and extra time to interact with their teachers (Sgro 2006). However, due to the distance between home and school, parents were not able to actively participate in formal home-school partnerships, e.g. Taking part in the school's decision-making process, serving on the board of trustees, or volunteering at the institution.

According to the literature, boarding schools' parents, particularly in Indonesia, have lower participation than parents in public schools due to the distance and their lack of understanding of PI (Azizah 2013; Parker & Raihani 2011). For example, four Islamic schools (madrasas) in West Sumatra and two madrasas in Yogyakarta still lacked PI due to a lack of understanding of parents' roles in schools (Parker & Raihani 2011). In addition, Parker and Raihani (2011) found that PI was limited to paying school fees and other related educational expenses.

In addition, previous studies on PI in Indonesian boarding schools also commonly have not used any PI framework. The parental participation framework is a paradigm that parents and teachers use to help their children with their education. As a result, analyzing the level of parental engagement was difficult to do, due to the lack of a measurement. Using Epstein's parental

*Corresponding Author: dionefrijum@uinjambi.ac.id

 DOI: 10.1201/9781003322054-26

engagement framework, this study will look at how parents and teachers collaborate to engage parents in school events and programs (2009).

1.1 *Islamic boarding school in Indonesia*

Boarding school in Indonesia is considered the oldest educational setting. Islamic boarding schools existed in Indonesia even during the colonial era (Azizah 2013). The trend of sending children to Islamic boarding schools is currently expanding due to parents' desire to provide their children with not only a standard curriculum but also an Islamic curriculum (Subianti 2007). Students will consequently learn the standard subjects as they would in a typical school setting during the day, and will continue to study Islamic courses in the evening. Due to the evening programs that the schools offer, the students need to reside at schools.

The fact that students are isolated from their parents is one of the biggest issues and obstacles for boarding schools. When parents are separated from their children, their participation in school programs and activities will usually decline. In the meantime, research shows that parents play an important role in their children's education (Castro et al. 2015; Khalifa 2012; Sgro 2006 Wilder, 2014). This is because parents are both caregivers and teachers for their children for the first few years of their life. No one understands children better than their parents, therefore teachers must learn from them.

Parents who reside close to the school and are aware of the need for PI will get in touch with the institution to talk about their kids' academic development. This would make it simple for parents and teachers to collaborate on raising their kids' academic achievement.

1.2 *Parental involvement and Epstein's framework*

Epstein's Concept of Six Forms of Participation has been used as a model in the education system of the United States and many others. In Epstein's model, the first form is parenting. Parenting is defined as efforts to assist all families in creating learning-supportive home settings for children (Epstein 2009). This form is defined as "help[ing] all families establish home environments to support children as students" (p. 16). Adults' support in preparing anything students need to get ready for school is an instance of parenting. The second form of Epstein's model is communication. Epstein (2009) recommended various activities or programs targeting parents and schools, for instance regular parent meetings, inclusive support for diverse families, regular reports to parents, in person and online communication between schools and parents, and supportive environment creation to facilitate teacher-parent communication. The third form proposed by Epstein is parent volunteering. An example of this form is the parent liaison assignment to make sure students are learning at school. Learning at home is the fourth form in Epstein's model. Parents are expected to provide students necessary assistance on their school assignments, projects, and other activities. This effort is important since it encourages students to enthusiastically learn at home.

The fifth form is making decisions. Among the many activities or programs recommended are regular conferences involving parents and teachers, follow ups to parents who are less engaged in school meetings, network development between schools and parents, and district council establishment to facilitate community engagement. Finally, collaboration is the sixth form. Epstein (2009) provided numerous examples of community engagement, such as disseminating knowledge of local resources, involving the community and families in school activities, involving the community and families in sponsoring specific school events, offering community services, and organizing alumni activities and events. In this study, this framework is used as a guided theoretical framework in identifying PI in the boarding school context

2 METHODS

The purpose of this research is to investigate parental engagement in an Islamic boarding school (RBS, pseudonym) in Indonesia. The boarding school educates students from kindergarten through high school. This study focuses primarily on high school settings. The qualitative design was used in this study because the data are kept in a descriptive narrative including notes and audio recordings (Lemmer & Van 2004), and the research was conducted in natural settings to study a contemporary phenomenon in a real-life context (Yin 2018). The data was gathered through interviews with six different teachers who differed in terms of their gender and the subjects they teach. The six teachers included a principal, an assistant principal, and two school owners. In addition, seven parents were asked about the following criteria: distance, gender, parental status, and economic background. The interviewee's profiles are as follows:

Table 1. Interviewees profile.

No	Name (Pseudonym)	Job/ Profession	Distance home to school (approximate)
1	Mr. Munas	Parent (a teacher at another district's school)	240 Miles
2	Mr. Gunadi	Parent (Rubber farmer)	50 Miles
3	Mr. Karyono	Parent (Rubber farmer)	60 Miles
4	Mr. Sukarno	Parent (Rubber farmer)	40 Miles
5	Mr. Safik	Parent (businessman)	30 Miles
6	Mrs. Marliyah	Parent (run agribusiness entrepreneurship)	30 Miles
7	Mrs. Yatno	Parent (single parent, farmer)	30 Miles
8	Mr. Badru	Teacher (teach Quranic Studies)	Live on campus
9	Mr. Jamil	Teacher, Founder, Owner	Lives on campus
10	Mr. Suyat	Teacher/Principal	Lives on campus
11	Mrs. Jamil	Teacher/Founder/Owner	Lives on campus
12	Mrs. Suryani	Teacher (Indonesian Language)	7 Miles
13	Mr. Hendra	Assistant Principal	30 Miles

Prior to data collection, the principal was contacted. The data collection process began with the support of the school principal, who ranked the participants to be interviewed based on the given criteria. The interview procedures were designed in accordance with Epstein's PI model. Each interview lasted about an hour and was digitally recorded and kept in a notebook. To enable participants to convey their perspectives properly, the interviews were done in their native language. One of the researchers of the study conducted the interviews both inside and outside the school.

Inductive and deductive analyzes were used to interpret the interview data according to the study framework (Patton 2002). Using Epstien's PI framework, the interviews were transcribed, interpreted, and coded for further verification. We use member validation to ensure data validity and reliability (Creswell 2012). In order to present a thorough and rich summary of the findings, verbatim quotes from the interviews were also included (Creswell 2003, p. 196).

3 FINDINGS AND DISCUSSION

This paper challenges the narrative of PI in Indonesia. We had a presumption that the Epstein (2009) parental participation models would be difficult to observe in this boarding school. On the other hand, the school has adopted five of Epstein's six models unwittingly. According to the results of the interviews, collaboration between school and home is valued by both teachers and parents. In front of the community, the school owner also stated that his school belongs to the community, not to him or his family. Therefore, the parents and the community have the full right to help the teachers in educating the students. The school owner's oath indicated that the school are opening its door to both parents and the community.

The study's findings are in opposition to earlier Indonesian research on parental participation. A prior study (Fitriah et al., 2013) found that PI is low in Indonesia because of parents' poor understanding of both the home and school environments. The current research did not find evidence of this phenomenon. The school successfully involved parents in its activities and programs, even though the vast majority of the families attending are middle-class and low-income. The majority of parents still participate in the school even when they are geographically separated from their children.

Our research also refutes Parker and Raihani's (2011) claim that teachers only have a responsibility to educate students in the classroom. When the pupils were home for the summer or other breaks from school, the parents and teachers actively worked together to collaborate in educational activities. For instance, when a child is at home during a school break, Mr. Sukarno realized that parents are the ones who teach them. "When she went home for a school break, it was our job to teach her in the same way she is taught at school," he said. Mr. Sukarno was also conscious of the necessity to take care of his daughter. "Relying only on the school is insufficient," he said. Meanwhile, The owner and administrator of the school, Mr. Jamil, made it clear to the parents that he no longer wanted to hear that their kids were spending long stretches of time sleeping during breaks from class or that they weren't engaged in academic or spiritual activities like five daily prayers and memorizing the holy Quran.

The school embraced all community members, as evidenced by the founders' declaration that "This school does not belong to me" (Mr. Jamil). Consequently, the community around the school is genuinely ready whenever it needs assistance in any situation. This includes not only parents who are actively involved in school activities and programs. For instance, Mr. Karyono stressed the significance of community involvement in educational programs: "Parents of other students also show up and offer assistance. Even those who didn't enroll their kids in school committed to volunteering at the institution."

Also in opposition to our findings was the claim (Majzub & Salim 2011) that PI in Indonesia is restricted to school funding. In this study, our observation revealed that parents helped with fundraising efforts and school fees in addition to taking part in a wide range of other school-based involvement programs. Additionally, the RBS a special fund was set out for refreshments for parents that visit the school in order to maximize the number of parent visits.

In light of the fact that RBS is a private Islamic school to which parents voluntarily choose to send their children, this study admits that parents' reasons for doing so may have an impact on their involvement in their kids' education despite the boarding school setting. Additionally, the school demonstrated that parents and the school can still actively participate in their child's education even if pupils live distant from their parents. The other five models were well-executed even though decision-making was not seen during the analysis.

According to the study, the family is a full partner with the school in making decisions that will improve the student's performance (Epstein 2009). Nevertheless, it has been challenging to observe the parental engagement decision-making process. In our interview, every parent affirmed that they had not yet participated in the decision-making process at the institution. According to the teachers, parents were not invited to the school's decision-making conference. There is no school committee at the RBS, in contrast to nearly all Indonesian schools (Karsidi, et al. 2013). There were some parent meetings at the school, but neither the school committee nor the parent teachers' association were in charge of planning them (PTO). The fact that most parents reside distant from the school could be the root of the problem. As a result, it was difficult to persuade parents to join a school committee. The absence of a school committee should not, however, be used as an excuse going forward. Asynchronous teleconference sessions may also be used by the school to allow parents to participate in the decision-making process.

4 CONCLUSION

This study has added to the corpus of knowledge on parental involvement in Indonesian boarding schools. According to the current study, parents can still take part in school activities and services despite their distance. RBS makes it possible for parents and the school to be actively involved in a student's education, even if they live a long way from one another. However, RBS also has a long way to go in terms of including parents in the school decision-making process. Parents' participation in decision-making is significant because it affects their sense of ownership to the school. More precisely, the school would be better able to end patterns of exclusivity, segregation, and injustice if parents were involved in decision-making since the school would value their opinions.

The study also paves the way for future research. First off, while we agree that if the research had been conducted in a wider range of institutions, it would have more consequences for both theory and practice, it was specifically centered on a background in an Islamic Boarding School. Our research focused on a private Islamic school where parents self-select their children to attend. Future research, in our opinion, would be far more effective if it were conducted in various boarding school settings. Instead, a prospective study may contrast Christian boarding schools in rural, suburban, and urban settings with non-religious boarding schools or with Islamic boarding schools. Regardless, comparisons between contexts may be useful in future study on parental participation.

REFERENCES

Azizah, Nur. (2013) Dukungan orangtua bagi anak yang belajar di pesantren. Parental involvement for boarding school students. *A Proceeding. National Conference on Parenting.*

Castro, M., Exposito-Casas, E., Lopez-Martin, E., Lizasoain, L., Navarro-Asencio, E., & Gaviria, J.L. (2015). Parental involvement on student academic achievement: A meta-analysis. *Educational Research Review*, 14, 33–46.

Creswell, J. (2012). *Educational Research: Planning, Conducting and Evaluating Quantitative and Qualitative Research* (4th ed.). Pearson Education, Inc.

Creswell, J. (2003). *Research Design: Qualitative, Quantitative, and Mixed Methods Approaches* (2nd ed.). Sage.

Crotty, M. (1988). *The Foundations of Social Research*. 1st ed. London: Sage.

Epstein, J.L. (2009) *School, Family, and Community Partnerships*. Corwin Press: Thousand Oaks, California.

Fitriah, A., Sumintono, B., Subekti, N.B., Hassan, Z. (2013). A different result of community of participation in education: an Indonesian case study of parental participation in public primary schools. *Asia Pacific Education Revision*. (14) 483–493.

Karsidi, R., Humona, R., Budiati, A.C., & Wardojo, W.W. (2013). Parent involvement on school committees as social capital to improve student achievement. *Excellence in Higher Education*. Vol. 4(2013), 1–6

Khalifa, M. (2012). A *e*-new-*ed* paradigm in successful urban school leadership: Principal as community leader. *Educational Administration Quarterly*.

Lemmer, E.M. & Van Wyk, J.N. 2004. Schools reaching out: comprehensive parent involvement in South African primary schools. *African Educational Review*, 1(2):259–267.

Majzub, R.M., & Salim, E.J.H. (2011). Parental involvement in selected private preschools in Tangerang, Indonesia. *Social and Behavioral Sciences*. Vol. 15(2011), 4033–4039.

Miles, M. B., Huberman, A. M., & Saldana, J. (2014). *Qualitative Data Analysis: A Methods Sourcebooks* (3rd ed.). California: SAGE Publications, Inc.

Patton, M. Q. (2002). *Qualitative Research and Evaluation methods* (3rd ed.). Sage.

Parker, L. & Raihani, R. (2011). Democratizing Indonesia through education? Community participation in Islamic schooling. *Educational Management and Leadership*. 39(6) 712–732.

Sanders, M. G., & Epstein, J. L. (1998). School-family-community partnerships and educational change: International perspectives. In *International Handbook of Educational Change* (pp. 482–502). Dordrecht: Springer Netherlands. http://doi.org/10.1007/978-94-011-4944-0_24

Sgro, A.H. (2006). *The Perception of Parents of the Appropriate Degree of Parental Involvement in an Independent Boarding School: A Matter of Trust*. Unpublished Dissertation. University of Pennsylvania.

Subianti, D. (2007). *Peranan Pondok Pesantren Terhadap Prestasi Belajar Pendidikan Agama Islam di Pondok Pesantren at-Tanwir Desa Talun Kecamatan Sumberejo Kabupaten Bojonegoro (The Role of Parents at the Islamic Boarding School at Pondok Pesantren at-Tanwir Desa Talun Kecamatan Sumberejo Kabupaten Boojonegoro*. Unpublished Thesis. Universitas Islam Negeri Malan.

Tan, Charlene. (2014). Educative tradition and Islamic schools in Indonesia. *Journal of Arabic and Islamic Studies*. p. 47–62

Tomlinson, H.B., & Andina, S. (2015). *Parental education in Indonesia: Review and recommendations to strengthen program system*. The World Bank: World Bank Group.

Wilder, S. (2014) Effects of parental involvement on academic achievement: A meta-synthesis, *Educational Review*, 66(3), 377–397.

Yin, R. K. (2014). *Case Study Research: Design and Methods* (5th ed.). Sage Publications.

Yin, R.K. (2018). *Case Study Research and Applications: Design and Methods*, 6th ed. Los Angeles: Sage.

Yulianti, K., Denessen, E., & Droop, M. (2019). Indonesian parents' involvement in their children's education: A study in elementary schools in urban and rural Java, Indonesia. *School Community Journal*, 29(1), 253–278.

Yulianti, K., Denessen, E., Droop, M., & Veerman, G. J. (2020). School efforts to promote parental involvement: the contributions of school leaders and teachers. *Educational Studies*, 1–16.

Religion, Education, Science and Technology towards a More Inclusive and Sustainable Future – Rahiem (Ed.)

The influence of Jambi Malay language use toward the implementation of Islamic *aqidah* educational values at Orang Rimba

D. Rozelin*, D. Mustika, M. Arifullah, Musli, Mailinar & U. Azlan
UIN Sulthan Thaha Saifuddin Jambi, Indonesia

ABSTRACT: The Jambi Malay language and Kubu dialect have many differences in their vocabulary. To ensure that the principles of Islamic *aqidah* education are thoroughly understood so that what is conveyed can be understood well, *ustadz* (Islamic preachers) should employ specific strategies when delivering their lessons. The purpose of the research was to find out the effect of using the Jambi Malay language in describing Islamic *aqidah* educational values for OR (Orang Rimba) and to find out a strategy for describing it to OR at Lubuk Kayu Aro. The research used a qualitative method – observation, interviews, and documentation – for collecting data. The results were (1) The influence of using the Jambi Malay language was quite strong because of the intensity of *ustad*-conducted religious lectures in the OR group; (2) the occurrence of code mixing carried out by OR when they converse with *ustad*; and (3) The strategy used by the *ustad* in conveying Islamic *aqidah* was a personal approach consisting of (a) exemplary behavior; (b) habituation; and (c) advice.

Keywords: Jambi Malay language, strategy, Islamic *aqidah* educational values

1 INTRODUCTION

Marginal groups or groups living in rural areas can be found in all countries in the world. They use language as a means of communication where the language is sometimes not understood by other community groups. A good language will be received well and understood by the interlocutor. Jambi Malay is a language used by most people in Jambi City, while people in the regency use different Malay dialects. Likewise, the Orang Rimba (OR) group in Lubuk Kayu Aro village uses the dialect Malay, which is different from the Jambi Malay language. The Islamic preachers (*ustadz*) in conveying Islamic religious teaching try to convey what can be understood by the OR group. Of course, it is not easy to provide an understanding of new teachings to others, especially to OR groups who previously had animism and dynamism beliefs.

The existence of this isolated tribe has been known since the 19th century. OR live in groups, and they do not care about education. Since 2012, they began allowing their children to study. In the early stages, the OR only allowed their children to study on palm plantations, a few years later, they began to allow their children to study in public schools (Rozelin & Fauzan 2020). This happened because their view of life and view of the world were not the same as other people. Muhaimin (2016) analyzed the pattern of communication and social integration of OR. Their pattern of communication was not the same as other community groups, where the communication pattern still uses the traditional way with cultural patterns of symbols and direct face-to-face communication. Mailinar and Bahren (Nurdin 2013) said

*Corresponding Author: dianarozelin@uinjambi.ac.id

DOI: 10.1201/9781003322054-27

that this group was also known as the isolated community, which was often identified as a community with low levels of education, economy, and health. This research is not the same as previous research. The findings of this article are related to the effectiveness of using the Jambi Malay language in preaching to the OR community and the appropriate strategies for teaching Islamic *aqidah*.

Based on the data, OR, who lived in Lubuk Kayu Aro embraced Islam, but their understanding of Islamic values is not very good. Ustad Asman Hatta said in carrying out Islamic teachings that the understanding condition of OR about Islam was not consistent yet. They profess Islam but do not perform prayers, fasting, or zakat. Even most of them return to consume foods that are forbidden in Islam.

Haryanti (2019), Octarina *et al.* (2018), and Waridah (2016) stated that language functions as a tool for integration, social adaptation, and launching activities. When adapting to a social environment, a person will choose the language depending on the situation and conditions they face. The intersection between the Jambi Malay language and the OR dialect has been going on for a long time. Both groups try to understand what the interlocutor wants, and sometimes the conversation is done slowly so that what is conveyed can be well received.

In education, especially in the learning aspect, there are the terms strategy, method, and technique. The three terms have conceptual differences. According to Rianto (2010), the method is a way of delivering learning materials to achieve curriculum goals. The definition confirms that learning methods are ways of delivering learning materials in an effort to achieve curriculum goals.

Ahwan (2014); Siahaan, n.d.; and Maragustam (2016) stated that learning strategies are methods and procedures adopted by students and teachers in the learning process to achieve instructional goals based on teaching materials. Therefore, strategy is a method with an orientation that emphasizes student participation, so that both method and strategy contain procedural dimensions. Learning strategy is a method used by a teacher to deliver learning material so that it will make the students accept and understand the subject. The goal is that learning is mastered at the end of the activity.

Several opinions (Ansori n.d.; Cahyono 2016; Hakim 2016); Hidayat and Wakhidah, n.d.) are related to Islamic education in character building. Islamic education is an effort to guide, direct, and foster students, which is carried out consciously and in a planned way so that a personality is developed in accordance with the values of Islamic teachings. There are several types of strategies that can be used by *ustadz* in conveying religious values: (a) Exemplary behavior: good behavior can be imitated by others. The process of developing and implanting Islamic educational values can be done by giving examples to students, not only in the form of words but also actions; (b) Habituation: This method invites students to do something repeatedly so that it becomes a habit. In implanting the values of Islamic education, this method is very effective because it will train students with good habits. This habituation method has deeper implications than implantation values through how to act and speak; and (c) Advice: In its application, this method is more flexible and can be done anywhere and anytime. However, in the context of implanting religious values, the advice method is carried out with due regard to various things, including the method of delivery, grammar, place, time, and material delivered, so that it can be well received by students.

Ustad Asman used the Jambi Malay language when communicating and conveying Islamic teachings to the OR group. However, the OR has not been consistent in implementing Islamic teachings. One of the problems here is that there is no permanent *ustad* who can routinely teach and guide them in implementing Islamic teachings.

Quite a lot of research related to the values of Islamic education has been conducted by previous researchers. Some of these studies include a journal article by Ahmad Bukhori Muslim, entitled *Disadvantaged but More Resilient: The Educational Experiences of Indigenous Baduy Children of Indonesia*. The results of this study reveal that although the local government has issued regulations regarding the necessity of attending school, most parents and children from the Baduy community reject them and tend to ignore formal education.

Mailinar and Bahren Nurdin's article entitled *Religious Life of the Anak Dalam Tribe in Senami III Hamlet, Jebak Village, Batanghari Jambi Regency*. This finding reveals that the OR community who lives in this area has become Muslims, but their view of Islamic teachings is still lower. Their implementation of worship still combines Islamic teachings with their traditions. This can be seen in the *basale* and *tahlilan*. Based on several references that have been stated previously, it appears that the problems studied are more focused on aspects of the implementation of Islamic education in educational institutions, as well as religious life among OR, while the influence of Jambi Malay language use and the strategy of *ustadz* in teaching Islamic education have not been touched. As such, the researchers wish to conduct in-depth research on this matter.

Previous research related to OR, especially *Strategy Islamic Education Teaching*, had been conducted by Rozali (2021). He found that the supporting factors that are related to Islamic education include: (a) OR has long embraced Islam; (b) the existence of operational assistance for religious teachers; (c) the construction of public facilities and infrastructure; and (d) emotional closeness due to family relationships. Inhibiting factors were: (a) the unavailability of transportation assistance for religious teachers; (b) lack of appreciation from the community; (c) a lack of parental support for religious education; and (d) a lack of motivation to learn religion. Zakiyah Darajat (2004) revealed that most Islamic education is aimed at improving mental health, which is manifested in the form of actions, both for oneself and for other parties. In addition, Islamic education also combines theory and practice, so that there is a combination of faith education and charity education, as well as personal and community education.

Based on the foregoing discussion, this paper will examine more about the effect of using the Jambi Malay language in implanting the values of Islamic *aqidah* education in the OR group and explore the strategies used in implanting Islamic *aqidah* values in the OR group at Lubuk Kayu Aro.

2 METHODS

This research was qualitative and used the descriptive method, trying to describe the object or subject that is studied in depth, broadly, and in detail. Qualitative research, according to Moleong (2016); Azwar (2014a); and Sugiono (2015), describes events systematically, realistically, and in accordance with facts related to a particular field and is not intended to test hypotheses, produce predictions, or study impacts that occur. In selecting the sample, this study used a purposive sampling technique consisting of 1 *ustad* and 3 members of the OR. Data collection techniques used observation, interviews, and documentation, while in data analysis, the writers used a descriptive technique. The research location is a group of ORs who have settled in Lubuk Kayu Aro village.

3 RESEARCH FINDINGS

Jambi Malay is a language that is not much different from Indonesian, so that the interlocutor can understand the outline conveyed. Ustad Saman Hatta used the Jambi Malay language when communicating with OR but sometimes mixed (code mixing) the vocabulary spoken between the Jambi Malay language and OR dialect. The OR community, in Lubuk Kayu Aro village, is a descendant of Palembang.

According to Rozelin (2014), the OR in Jambi Province comes from three descendants, namely the Padang, Palembang, and Bathin group. Based on the vocabulary collected, it is certain that the OR in Lubuk Kayu Aro are descendants of Palembang. When researchers asked how many times a Muslim should pray in a day, they answered: *ade lime waktu, subuh, juhur, ashar, magrib, isya* (there are five times, *subuh, dzuhur, ashar, maghrib, isya*). One of the characteristics of the dialect of the OR from Palembang is the syllable /E/-/limE/, while the OR from Jambi and Padang syllables that appear is /o/-/limo/.

In the beginning, Ustad Saman used Jambi Malay speech in the OR group, and then he tried to learn the Kubu dialect, mixing Jambi Malay vocabulary and OR dialect (code mixing). A few months later, the *ustad* was able to master the OR dialect well. It helped him convey Islamic teachings to the OR group. Of course, this adjustment takes a long time, but continuous communication and frequent meetings will help establish a good understanding between them.

The effect of using the Jambi Malay language at the beginning of preaching did not have a negative effect on the OR people. The researchers asked them: *Did any misunderstanding ever happen when the ustad gave a lecture using the Jambi Malay language?* The answer from one of the members of the OR group was, *Alhamdulillah, setau kami tak pernah, kami tentu ape yg di sampai ke, walaupun sebahagian* (Alhamdulillah, as far as we know, never; we certainly know what he explains, although some of them). Based on that explanation, it can be concluded that the use of the Jambi Malay language when preaching in the OR group can still be understood because their vocabulary comes under the same umbrella as the Malay language.

The vocabulary used by the *ustad* by mixing Jambi Malay Language and OR dialect, for example: *Kalo ada masalah kite serahkan semuanya pada Allah karena hanya Allah yang bise* ngaseh jalan, *sedangkan manusie cuma bise bantu* (If there is a problem, we leave it to Allah because only Allah can give the way, while humans only help). In this sentence, it can be seen that the beginning of the sentence uses the Jambi Malay language, then it is mixed with the OR dialect. This shows the effort of the *ustad* to understand and get closer to the OR group.

The researchers also asked the OR: Which one would you like when the *ustad* is preaching? Jambi Malay language or mixing between Jambi Malay language and OR dialect? *Kami tentu ape yang disampaikan ustad kalo memakai bahasa Jambi walau separoh, tapi bile campur dengan bahase kami, kami senang, lebih bagus kalo pakai bahase kami* (We understand what is being said when the *ustad* uses Jambi language even though half of it, but if it is mixed with OR language, we are happy, and it is better if the *ustad* uses our language). This explanation shows that the OR society is more comfortable if the *ustad* speaks in their language than the Jambi Malay language. To teach new beliefs to marginalized groups, *ustadz* must use a special approach that is not the same as the approach taken in rural or urban communities.

The delivery of Islamic teachings by the *ustad* includes the values of *aqidah* education, which is faith or belief. *Aqidah* in Islam is reflected in the pillars of faith. There are 6 pillars of faith, namely, Faith in Allah, Faith in Angels, Faith in Holy Books, Faith in Rasul, Faith in the Last Day, and Faith in Qada and Qadar. Submission is done slowly by giving examples and discussing while drinking tea. This was done because the level of OR education was not high, so the *ustad* had to use a suitable strategy so that what was conveyed could be understood well by them This situation not only happened with OR but also with another group called the Badui community (Muhammad Hakiki 2015; Muslim 2021). The table below shows the condition of OR education in Lubuk Kayu Aro village.

Table 1. OR education in Lubuk Kayu Aro village.

No	Education	Male	Female	Total
1	Elementary	15	10	25
2	Junior High School	4	6	10
3	Senior High School	–	2	2
4	Bachelor	–	–	–
5	University	–	–	–

Furthermore, efforts to describe Islamic educational values begin by teaching *tauhid* first. In this case, the OR is introduced to the pillars of Islam and the pillars of faith first. They are invited to know, understand, and believe in the existence of Allah, angels, the Holy Book, prophets, the last day, *qadha* and *qadar*. After that, they are taught to perform ablution, mandatory bathing, prayer, fasting, and *zakat*. This effort is carried out in stages. Since 2010, the OR has been carrying out worship, albeit gradually. In addition, they were also taught how to read the Qur'an, *hijaiyah* letters, and then to read *Iqro`*.

This activity is routinely carried out through the *pengajian* groups, not only for men but also for women. Recitation or *pengajian* for men is held every Friday night, while for women it is taught through *Majelis Ta'lim*. Although at first, many people attended this routine recitation, but it gradually began to decrease. According to Hermanto, this situation happens because they feel ashamed in terms of age, find it difficult to study, have no time because they have to work in the garden, and so on. This activity is held in the form of a direct approach to OR by providing explanations and solving problems related to the practice of Islamic teachings. Thus, they feel guided and finally willing to carry out the rules of Islamic law with pleasure.

This is in line with the *ustad*'s opinion that the main strategy of da'wah (especially for isolated communities) is problem-solving, which is a democratic effort for the development and improvement of the quality of life as part of empowering humans and society in solving various objectives of life problems. The concept and strategy of da'wah, which is directed at problem-solving or liberation of various problems in people's lives, will in turn, be positive (Azwar 2014b; Djamarah 2020; Suryana 2015; Uno 2011).

The strategy that was used by the *ustad* in conveying the values of Islamic *aqidah* education among the residents of OR Lubuk Kayu Aro was a personal approach consisting of exemplary behavior, habituation, and advice. We described the values of Islamic *aqidah* education to the OR by giving an example first: How to pray properly and correctly; How to perform ablution; What is halal and good food; and so on. So, they can easily understand through the examples we teach. Educators set an example through good exemplary stories, such as the history of the Prophets, friends of *Rasulullah*, and heroism stories in Islam. The goal is for the OR to make these figures as examples in their lives. As stated by the *ustad*: "to make them understand the values of Islamic *aqidah* education is not short; it takes two to three years for them to understand and implement it properly and correctly".

Habituation was done to ensure OR are accustomed in thinking and acting in accordance with the provisions of Islamic teachings. The essence of habituation is an activity that is done repeatedly so that it becomes a habit. The learning process through habituation involves the formation of new habits or the improvement of previously existing habits. Learning through habituation can be used in various ways, either in the form of orders, role models, special experiences, punishments, or rewards.

4 CONCLUSIONS

The delivery of Islamic *aqidah* educational values using the Jambi Malay language to the OR group initially encountered a few obstacles. Then, the *ustad* tried to understand and learn the OR dialect and finally did code mixing in communicating. Material delivery that is related to Islamic *aqidah* educational values is carried out using *a personal approach strategy* consisting of exemplary behavior, habituation, and advice. The habituation strategy was carried out because previously the OR were animist and dynamist, so it was not easy for them to leave their old beliefs.

This research had limited time and only covered the influence of the use of the Jambi Malay language on the implementation of Islamic *aqidah* in the OR community. As such, future research can examine the language shift, language maintenance, the religious life of the OR, their view of life, and their view of the world.

REFERENCES

Ansori, R.A.M., n.d. Strategi penanaman nilai-nilai pendidikan islam pada peserta didik, 19.

Azwar, S., 2014a. *Metode Penelitian*. Yogyakarta: Pustaka Pelajar.

Azwar, W., 2014b. *Sosiologi Dakwah*. Padang: IAIN Imam Bonjol Press.

B. Uno, R.C., 2011. *Model Pembelajaran Menciptakan Proses Belajar Mengajar yang Kreatif dan Efektif*. Jakarta: Bumi Aksara.

Cahyono, H., 2016. Pendidikan karakter: Strategi pendidikan nilai dalam membentuk karakter religius. *Ri'ayah: Jurnal Sosial dan Keagamaan*, 1 (02), 230.

Daradjat, Z., 2004. *Ilmu Pendidikan Islam*. Jakarta: Bumi Aksara.

Djamarah, S.B., 2020. *Strategi Belajar Mengajar*. Jakarta: Rineka Cipta.

Fanani, A., 2014. Mengurai kerancuan istilah strategi dan metode pembelajaran. *Nadwa: Jurnal Pendidikan Islam*, 8 (2), 171–192.

Hakim, R., 2016. Pola pembinaan muallaf di kabupaten sidrap provinsi sulawesi selatan. *Al-Qalam*, 19 (1), 85.

Haryanti, E., 2019. Penggunaan bahasa dalam perspektif tindak tutur dan implikasinya bagi pendidikan literasi. *Jurnal TAMBORA*, 3 (1), 21–26.

Hidayat, S. and Wakhidah, A.N., n.d. Konsep pendidikan islam ibnu khaldun relevansinya terhadap pendidikan nasional, 16 (1), 10.

Maragustam, M., 2016. *Filsafat Pendidikan Islam, Menuju Pembentukan Karakter Menghadapi Arus Global*. Yogyakarta: Kurnia Kalam Semesta.

Moleong, L.J., 2016. *Metodologi Penelitian Kualitatif*. Bandung: Remaja Rosdakarya.

Muhaimin, M., 2016. Melangun: Views of Life Orang Rimba (Remaining Local Wisdom). *Jurnal Gubernantia*, 1 (1).

Muhammad Hakiki, K., 2015. Aku ingin sekolah; potret pendidikan di komunitas muslim muallaf suku baduy banten. *Islam Realitas: Journal of Islamic and Social Studies*, 1 (1), 1.

Muslim, A.B., 2021. Disadvantaged but more resilient: The educational experiences of Indigenous Baduy Children of Indonesia. *Diaspora, Indigenous, and Minority Education*, 15 (2).

Nurdin, B., 2013. Kehidupan Keagam aan Suku Anak Dalam di Dusun Senam i Iii Desa Jebak Kabupaten Batanghari Jambi, 28 (2), 17.

Octorina, I.M., Karwinati, D., and Aeni, E.S., 2018. *Pengaruh Bahasa di Media Sosial Bagi Kalangan Remaja*, 1, 10.

Rianto, Y., 2010. *Paradigma Baru Pembelajaran: Sebagai Referensi Bagi Guru/Pendidik dalam Implementasi Pembelajaran yang Efektif dan Berkualitas*. Jakarta: Kencana.

Rozelin, D., 2014. *Dialek Melayu Orang Rimba di Provinsi Jambi: Kajian Dialektologi*.

Rozelin, D. and Fauzan, U., 2020. Education and proto language maintenance at Orang Rimba in Jambi province. *IJELTAL (Indonesian Journal of English Language Teaching and Applied Linguistics)*, 5 (1), 177.

Siahaan, A., n.d. *Strategi Pendidikan Islam dalam Meningkatkan Kualitas Sumber Daya Manusia Indonesia*, (1), 20.

Sugiono, S., 2015. *Memahami Penelitian Kualitatif*. Bandung: CV. Alfabeta.

Suryana, Y., 2015. *Metode Penelitian Manajemen Pendidikan*. Bandung: Pustaka Setia.

Waridah, W., 2016. Berkomunikasi Dengan Berbahasa Yang Efektif Dapat Meningkatkan Kinerja. *Jurnal Simbolika: Research and Learning in Communication Study*, 2 (2).

Religion, Education, Science and Technology towards a More Inclusive and Sustainable Future – Rahiem (Ed.)

Analysis of the utilization of instructional media and technology during instruction of Arabic language at Umma University

O.R. Omukaba*
Arabic Department Umma University, Kenya

ABSTRACT: Technology advancement has become integral in all spheres of human life, including the education sector. In that respect, we cannot separate instructional media and technology (IMT) from education. This study analyzes the available IMTs vis-a-vis utilizing them for instructing the Arabic language at the Umma University, Kajiado Campus. The research was driven by a lack of adequate IMT in many institutions, and sometimes, if provided, they were underutilized either due to a lack of proficiency or the feeling that they were not important. The study is guided by the cognitive theory of multimedia learning developed by Richard E. Mayer (Mayor 2005). A descriptive survey design was employed to collect data through questionnaires and interviews, and a sample size of 30 participants was drawn from the target population of 80 lecturers and students by purposive sampling. Qualitative data was analyzed and presented thematically in narrative form, and quantitative data was analyzed using SPSS version 22.0. The study findings revealed that IMT is essential in enhancing the instruction of the Arabic language at Umma University, but its availability is minimal and it is underutilized. Thus, it underscores the need for increased availability of IMT and establishing initiatives that will ensure more of its utilization. It recommends that the administration ensures provisions of IMT and monitor its utilization by requiring that all lecturers use IMT in teaching.

Keywords: instructional media, technology, instructing, learning, Arabic language

1 INTRODUCTION

There is strong and growing evidence from educational practitioners that the teaching and learning process has tremendously gained power with the innovation of instructional media and technology. The history of educational technology goes back to the Stone Age, when humans started scratching figures and images on the surface of rocks as a medium of communication (World Encyclopedia 2001). As man's brain developed further, he started using the bark of trees and textiles as a medium of writing. Interest in instructional media and technology (IMT) increased, and by 1950, the utilization of computers started. In 1980, there was an increase in the use of computers as a medium of instruction, which led to the development of a more advanced IMT. Some of the modern IMTs that have emerged to advance the field of education include: Distance learning, learner-centered learning, web tools, the internet and virtual environments (such as Second Life, wikis, and blogs) (Center for Social Organization of schools 200)

Instructional media refers to the physical means through which instruction is conveyed to learners (Reiser & Gagne 1983). It applies to all materials and resources that a teacher may employ in education to achieve learning objectives (Scanlan 2003). This includes materials such as black and whiteboards, printed media (handouts, books, and worksheets), display boards, charts, slides, overheads, real objects, and video tapes or films, as well as the latest

*Corresponding Author: omukaba@umma.ac.ke

DOI: 10.1201/9781003322054-28

media such as computers, models, DVDs, CD-ROMs, smartboards, the internet, and interactive video conferencing, etc. (Talabi 2001).

A study conducted in the United States revealed that media and technology employed for instructional purposes engage learners strongly in the learning process (Mohan *et al.* 2001). Dr. Chidi E. Onyenemezu and E. S. Omulati of Nigeria indicated in their study that, regardless of the subject teacher and the level of learners, the role of educational media is paramount. Naomi Kutto Jebungei of Kenya, in her study in Eldoret, concluded that, from all indications, students taught with IMT absorb more knowledge from their interactions with the resources. They understand what is taught better and faster than with the use of textbooks, chalkboards, and lectures.

It is against this background that the study opts to suggest effective ways of improving the teaching and learning of the Arabic language at Umma University, located in Kajiado County, Kenya. Arabic is a foreign language to Kenyans and is taught in very few schools. It is not spoken on the streets, which makes its learning very difficult. Therefore, this calls for Arabic instructors to employ a variety of techniques to teach the language. In that respect, IMT plays a vital role in learning this language. Umma University is among the few institutions that offer Arabic in the country. It is important to be taught effectively. Moreover, the Kenyan government through the Ministry of Education, has a policy of higher education (2006), with the vision to make ICT a universal tool in education and training. However, educational technology is not effective yet due to a lack of adequate resources and the insufficient qualifications of teachers. This drives the need to determine the IMT available at Umma University and the extent of its utilization.

1.1 *Statement of the problem*

With the observation and criticism of education technology, which has been cited as heavily affecting the teaching and learning process, several studies have proved that IMT is a significant factor in effective learning and teaching processes as it bridges the gap between teaching and learning at all levels of education. Different studies have shown that the use of several teaching media and appropriate teaching methods is far better than lengthy explanations. It enhances learning for students with different learning styles (Montgomery 1995).

Responding to that, Kenya as a country came up with a policy of higher education in 2006, with the vision to make ICT a universal tool in education and training. Based on that perspective, the overall problem addressed in this study is to find out if Umma University is adhering to this policy by providing a variety of IMTs to lecturers and students for teaching and learning Arabic. Provision alone is not enough; it is also important to find out if the available IMTs are fully utilized. Nevertheless, there are challenges in the field of educational technology, such as inadequate provision of instructional resources, a lack of skills and creativity in using them effectively, and some of the lecturers and students lack innovation skills to come up with substitutes that can make the teaching and learning processes interesting, and above that lack of interest or attitude, which calls for motivation.

An important point to note is that we should not only provide IMT to institutions, but efforts should be made to make sure that they are adequately utilized. Studies on the types of IMT used for teaching several disciplines and how to utilize them are available. Moreover, countries where the Arabic language is their instructional medium have studies on IMT applicable for teaching and learning, but in Kenya, none exists. This is a gap that needs to be filled.

It is against this background that the study aimed to investigate the availability and utilization of IMT during the instruction of the Arabic language at Umma University, Kajiado Campus, Kenya.

1.2 *Objective of the study*

1. Identify the IMT available at Umma University for instructing Arabic

2. Evaluate the extent to which lecturers and students utilize the media and technology available for instructing and learning the Arabic language at Umma University

1.3 *Theoretical and conceptual framework*

The study was guided by the cognitive theory of multimedia learning developed by Richard E. Mayer and his colleagues in the year 2000. The theory operates on the fundamental principle that the way the human brain learns is supported by multimedia (Mayer 2005). It says that people learn more using pictures and words than they learn from words alone. This conquers our study, which asserts that IMTinfluence the learning process. According to the cognitive theory of multimedia learning, the main goal of multimedia is to encourage the learner to construct new knowledge by building a wide mental presentation from the material presented as an active participant.

2 METHODS

The study employed a descriptive survey design and purposive sampling techniques. Questionnaires and interviews were used to collect the data. A sample of 30 participants was drawn from the target population of 80 lecturers and students. It was composed of diploma and certificate students since they are the ones pursuing the Arabic language at Umma University. A selected sample of 17 diploma students, 7 female and 10 male, were picked out of the total 52 diploma students. For the certificate level, 10 students, 5 female and 5 male, were selected from a population of 25 certificate students, and all 3 lecturers were involved. This was credible to provide reliable results that could be representative of the whole department. In addition, the sample size of 30 respondents in a population size of approximately 80 people yields results that have a 95% confidence level and a 15% confidence interval.

This gave rich data, which assisted in finding out the available IMT at Umma University and the extent of its utilization. Data of qualitative nature was analyzed and presented thematically in a narrative form, while quantitative data was analyzed using SPSS version 22.0

3 FINDINGS AND DISCUSSIONS

Table 1. Instructional media and technology available at Umma University for instructing Arabic.

Resource material	Lecturers	%	Students	%	HOD	Observation
Textbooks	3	100	22	85	Yes	Yes
Language laboratory	–	–	–	–	–	–
Whiteboards	3	100	22	85	Yes	Yes
Computers	–	–	15	58	Yes	Yes
Smartboards	2	66	10	38	Yes	Yes
Overhead projector	1	33	10	38	Yes	Yes
Charts	2	66	15	58	Yes	Yes
Journals	2	66	10	38	Yes	No
Handouts	4	80	14	52	Yes	Yes
Magazines	2	40	10	37	Yes	No
Laptops	3	100	–	–	Yes	Yes
Pamphlets	–	–	3	12	No	No
TV	–	–	–	–	No	No
Internet connectivity	3	100	22	85	Yes	Yes

Regarding the availability of IMT, the study found that internet connectivity and textbooks were most accessible at Umma University, as almost all the respondents confirmed their availability. Based on the results, the availability of whiteboards is second, followed by smartboards, overhead projectors, charts, journals, handouts, computers, and magazines. The university lacked a language laboratory, laptops, pamphlets, and TVs. The results on the availability of IMT nearly correspond with the findings on its utilization. For instance, all the lecturers confirmed that they often use textbooks, whiteboards, handouts, and the internet when delivering lectures. Most of the lecturers occasionally deliver lessons using charts, projectors, smart boards, and laptops. The results also indicate that TVs, journals, pamphlets, and language laboratories are never used by all the lecturers during classes. The results on the utilization of IMT by students also correspond to the findings on availability. For example, a majority of the students (at least 60%) stated that they often use whiteboards, textbooks, smartphones, and the internet for learning. Computers, charts, magazines, journals, projected media, and smart boards are normally used by students for learning. The least used IMT by students includes pamphlets and journals.

The relative comparison of availability and utilization of IMT by both learners and lecturers indicates that an increase in the usage of IMT at Umma University would be initiated if the availability was improved. In addition, though all media are important, there is an urgent need of increasing the availability of projected media because it enables students to learn through videos and PowerPoint presentations, which enhance the learning environment and understanding of the courses. The projected media can substitute for other IMTs, such as charts and TVs if the lessons can be converted into appropriate soft copy format. The language laboratory is essential for learning language, which is missing at Umma University (Curzon 2001). It is therefore recommended that the university step up and provide the missing IMT to improve the learning and teaching of the Arabic language. Despite the availability of computers, the results also indicate that a few learners often use the technology for learning purposes. Instead, the students use smartphones and the internet to learn. Computers have large screens and have more applications that can ease learning compared with smartphones.

4 CONCLUSIONS

The study has potential limitations: (1) The unwillingness of respondents to participate in the study at the time of data collection as they were preoccupied with lessons: to overcome this, interviews were carried out during break time, lunchtime, or when the respondents were free to avoid interruption with lessons. (2) The unwillingness of the respondents to disclose some information, especially lecturers and students, for fear of disciplinary action from the administration: the respondents were assured of the confidentiality of any information they provided. (3) Dishonest response from the university management in case of shortage in procuring IMT for the Arabic language department: the researcher convinced them that the information collected was only meant for educational purposes.

In conclusion, the results show that generally internet connectivity, textbooks, and whiteboards are most accessible at Umma University, Kajiado Campus, followed by computers, and then charts, journals, handouts, and magazines. The university lacks laptops, smartboards, overhead projectors, language laboratories, TVs, and pamphlets, which are essential for teaching and learning any language. The results on the availability of IMT are generally consistent with the findings on their utilization in the institution. The results also indicate that lecturers and students lack training in the utilization of IMT. Both instructors and students had positive opinions regarding learning using IMT. The study, therefore, makes a significant contribution to the literature about utilization of IMT in classrooms. Most of the findings in this study are consistent with results from other related research (Green et al. 2015; Ma & Au 2014; Seok et al. 2010).

The study shows that IMT is essential in enhancing learning, but its availability is minimal and it is underutilized. Thus, this study underscores the need for increased availability of IMT in learning institutions and establishing initiatives that will ensure more of its utilization.

5 RECOMMENDATIONS

The main findings suggest that several essential IMTs, such as overhead projectors, language laboratories, journals, laptops, pamphlets, and TV, among others, have limited availability and are underutilized at Umma University. The results also indicate both lecturers and students have positive attitudes toward utilizing IMT for learning. Based on the obtained results, it is recommended that the administration introduce more IMTs at Umma University, Kajiado Campus. The administration can achieve that by developing a funded program through which more types of IMT will be purchased and human resources to train both lecturers and students on the use of IMT. This may involve writing proposals for funding to other stakeholders, such as the sponsors and the government, indicating the need for purchasing more IMT in the institution. The administration can ensure that IMT is utilized in the institution by requiring that all lecturers use IMT in teaching. The lecturers can be required to keep a record of lessons delivered and the IMT that they used, and then submit the data to the administrators. To ensure that the students become proficient in using IMT, they can be required to submit assignments that involve IMT, such as completed assignments submitted via email, presenting PowerPoint slides, and recording and decoding audio. Learning institutions in developed countries have made major advances in the use of IMT because students can opt for distance learning online and through video and audio conferences (Watson & Pecchioni 2011). The stakeholders of Umma University, Kajiado Campus, need to realize the potential benefits of IMT and initiate activities that will contribute to its utilization in the institution.

REFERENCES

Amutabi, M. (2004). Challenges facing the use of ICT in Kenyan Universities. *UNESCO Forum Mollegiums on Research and Higher Education Policy 1–3 December 2004*

Bogdan, R. & Biklen, K. (2003). *Qualitative Research FOR Education. An Introduction to Theories and Methods*. Fourth Edition, New York

Dahiya, S. (2004). *Education Technology towards Better Teacher Performance*. Shipra, New Delhi

Duffy, J. Johassen D, and Lowyck, J. (1993). *Designing Constructivist Learning Environments*. Springs- w.w.w. zu.ac.ae/the. http://groups.yahoo.com/LTHE

Green, A. J., Chang, W., Tanford, S., & Moll, L. (2015). Student perceptions towards using clickers and lecture software applications in hospitality lecture courses. *Journal of Teaching in Travel & Tourism*, *15*(1), 29–47. doi:10.1080/15313220.2014.999738

Gu, P., & Guo, J. (2017). Digital case-based learning system in school. *PLoS One*, *12*(11), 1–15. doi:10.1371/journal.pone.0187641

Kimamo, G. (2012). *Educational Communication & Technology*, CUEA, The Catholic University of East Africa.

Moore, A.H; Moore, J.F. Bofern, P; H.R (2003). *Information Technology Review*, Zayed University Dubai, United Emirates. www.zu.ac.ae/the

Muriithi, P. (2005). *A Framework for Integrating ICT in the Teaching and Learning process In Secondary Schools: School of Computing and Informatics*. University of Nairobi. eomwenga@uonbi.ac.ke

Omariba, A. (2012). "Challenges facing teachers and students in the use of instructional technologies: *A Case of Selected Secondary Schools in Kisii County, Kenya*". Unpublished Thesis, Kenyatta University.

Omwenga, I.E. (2008). *Pedagogic Issues and E-Learning Cases: Integrating ICTS into Teaching-Learning Process. School of Computing and Informatics*. University of Nairobi, eomwenga@uonbi.ac.k

Seok, S., DaCosta, B., Kinsell, C., & Tung, C. K. (2010). Comparison of instructors' and students' perceptions of the effectiveness of online courses. *Quarterly Review of Distance Education*, *11*(1), 25–36.

UNESCO, (2002). *Information and Communication Technologies in Teacher Education: A Planning Guide*. Division of higher Education, France (Paris): UNESCO.

UNESCO, (2005). *Toward the Knowledge Society: UNESCO World Report*: Paris: UNESCO. Available online, www.unesco.org/en/worldreport

Watson, J. A., & Pecchioni, L. L. (2011). Digital natives and digital media in the college classroom: Assignment design and impacts on student learning. *Educational Media International*, *48*(4), 307–320.

Religion, Education, Science and Technology towards a More Inclusive and Sustainable Future – Rahiem (Ed.)

The mythology of Putri Mandalika in the Sasak Islamic tradition in Lombok

F. Muhtar*, S.A. Acim, Ribahan & A. Fuadi
UIN Mataram, Indonesia

ABSTRACT: The context of this essay is the prevalence of the mythology of Putri Mandalika, who changed into Princess Nyale. The Mandalika myth is inextricably entwined with Sasak culture. For Sasak women, the myth of the Mandalika princess is to represent enthusiasm. This study aims to investigate religious practices in which Princess Mandalika is a mythical figure. Feminist and phenomenological approaches were applied in this investigation. The findings of this research demonstrated that the animism and vibrancy of Lombok society before the coming of Islam were the roots of Princess Mandalika mythology. In addition, the existence of Princess Mandalika is seen as a symbol. The mythology of Mandalika is easier to comprehend from the perspective of Wasathiyah Islam. This study also demonstrated how Mandalika mythology in Lombok has evolved into a representation of feminine energy and has remained contemporary as a result of people's unwavering adherence to Islamic beliefs.

Keywords: mythology, Princess Mandalika, Islamic religion, Sasak tradition, Islamic moderation

1 INTRODUCTION

According to Azyumardi Azra, religion has myths; therefore, some religious people still believe in myths (Ramakrishna & Tan 2003, p. 39). One such myth is the myth about Princess Manda Lika that developed among the Sasak Islamic community in Lombok, which is religious and famous for the island of a thousand mosques.

The legend of the Sasak people describes Princess Manda Lika as a person who never gives up, has an attractive face, and has a positive outlook on life. The community around Seger mandala Beach performs traditional Sasak rituals and ceremonies in memory of Princess Mandalika (Sujarwo 2019; Wahidah 2019). Mandalika has become one of the ten government national economic zone projects and tourist attractions. Mandalika, in the tradition of the Sasak people, is considered a sacral tradition.

According to Robert Graves, the basis of myth is the desire of a person to understand the universe and to view the universality of the existence of myth as an interesting medium of justification throughout the world. Mandalika mythology is considered the most appropriate name for naming important places among some of the Sasak people (Sellers 2001).

The transformation of one person into another is the subject of many ancient mythologies. For instance, several of Cinderella's most well-known tales depend on our capacity to accept that the transformation was feasible: A soiled hedgehog can turn into a beautiful princess (Lies 1999, p. viii).

The myths we now refer to most likely developed when the ancients told these tales to one another in an effort to explain the world and its existence. Even though myths have gods and

*Corresponding Author: fathurrahmanmuhtar@uinmataram.ac.id

 DOI: 10.1201/9781003322054-29

goddesses, they shouldn't be confused with traditional faiths. The feminine Mother Earth, who gives life and bears fruit, is the starting point of mythology (Lies 1999, p. vii). Compared with men, women make up 60%–80% of those who contribute to food security and variety (Doss et al. 2018).

It may seem more logical to think of life as an inflow and outflow of transformations before these discoveries. According to certain metamorphosis myths, a weaver transforms into a spider that spins its web continuously to create the universe. Another example illustrates the difference between gods and humans: A deity disguises a girl as a cow, so his wife cannot find her. The significance of this research is to show how the Princess Mandala legend becomes a metaphor for ideals that might elevate the status of women. The metamorphosis of the Mandala princess into a "Yale" who can produce food for the people illustrates it.

This study contributes to the findings about how mythology can be used as a guide in daily life, since myths include ideals rooted in ancient religious traditions. Regarding equal rights to public expression, women in Islam are on par with men. Ibn Rush, an Islamic philosopher who lived in the 11th century AD, provided a description and commentary by stating that, in his opinion, women and men share the same basic biological makeup and primary function as humans (Al-Jabiri 2003, p. 256).

When a man proposes to a woman in Sasak society, he must go through a lengthy traditional procession, and incur significant costs to fund the various stages of the marriage tradition (Sumadi 2013, p. 49). Islam, with its teachings of religious moderation, is accommodating to the religious traditions of the Sasak community. Religious moderation contributed to preserving and maintaining the traditions of the Sasak people, although syncretism in those traditions began to erode with moderate Islamic teachings.

Bianca Smith once researched women mythologies – the relationship between a Princess Goddess Anjani who is the "Queen of Jinn," residing in Mount Rinjani and considered to be the guardian of Allah. She led Lombok from Mount Rinjani, which influenced an Islamic female figure in Lombok (Smith & Woodward, n.d., p. 207). The daughter of goddess Anjani is very popular and is a symbol of mystical power that influences Sasak female leaders in Lombok. This paper discusses the mythology of Mandalika in the traditions of the Sasak people and the interpretation of the mythology of Mandalika from the perspective of religious moderation.

2 METHODS

This research is qualitative, with several data collection methods: in-depth interviews with ten respondents. The topic of the interview consists of life history, experiences in detail, and meaning (Seidman 2006, p. 17). Direct observation of various religious tradition research and documentation-related activities constitutes an observation. The close relationship between phenomenology and ethnography serves as a foundation for understanding reality in the context of religious life and behavior. According to Mircea Eliade, the phenomenon of religious people has always been associated with the sacred (Abdullah 1999, p. 35). In addition to phenomenology, this study makes use of an ethnographic method and Van Kaam's theory of analysis to comprehend the beliefs that people develop throughout the data analysis stage of their lives (Kuswarno 2009, p. 69).

3 RESULTS AND DISCUSSION

3.1 *Mandalika mythology in the tradition of the Sasak people*

The first investigation reveals the Mandalika princess legend in the Sasak tradition has doctrinal principles concerning Sasak women's persistence in battling life hardships. Mandalika Mythology is employed in place names such as Mandalika Circuit, Mandalika

University, and Mandalika Terminal. Religious moderation helped preserve Mandalika mythology's principles.

Mandalika represents Sasak women in public arenas including school, work, and work in general. Women have always induced negative things, such as the narrative of Siti Hawa, a metaphor for humanity's mother, and Israel's portrayal of the people (Israilliyat). "When the serpent went to heaven, the devil sprang from his gut," says Ibn Jarir. He gave Eve the forbidden fruit from Adam and his wife. Eve ate the fruit. (M.I.M, 2016, p. 177)

After Adam ate the fruit, he went to him and told him about what the devil had said. In the Qur'an or the Hadith, Eve is not explained in depth. According to Habib Ziadi, in the cultural realm of Lombok Sasak, the narrative about Princess Mandalika is a legend story told for generations, which has become the belief of some people that Mandalika has transformed into "Putri Nyale," which is a sea worm that is used as a meal for the surrounding community (Habib Ziadi, personal communication, July 1, 2022).

The researcher observed that "Bau Nyale on the Seger beach" caught the sea worms at a party. Tens of thousands of people thronged the beach to catch the Nyale, which appeared in the sea on the fifth day of the full moon of the 10th month of the Sasak calendar, February 19–20, 2022. People stayed in setting up tents along the beach, catching Nyale at night. (Observation, February 19, 2022)

The myth of Putri Mandalika, a heredity belief that became the local culture of Sasak where Princess Mandalika threw herself into the sea so that she became Nyale, is a symbol that encourages women to be more advanced and independent (Lalu Zulkifli, personal communication, July 4, 2022). Wasathiyah Islamic teachings that are assimilative and accultirative play a role in preserving the positive values of the tradition of celebrating Mandalika rituals (Ali Sukma Jaya, personal communication, July 1, 2022) According to Ahmad Suja'i, Islamic teaching is acculturative in addressing this culture (Ahmad Suja'i, personal communication, July 1, 2022). Salimul Jihad also stated the existence of Wasathiyah Islam as a link and interaction between Islam and culture (Salimul Jihad, personal communication, July 4, 2022).

The myth of Mandalika, according to Nasihun Badrin, Secretary General of the Nahdlatul Wathan Diniyah Islamiyah (NWDI) must be viewed from the aspect of Islamic teachings that maintain good old ways and accept new, better things (*al-muhafadza ala qadimis sholih wal akhdzu bil Jadidil Ashlah*) (Nasihun Badrin, personal communication, July 4, 2022). Islam is here to care for culture, not to destroy culture, What is the proof? When the Messenger of Allah was present never to damage the Arab culture, his presence in the Arab culture still exists (TGH, Mythology about women in the Islamic tradition has existed since the story of Isra'iliyyat; the Islamic tradition extensively addresses gender misunderstandings. Some legends come from divine religions before Islam, including Judaism and Christianity. It is possible that former believers of both religions who later converted to Islam provided these anecdotes. Additionally, it is conceivable that adherents of both faiths do this to obstruct Islam's teachings (Umar 1999, p. 286). Eve's portrayal as a temptress in the Bible hurt Judaeo-Christian women. All women inherited Eve's guilt and cunningness. The characteristics of women are summed up as unreliable, immoral, and evil. Menstruation, pregnancy, and childbearing are just punishment for the female sex ('Abd al-'Aẓīm, n.d., p. 7).

The interpretation of gender verses in the Qur'an makes extensive use of tools from Jewish-Christian history due to the "customary law" that predominated in Medina society—where the legal verses of the Qur'an were revealed—and the prevalence of hadith in the Jewish tradition. Not to mention that during Islam's formative years, the early Jewish communities served as the means of spreading (futuhat) the religion. The Prophet avoided some Jewish traditions because they were thought to go against the fundamentals of "absolute monotheism" (*tawhid*) (Ahmed 1992, p. xv).

The Qur'anic genesis story contains tales about the existence of women, such as the Israilliyyat myth. The Old Testament contains tales that are typically thought to portray women negatively, such as feminist encounters where women are present to meet some of the men's demands (2:20). Because she was made from a man's rib, women are viewed as the second creation and as inferior to men (2:21–22). Women must take a larger risk when considering the idea of heredity since they are held responsible for the cosmic drama that led to humankind's falling to Earth (3:12). (3:12). The Talmud, a text that examines Old Testament verses, goes into great detail to interpret these verses (Umar 2014, p. 9).

Moslem philosopher Al-Biruni was the most knowledgeable medieval Muslim writer regarding mythology in Greek gods. Bruni cites "the mythologists" (*ashab al-amthal*), "the well-known histories" (*al-tawarikh almashhura*), and "the historians" (*al-mu'arrikhun*) as his sources. He references the Phaenomena of Aratus, an epic poem on stars and mythology. "Whatever his sources were – perhaps Christians – they were not generally known; Greek mythology is largely unknown to other authors (Walbridge 2001, p. 27).

The scholars of the classical era disagreed about the strategic role of women in Islam, just as they disagreed about whether women could be prophets. The great Andalusian scholar Abu Bakr Muhammad Ibn Mawhab al-Tujibi al-Qabri (d. 406 H/1015 AD), who lived in the second part of the 4th century AH/10 AD, made a contentious claim that it was acceptable for women to be prophets and to receive revelations from the prophets. According to Allah, Maryam, the mother of the prophet Isa (as), was one of them. Other scholars, such as Abu Muhammad Abdullah ibn Ibrahim ibn Muhammad ibn Abdullah ibn Ja'far al-Ashili (d. 392 H/1001 AD), instantly rejected this viewpoint. He contends that Maryam is just a *siddiqah*, as described in the QS, and not a prophet. Al-Maidah [5]: 75, *wa ummu hu siddiqah* (and her mother was a very righteous person) (Umar 2014, p. 9).

3.2 *The myth of Mandalika in the interpretation of religious moderation*

All great religions and civilizations emphasize moderation. Graeco-Judaic and Christian creeds call it the "golden path," whereas Confucianism and Islam call it Chung Yung and Wasathiyah (Kamali 2015, p. 1). Azyumardi Azra describes Islam Wasathiyah as moderate, inclusive, and tolerant. Wasathiyah Islam is sometimes termed justly balanced Islam. The Islamic term Wasathiyah is from the Qur'an and means "middle folk" (Qur'an 2:143). Wasathiyah is the ideal position, according to a hadith of Prophet Muhammad SAW (Azra 2020, p. x). Religion is depicted in stories and myths that show how man must translate abstract concepts based on culture for religion to be accepted by his followers (Nimer 2010).

The Islamization of Hindu doctrine caused Muslims and Hindus in Lombok to have the same cultural characteristics. Mandalika mythology combines religion and culture. Religious traditions involve rituals of respect and service involving sacred symbols. Culture influences personal and social behavior (Liliweri 2016, p. 59). Kuntowijoyo said Indonesian rituals show Islamic influence. Islam influences our popular sociocultural ceremonies (1999, p. 235). Haidar Baqir said that cultures must interact/acculturate so that the culture is rich with various cultures that influence it, as long as the culture contains goodness, brotherhood, and humanity (Baqir 2018, p. 28).

Mandalika became famous when the Mandalika circuit was established as the track for the Formula 1 Grand Prix motorcycle event. The concept of religious moderation is well known from an Islamic perspective, and one of the movements for religious moderation is to respect the local culture. Religious moderation in practice can accommodate local cultures and religions. Some people still uphold and preserve culture, which is the result of human creativity and revelation from Allah (Abdul Aziz 2019, p. 16).

4 CONCLUSION

In West Nusa Tenggara, the legend of Mandalika has also turned into a popular tourist destination. Because the Mandalika University of Education in Mataram, which uses the name as both a college and the Formula 1 Grand Prix motor track became well-known,

Mandalika gained notoriety. The Princess Mandalika legend is used to support the sustainability of Lombok residents' way of life by serving as a symbol of women's emancipation. The animism and dynamism-related beliefs of the Lombok people's ancestors are the source of the Mandalika story. Islamic teachings react to these concepts in an acculturative, assimilative, and adaptive manner. This belief evolved as a result of the accommodating nature of Islamic teachings as expressed in the Islamic Wasathiyah.

The shortcoming of this study is that field observations are still lacking, and this is because events related to religious ritual activities are temporary. However, this study recommends that we improve the understanding of religious moderation in society to develop a dialectic of religious moderation to maintain the values contained in religious traditions.

REFERENCES

Abdul Aziz, A., 2019. *Implementasi Moderasi Beragama dalam Pendidikan Islam*. Kementerian Agama, Jakarta.

Abdullah, A., 1999. *Studi Agama, Normativitas atau Historisitas*. Pustaka Pelajar, Yogyakarta.

Ahmed, L., 1992. *Wanita dan Gender dalam Islam, Akar-Akar Historis Perdebatan Modern*. Lentera, Jakarta.

Al-Jabiri, M.A., 2003. *Tragedi Intelektual Perselingkuhan Politik dan Agama*. Pustaka Alief, Yogyakarta.

Azra, A., 2020. *Relevansi Islam Wasathiyah, dari Melindungi Kampus Hingga Mengaktualisasikan Kesalehan*. Kompas, Jakarta.

Baqir, H., 2018. *Islam Tuhan, Islam Manusia, Agama dan Sprituallitas di Zaman Kacau*. Mizan, Jakarta.

Doss, C., Meinzen-Dick, R., Quisumbing, A., Theis, S., 2018. Women in agriculture: Four myths. *Global Food Security* 16, 69–74. https://doi.org/10.1016/j.gfs.2017.10.001

'Abd al-'Aẓīm, Š.M.-, n.d. *Women in Islam_Versus Women in the Judaeo-Christian Tradition_The Myth and the Reality-Conveying Islamic Message Society.pdf*. Conveying Islamic Message Society, Canada.

Kamali, M.H., 2015. *The Middle Path of Moderation in Islam: The Qur'anic Proncipie of Wasatiyyah*. Oxford University Press, New York.

Kuntowijoyo, 1999. *Paradigma Islam Interpretasi Untuk Aksi*. Mizan, Jakarta.

Kuswarno, E., 2009. *Metode Penelitian Komunikasi, Fenomenologi, Konsepsi, Pedoman dan Contoh Penelitiannya*. Widya Padjajaran, Bandung.

Lies, B.B., 1999. *Earth's Daughters: Stories of Women in Classical Mythology*. Fulcrum Resources, Golden, Colo.

Liliweri, A., 2016. *Konfigurasi Dasar Teori-Teori Komunikasi Antar Budaya*. Nusa Media, Bandung.

M.I.M, A.S., 2016. *Israiliyyat dan Hadits-Hadits Palsu 176. Tafsir Al-Qur'an, terj. Mujahidin Muhayyan*. Keira Publising, Depok.

Nimer, M.A., 2010. *Nirkekerasan dan Bina Damai dalam Islam: Teori dan Praktik*, Terj, M. Irsyad Rhafsady. Pustaka AlFavet, Jakarta.

Ramakrishna, K., Tan, S.S. (Eds.), 2003. After Bali: the threat of terrorism in Southeast Asia. *Presented at the Workshop on "After Bali: The Threat of Terrorism in Southeast Asia,"* World Scientific; Institute of Defence and Strategic Studies, Singapore; River Edge, N.J.: Singapore.

Seidman, I., 2006. *Interviewing as Qualitative Research: A Guide for Researchers in Education and the Social Sciences*, 3rd ed. Teachers College Press, New York.

Sellers, S., 2001. *Myth and Fairy Tale in Contemporary Women's Fiction*. Palgrave, Houndmills, Basingstoke, Hampshire; New York.

Smith, B.J., Woodward, M., n.d. *Gender and Power in Indonesian Islam: Leaders, Feminists, Sufis and Pesantren Selves* 207.

Sujarwo, W., 2019. *Sasak Traditional Villages: A Potential Tourism and Portrait of Conservation Efforts for Culture and Plants* 21, 18.

Sumadi, I.W.S., 2013. *Tradisi Nyongkol dan Eksistensinya di Pulau Lombok*. Ombak, Yogyakarta.

Umar, N., 2014. *Islam Fungsional, Revitalisasi dan Reaktualisasi Nilai-Nilai Ke Islaman*. Gramedia, Jakarta.

Umar, N., 1999. *Argumen Kesetaraan Jender Perspektif al-Qur'an*. Paramadina, Jakarta.

Wahidah, B.Y.K., 2019. Mitologi putri mandalika pada masyarakat sasak terkait dengan bau nyale pada pesta rakyat sebagai kearifan lokal tinjauan etnolinguistik tahun 2018. *Jurnal, Pendidikan* 4. https://doi.org/10.36312/jupe.v4i5.1297

Walbridge, J., 2001. *The Wisdom of the Mystic East, Suhrawardi and Platonic Orientalism*. Sunny Press.

Society and humanity

Religion, Education, Science and Technology towards a More Inclusive and Sustainable Future – Rahiem (Ed.)

Seloko adat as an identity and existence of the Jambi Malay society in Indonesia

M.I. Al Munir*, M. Habibullah, P. Abbas, M. Rusydy, S.R. Jannah & Masiyan
UIN Sulthan Thaha Saifuddin Jambi, Indonesia

ABSTRACT: This article is based on the context of the Jambi Malay *seloko adat*, which still survives in the midst of the global culture that characterizes Indonesian culture. *Seloko adat* was able to survive because it was seen as an inseparable part of the existence of the Jambi Malay society. In addition, *seloko adat* is also used as a guide in their daily life. The main focus of this article is to discuss the existence of *seloko adat* and its implementation in the life of the Jambi Malay society. This article uses interview, observation, and documentation techniques for data collection and uses data analysis of Aristotle's theory of cause and Søren Kierkegaard's existentialism. This research finds that *seloko adat* is the embodiment of the identity of the Jambi Malay society, as well as the basis for their existence. Based on these research findings, this article argues that the Jambi Malay *seloko adat*, in addition to functioning as the Malay identity of the Jambi people, is also a means of proving their existence.

Keywords: Jambi Malay, *seloko adat*, identity, existence

1 INTRODUCTION

Seloko adat are expressions containing messages that have ethical and moral values to be obeyed by society. The existence of *seloko adat* has spread well in the Jambi Malay society, the majority ethnic group, residing along and around the outskirts of the Batanghari River in Jambi, Indonesia. The spread of *seloko adat* is not only in homogeneous rural areas thick with Malay culture, but also in heterogeneous urban areas. *Seloko adat* is always heard orally in many official events in government offices and among ordinary people, such as at wedding receptions. In general, the society understands that in the *seloko adat* there are values and norms whose existence goes hand in hand with the existence of the Jambi Malay society. According to Cassirer (1990) and Koentjaraningrat (2002), the existence of culture is necessary and important because it exists along with the existence of humans. Furthermore, *seloko adat* was born and developed in the Jambi Malay society and serves as a guide in their lives. *Seloko adat* is a part of the Jambi Malay customs that continues to be used by the society, as well as the teachings of Islam that color it.

Many studies that have been carried out related to the Jambi Malay *seloko adat*. These studies have several trends. First, studies that tend to examine the legal aspects of *seloko adat* (Nauli 2014; Ulum et al. 2020). Second, studies that tend to examine aspects of education (Rahima 2018; Putriyanti et al. 2019; Syuhada 2020). Third, studies that tend to examine aspects of religion (Ahmad & Amin 2015; Rahima & Ridwan 2016; Noor 2019). Fourth, studies that tend to examine cultural aspects (Kurniadi & Zulkarnain 2021). Fifth, studies

*Corresponding Author: m.iedalmunir@uinjambi.ac.id

DOI: 10.1201/9781003322054-30

that tend to examine aspects of language and literature (Irpina et al. 2020; Suhardianto & Fitrah 2018; Zahar 2018). Sixth, studies that tend to examine aspects of philosophy (Munir & Ja'far 2013). These trends show that the philosophical aspects of the Jambi Malay *seloko adat* are still under-studied, especially in relation to the subaspects of metaphysics and existentialism. Whereas the philosophical approach is important in an effort to study culture (Lubis 2016; Noerhadi 2013; Sugiharto 2019).

This study aims to complement existing studies related to Jambi Malay *seloko adat* by presenting the existence of *seloko adat* with a philosophical approach with subaspects of metaphysics and existentialism. The main question of this research is: Why is *seloko adat* able to survive in the life of the Jambi Malay society? This question is broken down into the following two questions: (1) How is *seloko adat* positioned as a self-identity by the Jambi Malay society? (2) How is *seloko adat* implemented as their self-existence?

This research is based on the argument that the Jambi Malay *seloko adat* does not only function as the Malay identity of the Jambi people, but also as a means of proving their existence.

2 METHOD

This is a qualitative research. Data was collected by interviewing informants consisting of traditional leaders from the Jambi Malay Traditional Institute, both in the city, district, and province. Interviews were also conducted with academics from universities in Jambi. In addition, data is also generated from observations of the procession of delivering *seloko adat* orally in certain events in the society, and is equipped with various documentation related to *seloko adat*. The collected data was then analyzed using Aristotle's theory of cause and Kierkegaard's existentialism. Aristotle's theory of cause is used to analyze the identity of the Jambi Malay society, while Kierkegaard's existentialism is used to analyze their existence.

3 FINDINGS AND DISCUSSION

In this section of the findings and discussion, the writer will present the findings of the research, which consist of two subsections, namely *seloko adat* as a Malay identity and *seloko adat* as a means of proving the existence of Malays. In the first subsection, it is explained about the Malay identity resulting from *seloko adat* with the approach of Aristotle's theory of cause. Meanwhile, the second subsection describes the use of *seloko adat* as a means to prove the existence of the Jambi Malays using Aristotle's theory of cause and Kierkegaard's existentialism. Next, the findings section will be complemented by a discussion section where the authors compare the findings with the findings of previous studies.

3.1 *Seloko adat as Malay identity*

Aristotle, in Physics II 3 and Metaphysics V 2, offers four causes for everything that requires explanation or detail, including the results of art and human action. Aristotle describes four causes that can be given as an answer to the question of why something exists. First, the material cause, which is the answer to the question of what it comes from. For example, the statue comes from bronze. Second, the formal cause, which is the answer to the question: In what form? Or what will happen? For example, the shape is that of a statue. Third, efficient cause, which is the answer to the question: Who is the main actor? For example, the sculptor. Fourth, the final cause, which is the answer to the question: For what does something exist? For example, health is the result of exercise, weight loss, hygiene, and medication (Falcon 2022). The theory of cause is what the author uses to examine the existence of *seloko adat* as the identity of the Jambi Malay society.

From the results of the interview, the three causal theories above, when applied in the Jambi Malay *seloko adat*, will produce the following results. First, the material cause of *seloko adat* is the two main sources of law in Islam, namely the Qur'an and Hadith, in addition to being equipped with existing Malay customs that have been passed down from generation to generation. Second, the formal cause of *seloko adat* is not standardly patterned like rhymes or poetry. *Seloko adat* is flexible; there are *seloko adat* that consist of two or four lines or even more. Third, the efficient cause of *seloko adat* is the ancestors and traditional leaders, who are considered the most authorized parties in maintaining, developing, and inheriting *seloko adat*.

Based on interviews, one example of a *seloko adat* that is closely related to this explanation is as follows:

"Adat bersendikan syarak, syarak bersendikan kitabullah;
Syarak mengato, adat memakai;
Syarak berbuhul mati, adat berbuhul sentak."

"Custom based on sharia, sharia based on the Qur'an;
Sharia says, custom uses;
Sharia has fixed knots, custom has jerky knots".

The explanation above shows that for the Jambi Malay society, *seloko adat* is their Malay identity because the source comes from the Qur'an, Hadith, and *adat*. The form is flexible and simple, like their general mindset. The composers are also their ancestors and authorized traditional leaders, so they must be preserved and guided.

3.2 *Seloko adat as a means of Malay existence*

In line with the explanation in the previous subsection, which consists of three causes in the Jambi Malay *seloko adat*, this subsection leaves one theory of cause, the final cause, namely why the *seloko adat* exists. From the interview, it was found that the final reason for the Jambi Malay *seloko adat* was to be a companion to the main sources of law in Islam, namely the Qur'an and Hadith. *Seloko adat* is an additional way of life for the Malay society. As a way of life, the *seloko adat* is also a means for the Malay society to show their existence consciously.

Based on interviews, examples of *seloko adat* explaining this case are as follows:

"Beruk di rimbo disusukan, anak di pangku diletakkan;
Tibo di mato jangan dipicingkan, tibo di perut jangan dikempeskan;
Lurus benar dipegang teguh, kato benar diubah idak."

"Monkey in the jungle is fed, child is placed on the lap;
Don't squint it in the eyes, don't squeeze it in the stomach;
Righteousness is firmly held, right words are not changed".

"Jalan berambah yang diturut, baju bejait yang dipakai;
Nan bersesap berjerami, bertunggul berpemareh, berpendam berpekuburan".

"Paved path that you follow, dirty clothes that you wear;
Come from a clear origin, have good parents, have a permanent residence"

The explanation above illustrates that *seloko adat* is a means for the Jambi Malay society to exist in life. They consciously make *seloko adat* a guide in today's life, either as individuals or as part of society. In the book Fear and Trembling, Kierkegaard explains that a person's life becomes meaningful when he brings himself into a universal world that embodies his desires and natural tendencies as the purpose of the life in which he exists. Even though he loses his self because he is generally accepted, his actions become meaningful because they can be understood and regulated by a norm (Crowell 2020). The norms referred to by Kierkegaard in the understanding of the Jambi Malay society are *seloko adat*.

4 DISCUSSION

Many aspects of the approach can be used to understand the Jambi Malay *seloko adat*, so that many findings can be obtained from research on it. The legal aspect shows that *seloko adat* can be used as an alternative law other than positive law. The educational aspect shows that *seloko adat* is a source of character education in addition to formal education. The religious aspect shows that *seloko adat* is a perfect companion for religion as a guide for the life of the Jambi Malay society. The cultural aspect shows that *seloko adat* is a form of generic culture that is passed down from one generation to the next, although it can also be juxtaposed with popular culture that is developing today. Aspects of language and literature show that *seloko adat*, in its delivery, uses a lot of figurative language or metaphors, so that it must be able to be understood properly. However, the philosophical aspect shows that *seloko adat* holds certain values that are important to be explored and understood so that they can be used in the life of the Jambi Malay society. Especially in the philosophical aspect, there is still a lot of empty space for the study of *seloko adat*. In this study, the philosophical aspect enriches and completes the study of *seloko adat* related to its existence as a manifestation of the identity of the Jambi Malay society as well as a means for the Jambi Malay society to realize their existence. By analyzing *seloko adat*, using Aristotle's theory of cause, the result is that *seloko adat* is an inseparable part of the Jambi Malay society, while through the analysis of Kierkegaard's existentialism, the results show that all the existential manifestations of the Jambi Malay society cannot be separated from the *seloko adat* that they make as a guide in their lives.

5 CONCLUSION

Seloko adat is the Malay identity of the Jambi society, as well as a means to prove their existence. This answers the main question of this research that *seloko adat* is able to survive in modern life today because of its vital function as the identity and self-existence of the Jambi Malay society. One of the weaknesses of this research is that it has not been directed specifically at the younger generation of Jambi Malay society. Therefore, more intense efforts are needed from all parties so that the younger generation of Jambi Malay society is better able to understand *seloko adat* well so that they are able to color the global world with their Malay spirit. In addition, further research on *seloko adat* can take a more practical approach, such as environmental ethics or bioethics.

REFERENCES

Ahmad, H. and Amin, E., 2015. Integrasi ayat-ayat al-quran dalam seloko adat Jambi: Transformasi dakwah kultural. *Kontekstualita: Jurnal Penelitian Sosial Keagamaan*, 30 (1), 1–24.

Cassirer, E., 1990. *Manusia dan Kebudayaan: Sebuah Esei Tentang Manusia*. Jakarta: Gramedia.

Crowell, S., 2020. *Existentialism*. The Stanford Encyclopedia of Philosophy.

Falcon, A., 2022. *Aristotle on Causality*. The Stanford Encyclopedia of Philosophy.

Irpina, W., W, S.A., and Raudhoh, R., 2020. Literasi seloko adat masyarakat kelurahan sengeti. *Baitul 'Ulum: Jurnal Ilmu Perpustakaan dan Informasi*, 3 (1), 57–69.

Koentjaraningrat, 2002. *Manusia dan Kebudayaan di Indonesia*. 19th ed. Jakarta: Djambatan.

Kurniadi, M.D. and Zulkarnain, Z., 2021. The role of Jambi malay traditional institutions in the seloko oral traditions of the Jambi culture in maintaining the advancement of culture. *Budapest International Research and Critics in Linguistics and Education (BirLE) Journal*, 4 (1), 570–576.

Lubis, A.Y., 2016. *Postmodernisme: Teori dan Metode*. 1st ed. Jakarta: Rajawali Pers.

Munir, M.I. Al and Ja'far, M.H., 2013. Etika kepemimpinan dalam seloko adat melayu Jambi leadership ethic in traditional adage in jambi malay. *Kontekstualita*, 28 (2), 127–140.

Nauli, M., 2014. Pengaruh hindu dalam seloko melayu di hulu batanghari. *Jurnal Ilmu Hukum*, 4 (2), 105–114.

Noerhadi, T.H., 2013. *Aku dalam Budaya: Telaah Teori & Metodologi Filsafat Budaya*. Jakarta: PT Gramedia Pustaka Utama.

Noor, S., 2019. Local wisdom based Da'wah in the oral tradition of the jambi malay seloko adat. *Ilmu Dakwah: Academic Journal for Homiletic Studies*, 13 (2), 233–249.

Putriyanti, I., Mila, S.S., Ade, L., and Wahyu Dwi, 2019. *Teaching the Seloko Adat Jambi to Improve Morality and Social Control*, 365 (Icsgs 2018), 63–66.

Rahima, A. and Ridwan, S., 2016. Religious values in the theme structure of traditional seloko of jambi malay. *Ijlecr – International Journal of Language Education and Culture Review*, 2 (1), 82–91.

Rahima, A., 2018. Educational character values in seloko custom utterances of jambi malay society. *KnE Social Sciences*, 3 (9), 754.

Sugiharto, B., 2019. *Kebudayaan dan Kondisi Post-Tradisi: Kajian Filosofis atas Permasalahan Budaya Abad Ke-21*. Yogyakarta: PT Kanisius.

Suhardianto and Fitrah, Y., 2018. Seloko Adat Jambi: Kajian Struktur, Fungsi Pragmatik dan Fungsi Sosial. *Dikbastra: Jurnal Pendidikan Bahasa dan Sastra*, 1 (1), 79–97.

Syuhada, N.I.S., 2020. Seloko adat melayu to building character and multicultural of jambi society. *Criksetra: Jurnal Pendidikan Sejarah*, 9 (2), 194.

Ulum, B., Arifullah, M., and Fuhaidah, U., 2020. *Conserving Islamic Law and Seloko Adat Melayu Jambi in the Globalization Era*, (Icri 2018), 1045–1051.

Zahar, E., 2018. Analisis struktur majas seloko hukum adat sebagai bentuk ekspresi simbolik nilai-nilai religius masyarakat melayu jambi. *Jurnal Ilmiah Dikdaya*, 8 (1), 150.

Religion, Education, Science and Technology towards a More Inclusive and Sustainable Future – Rahiem (Ed.)

Challenges facing Muslim converts in the Republic of Kenya: A case study of Mumias-Kakamega county

H.Y. Akasi*
Umma University-School of Sharia and Islamic Studies, Kenya

ABSTRACT: Converts are a significant group within Kenya's Muslim population. If converts are handled in the right way, they may play a big role in the spread of Islam to non-Muslims. If converts are treated properly, they could have a significant impact on the propagation of Islam among non-Muslims. In Mumias, Kakamega County, Kenya, converts face so many challenges as they navigate through their new-found faith. The aim of this study was to investigate the problems that convert in Kakamega face. The data were collected using semi-structured interviews as primary data. For the interviews, a sample of 25 participants was selected. There were 15 men and 10 women. Converts face physical assault, derogatory language, neglect by family and friends, difficulty finding a scholar, social integration, difficulties acquiring Islamic knowledge, difficulties learning the Quran, humiliation, identity issues, marital problems, and other problems, according to the findings. The study recommends a systematic process of educating converts for a better understanding of Islam and providing them with self-efficiency projects that will enable them to be self-reliant.

Keywords: Islam, society, religious conversion, Muslim community

1 INTRODUCTION

Muslims view Islam's mission as a fundamental virtue found in its religious teachings. With wisdom and teachings, one of the methodologies used in Islamic missions is an invitation to all to the way of the Lord. Religious conversion and reversion are part of the mission of Islam, and new Muslims go through various phases. Religious reversion or conversion marks an entry into a new relationship that is characterized by Islamic brotherhood. During conversion into Islam, a new member has to say affirmations to Islam before an imam. The individual must be made out of free will. In the post-conversion phase, there are significant changes that happen in the life of an individual. This is the time when an individual learns how to live like a Muslim and practice the faith (Razick & Rushana 2018).

Convert persecution began in the early days of the Prophet Mohammad (Pbuh) in Mecca, and spread across East Africa before reaching the Arabian Peninsula. Africa was the first continent to receive Muslim converts fleeing the Arabian Peninsula (Zaki 1974). In 615 AD, they fled to the Christian kingdom of Aksum, which is now part of Ethiopia and Eritrea. Abyssinia took the lead in welcoming the first immigrants since it is the closest African country to the Arabian Peninsula. The Companions' migration, which included Prophet Muhammad's daughter Ruqayyah and his son-in-law Uthman ibn Affan, among others, and lasted fifteen years, was a divine will and prophetic guidance to honor the countries of Abyssinia, including East Africa, with this religion since the beginning of the revelation (Al-Mubarak furi 1996).

Studies conducted in Britain have shown that newly converted Muslims face rejection, starting from their immediate family members, with the increased visibility of their faith in public spaces. For instance, close relatives of the newly converted Muslims may even turn away from them. New converts have mixed feelings because they have the joy of practicing and learning the new faith

*Corresponding Author: hakasi@umma.ac.ke

 DOI: 10.1201/9781003322054-31

while at the same time experiencing anxiety while mingling with their community. They also have concerns about losing their previous social life (Brice 2010).

In Kenya, and particularly in Kakamega, the county is divided into 12 sub-counties, 60 wards, 187 village units, and 400 community areas. Furthermore, the county had a population of 1,660,651, comprising 800,896 males and 859,755 females, giving a population distribution of 48% male and 52% female. This population is projected to be growing at an annual growth rate of 2.5%. The county labor force is projected to be 1,033,512 in the year 2018, consisting of 485,383 males and 548,129 females. This population is projected to be 1,086,501 and 1,142,207 by 2020 and 2022, respectively. This high labor force implies that the county government should put appropriate policies in place to create employment and encourage the setting up of private enterprises to absorb this labor force (Kakamega County: 2022). The number of Muslims recorded in the county is 88,412 (City population: 2022), and the majority were found in Mumias. Conversion to Islam is a gradual process and mostly happens during the Ramadan period. In Kenya, initiation into Islam is associated with a warm welcome from the community members. After the welcome phase, the new converts are left on their own to navigate through the faith. This leaves the individuals burdened with obligations about the new beliefs (Sahad et al. 2013).

In general, the word convert refers to a person who changes from one religion to another (Alalwani 2012). In scholarly works, the word convert means someone who has converted to Islam based on the notion of "al-Fitrah." According to this concept, everybody is born a Muslim, but it is only the parents or environment that cause one to become a non-Muslim. The Prophet (Pbuh) said, "Every child is born with a true faith of Islam (i.e., to worship none but Allah Alone) and his parents convert him to Judaism or Christianity or Magianism, as an animal delivers a perfect baby animal. Do you find it mutilated?" (Bukhari 1385).

Converts have an important role to play in our society. For instance, converts are required to practice Islamic teachings and encourage others to do the same. Converts have made a profound contribution to the spread of Islam in Kenya and globally. In Kenya, conversion to Islam has a very long history and has happened alongside the spread of Islam (Kilonzo 2001). Although recent Muslim converts contribute toward the development of Islam, they face numerous challenges, ranging from discrimination to assault and neglect from their family members. The study by Brice in the UK addressed some of the reasons that make people join Islam. He states that people convert to Islam because of acceptance based on conscience and convenience. With the route of conscience, an individual is persuaded to join Islam based on the explanations provided about Islam and its benefits. For convenience, people join Islam because of individual needs that must be met. For instance, the need for marriage may push an individual to convert to Islam. A business contract may also be a pressing need that will cause some people to join Islam.

The work of Rashida Alhassan analyzes the reasons that Islamic converts had while trying to convert to Christianity. The study revealed that converts experienced challenges such as parents attempting to kill their children because of their decision to join Christianity. The study also reports some parents subjecting their children to torture for wanting to leave Islam. As the challenges continue to manifest among Muslim converts, formal structures provide no significant support.

1.1 *Problem statement*

New converts experience some dilemmas in balancing their religious practice with their cultural heritage. It is therefore a challenge for the new converts to navigate their way through the new religious faith communities. Perceptions about new Islam converts being at a high risk of joining extremist groups such as ISIS and Al-Shabaab also have a psychological impact on new Muslims. Within the prism of risk, it is perceived that new converts are a threat to public safety because of perceptions about engaging in acts of violence and bomb plots. Past studies have shown that Muslim converts are more likely to be attached to terrorism and extremism in media coverage. Little research exists on the support that mosques and Islamic organizations provide to newly converted Muslims. The lack of guidance for newly converted Muslims puts them at high risk of exposure to extreme ideas that are shared on the internet. The goal of this study was to look into the issues that new Islam converts in Mumias face and make recommendations on how they can be helped to strengthen their faith and protect themselves from extremist ideas spread on the internet.

2 METHODS

2.1 *Theological approach*

The challenges that convert experience can be understood from the theological approach, which is in accordance with Islamic teachings. Theology is more concerned with transcendence, doctrine, and faith interpretation. There are four perspectives that are associated with each tradition. Descriptive theology allows one to describe doctrine in a functional way. Systematic theory entails summaries of the theological doctrines of faith in a professional way. There is also a theory of dialogue in which there is an attempt to understand other people for their own sake. The descriptive theology was the best fit for this research.

A qualitative approach was selected because issues of religious conversion involve relationships between individuals and their families, friends, or members of the community and may be challenging to discuss in open forums. Participants would want to maintain confidentiality when discussing such matters. Therefore, qualitative techniques were the most appropriate. To achieve the objectives, the snowball or chain technique was used, and it allowed access to the participants that were difficult to reach. As alluded to by Creswell, the snowball technique is useful in identifying interesting cases from individuals who can be contacted to provide useful information. Purposeful sampling was also a reliable technique used in this study as it led the researcher to mosques and other Islamic institutions to collect data about the challenges facing converts.

Banerjee and Chaudhury (2010) defined a target population as all the subjects involved in the study. The sample size comprised three categories of participants. The first group consisted of converts, while the second group comprised Muslim converts who had gone back to their previous religion. There was interest in finding out the reasons for converting to their previous religion. The third group had Islamic religious leaders who had knowledge about the challenges that converts experience. The sample consisted of 15 males and 10 females aged between 18 and 40 years. The choice of this age was informed by the fact that it is regarded as the adult age group where individuals are expected to be responsible for their actions.

2.2 *Data collection*

A qualitative approach was used to study the problems faced by new Islam converts in Mumias. Data was collected using observations, interviews, and a review of secondary sources.

2.3 *Observations*

The study involved observations as one of the methods of collecting data. Selected Islamic preaching was observed with the aim of understanding the different ways in which non-Muslims are attracted to Islam. Muslim clerics provided an important source of information about the converts and the role they were supposed to play once they started practicing the new faith. More observations were made on some of the programs that are designed for participants in the mosques.

2.4 *Interviews*

Qualitative data was collected through informal interviews that were conducted among selected Muslim leaders from mosques in Mumias. Relevant questions about problems facing Islam converts were asked during the interview sessions. The leaders were first approached and informed about the objectives of the study, and they gave consent for interview sessions. The data from the informal interviews were categorized and analyzed under several headings based on the major themes that emerged during the discussions (Jessica & others: 2020).

2.5 *Secondary data sources*

The study also collected secondary data from the mosques within the area of study. Some of the data collected from the secondary documents included the age, gender, level of education, and previous religion of the converts. These variables have a big impact on the post-conversion life of converts.

3 RESULTS AND DISCUSSION

After a rigorous thematic analysis of the data, three themes emerged concerning the converts in Kakamega County. The first theme is the "outside challenges," and the second is the "inside challenges."

3.1 *Challenges from outside faith*

Physical assault: The result found cases of physical assault directed at the respondents because of their stand on Islam. The researcher interacted with Mr. F, who is a young man working as a construction worker. The young man described an incident where he was involved in an argument with other men at a construction site, and he had been physically assaulted by acts of pushing by his coworkers, who called him names and claimed that he was foolish to join Islam, yet most of his family members were non-Muslims. He explained that the incident had made him shy away from discussing his new faith with his coworkers.

Derogatory language: The use of derogatory language was found to be common among the new converts in Mumias. Some of the sentiments recorded from the participants include: "That religion you've joined is for terrorists" Do you want to become one?" This is what one of Miss J's male friends told her when he noticed that she had started wearing Muslim attire. Miss J, a student at a local computer college in Mumias, narrated how she had experienced derogatory language after joining Islam. She explained that she had joined Islam after completing her high school education, and she had one male friend who had invited her to join Islam. She also claimed that one of her friends told her that she had only gone into Islam to hunt for a rich boyfriend from the Arabs.

Neglect by family and friends: Converts often experience neglect from their family members and friends when they join Islam. In some cases, family members may stop supporting one of their own financially or socially once they realize that he has joined Islam. Mr. J, who is a college student, gave some insight into how his sister stopped supporting him while in college because of his joining Islam. He claimed that her sister is a born-again Christian, and she used to support him with pocket money and other materials during his college life. When she got word that he had joined Islam, she started avoiding him and had a negative attitude toward him. Mr. J claimed that he had once heard his sister say that she did not want to be converted to Islam, a faith that has extremists.

3.2 *Challenges from within Islamic faith*

a) *Difficulty in finding scholars:* The study observed that converts had difficulty getting scholars to teach them Islam. It was established that the Imams who had initiated converts into Islam had neglected them. As responded by Mr. T, who is a carpenter, he was converted while in the village and was left to grow on his own. He had not seen anyone to offer assistance in learning the basic principles of Islam. He had tried getting someone to assist him, but his efforts were fruitless. The lack of support had made this respondent and others unable to recite the opening chapter of the Quran. To make matters worse, he did not live near a mosque where he could go to pray and engage in some studies, even if he found an Islamic scholar.
b) *Social integration:* Social integration was identified by some converts as a major social problem that affected them when they joined Islam. It was stated that integration problems were felt from the mosques to the larger community. To some converts the environment in the mosque was new to them and they did not understand the norms about

standing and so on. Mr. K, who is a welder and a new convert, complained that he did not know what to do in the mosque and feared asking other worshippers. He claimed that his second day at the mosque had a lot of confusion and he felt like leaving. He also claimed that some of the people in the mosque were looking at him like he was in the wrong place. For Mr. N, language was a major problem that affected his integration within the Muslim community. Ethnic differences within the Muslims in Mumias made him not feel comfortable in the environment. He had problems communicating with fellow Muslims because of the differences in language. He expressed fear that he was going to have a very difficult time attending the community activities because of the communication problems.

c) *Difficulties in acquiring Islamic knowledge:* From the interviews, it emerged that participants do not have a sufficient understanding of Islamic principles. Some of the participants do not know how they will gain knowledge about Islam. For instance, Miss B. was a new convert who was recently married to a Muslim, but she expressed her lack of knowledge about Islam. She claimed that she was interested in learning the faith but did not have a clear way of understanding it. For some of the participants, they had tried to put in their own efforts to learn Islam, but they did not get support from their family, friends, or mosques.
d) *Difficulties in learning Al-Quran:* A few of the respondents in this study mentioned that they had problems understanding the Al-Quran. They understood that it was their responsibility to learn and adapt their lives according to the Quran. Some did not have an understanding of the basic principles and chapters in the Quran, and some were even ignorant about knowing the Quran. For some of the female participants, their husbands did not have good knowledge of the Quran, yet they depended on them to learn about Islam. The inability of the husbands to recite some important parts of the Quran was a major obstacle to the learning of their wives. Outside the home environment, some converts claimed that it was more difficult to study the Quran since they were preoccupied with work or studies. The converts did not give the Quran more priority compared with their work or studies.
e) *Humiliation*: From the sermons I observed, humiliation of fellow Muslims was one of the issues that was preached about. The preachers asserted to their worshippers that it was wrong to humiliate fellow Muslims. Examples of cases where converts were humiliated were highlighted in such preaching, and it was a behavior that was abhorred because it was against the principles of brotherhood in Islam. An observation was also made about Muslim events such as weddings, funerals, and other ceremonies. The converts invited to such events claimed that they felt lost and isolated from their community. It emerged that there is a differential treatment of Muslims by birth and those who have converted to Islam.

The results from interviews also showed that converts suffer from expressions of Islamic identity. There is difficulty among the converts in expressing their Muslim identity. This is because they regard Islam as a minority in the community. Some of the converts feared revealing their identity through their dress when they were in the company of other people in the community. The differences between Muslim dressing and traditional cultural practices were also making some converts fear expressing their identity. Some of the converts feared that expressions of their identity could result in their complete exclusion from the previous community.

3.2.1 *Marital problems*

The issues of marriage were prominent among the converts in Mumias. Some of the converts were worried about the strict rules that Islam has for marriage. Converts who are married and have Christian partners complained that it was sometimes difficult to strike a balance between the two diametrically opposed religious faiths. Some of the converts opened up and stated that they had gotten into marriage in complete opposition to Islamic practices. For instance, there are men who have married women who are non-Muslims, and they felt that they had gone against the teachings of Islam.

3.2.2 *Mixing Islam and traditional practices*
From the interviews, it emerged that converts had problems mixing Islamic practices with their previous practices. According to the teachings of Islam, a Muslim is not supposed to mix traditional practices with Islamic practices. The converts therefore find themselves in difficult situations when they are required to go back to their traditional practices when they are in the company of non-Muslims. When converts mix their traditional practices with Islam, they are likely to come into conflict with the Muslims who were born Muslims. They are also likely to be perceived as jokers in their faith.

4 DISCUSSION

A significant number of challenges were identified, and there are reasons that explain them within the region and within the larger Islamic faith. The finding that Muslim converts are vulnerable to rejection, assault, and neglect is consistent with other studies that have been done globally about conversion to Islam and its effects. In Malaysia, a study by Faisal Azni and Md. Yusoff (2019) concluded that misconceptions surrounding Islam as a religion contribute to the resentment and conflict that family members have toward converts. For instance, misconceptions exist about Islamic teachings and ideology, and they are present on social media networks (Rink & Kunaal 2018). When family members listen to misconceptions, they become negative and will resort to assaulting, neglecting, and resenting their members who have converted to Islam. Another finding was that converts experience family conflict and identity crisis, as has also been found in other previous studies. Being a Muslim is more than just wearing attire, going to mosques, and reading the Quran; but it is a deeper lifestyle change that an individual is supposed to go through, as indicated by Faisal Azni and Md. Yusoff (2019). Islam is the most widely misunderstood religion, which means that it is not an easy journey for converts. It is expected that clashes may happen anytime between the traditional practices of previous religions and those of Islam. Once an individual has converted to Islam, it is expected that a new lifestyle, different from the former, will be adopted. For instance, if an individual was involved in taking alcoholic drinks, then he or she is expected to abandon such a life and live a different one. (Shahid 2019).

The study also found that a lack of knowledge about Islam is a major problem for the converts. Previous studies in Ghana have also shown similar results among the Islamic communities. The lack of knowledge about Islam is an issue that troubles so many young people who convert to the religion, and they become easy targets for misinformation and radicalization by extremists (Rahman 2015).

There is keenness among the new converts to gain Islamic knowledge. When anxiety is accompanied by feelings of isolation and disappointment, the results are that converts become vulnerable to online resources. They are also predisposed to the slippery slope effect that can sweep them toward extremism, as underscored by Karagiannis (2012). Efforts should be made to meet the needs of the new converts by providing good scholarly courses that can help them gain significant religious knowledge and awareness about Islam and its cultural context.

Brice (2010) found that the lack of support networks is one of the most challenging problems that new converts face. The implication for the new converts, especially for those who are single, is that it becomes difficult to get married. This is further complicated by the fact that new converts are making efforts to try and follow their new religious values and instructions, while at the same time they are expected to have issues in their work and occupation. For instance, those who do not have alternative sources of finance may find it difficult to follow Islamic instructions. Women are more vulnerable to the cultural shift that happens, and they require counseling to manage their careers and get finances to support their faith in Islam. Career counseling is one of the needs that can help converts help themselves as they struggle to integrate into Islam.

Brice (2010) also demonstrated that new converts feel under pressure to abide by Islamic cultural norms. This is where they need more knowledge and help, and where they don't get it, they turn to the internet. Mosques are viewed as not being well equipped to provide converts with adequate support and guidance. It has been found that the scholars with knowledge are too busy to attend to the converts. There is also a lot of reference to online resources to get knowledge about Islam, but this tends to be confusing to the converts, and it is also difficult for the converts to find what is relevant and good for themselves.

5 CONCLUSIONS

Muslim converts face numerous challenges, both from their families and their newfound faith. From the interviews conducted, converts suffer from assault, rejection, humiliation, marital conflict, and neglect because of their involvement with Islam. The converts also face problems related to the lack of knowledge of their new faith, lack of social integration, lack of knowledge of the Al-Quran, identity problems, and mixing practices. According to the findings of the study, there are no adequate support structures in Mumias to assist converts and guide them through their newfound faith. Mosques and community organizations should keep good records of converts. Well-kept records are useful in following up on converts and finding out the individual problems they face and how they can be assisted to remain in their new faith.

The findings of this study show that converts are aware of the problems they face and the needs they have in their new faith. Organizations that are run by converts and are supported by the community should be established. This should be informed by the fact that converts know their needs and the journey they are going through. The study recommends the establishment of programs to support Muslim converts in Mumias. For instance, there should be programs to offer social, emotional, and psychological support to converts. The converts also need material support in terms of clothes and other attire that they can use to smoothly integrate into the Muslim community. Programs should also be established to provide Islamic literature to the converts so that they gain knowledge about their new religion and grow spiritually to spread Islam to more people.

REFERENCES

Alalwani, T.J., 2012. Apostasy in Islam: A historical and scriptural analysis. *International Institute of Islamic Thought (IIIT)*.

Brice, K., 2010. *A Minority within a Minority: A Report on Converts to Islam in the United Kingdom*. Swansea University.

Faisal Azni, N.L. and Md. Yusoff, Y., 2019. Nurturing an Islamic identity throughout familial conflict among new converts: A critical review. *Jurnal Sains Insani*.

Karagiannis, E., 2012. European converts to Islam: Mechanisms of radicalization. *Politics, Religion & Ideology*, 99–113.

Kilonzo, J., 2001. *Islam, Indigenous Traditions, and Adventism in Kenya: A Comparative Study to Determine Effective Approaches to Evangelize Kenyan Muslims*. Dissertation Projects.

Rahman, A.A., 2015. *An Analysis of the Challenges and Prospects of Convert to Islam in the Greater Accra Region*. Doctoral dissertation. University of Ghana.

Razick, A.S. and Rushana, A., 2018. *Problems Faced by Converts to Islam in Sri Lanka: A Study Based on Anuradhapura District*.

Rink, A. and Kunaal, S., 2018. The Determinants of religious radicalization: Evidence from Kenya. *Journal of Conflict Resolution*, 99–105.

Sahad, M.Z., Siti Aishah Chu Abdullah, and Suhaila Abdullah, 2013. Malaysian news report on Muslim converts' issues: A study on Malaysiakini. *International Journal of Humanities and Social Science*, 3 (13), 231–245.

Shahid, H., 2019. Forging a brazilian Islam: Muslim converts negotiating identity in São Paulo. *Journal of Muslim Minority Affairs*, 231–245.

Zaki, A., 1974. Al-Islam Wal-Hadhara Al-Islamiyyah Fī Sharq Africa. *Majala Tarekheya Al-Masriyyah*.

Religion, Education, Science and Technology towards a More Inclusive and Sustainable Future – Rahiem (Ed.)

Participation of elderly workers in Jambi province

A.S. Prasaja*, M. Soetomo, R. Ferawati & A. Halim
UIN Sulthan Thaha Saifuddin, Jambi, Indonesia

R.D. Ariani
CV. Geo Art Science, Yogyakarta, Indonesia

ABSTRACT: Improving the social welfare of the community can be done with development efforts. The indicators that can be used is the life expectancy figure. Where the high life expectancy has an impact both positive and negative. The purpose of this study is to analyze the factors that influence the elderly population in Jambi Province to work based on demographic characteristics consisting of age, education, gender, health level, and marital status. The data used in this study is secondary data obtained from SUSENAS in Jambi in 2020. The population in this study was all residents aged 60 years and over who had working status in Jambi Province, which was 1,067 people. The method that used is regression analysis. The analysis showed that the variables of age, education, health, marital status, and gender had a simultaneous effect on the work participation of the elderly population. Meanwhile, partially affecting the work participation of the elderly, all variables except the gender variable.

Keywords: work participation, elderly, labor, life expectancy

1 INTRODUCTION

The growth of social welfare aimed at developing and improving the community's standard of living can be done with development efforts (Turner 2021). One of them is population-based development which uses a development bottom-up planning strategy. The bottom-up planning approach aims to optimize the distribution of resources and potential owned throughout the region and carry out development based on the potential problems owned by the region. The meaning of population-minded development is a development that adapts to the population's potential and conditions to emphasize improving human resources quality. One of the indicators that can be used to measure the quality of human resources is the Life Expectancy Rate (Widiastuty 2019).

The elderly are 60 years old and over (Law (UU) on the Welfare of the Elderly 1998). The number of elderly people in Jambi Province continues to increase every census. Based on Figure 1, elderly people increased by 47% in 2010 and 60% in 2020. It indicates that the Life Expectancy level in Jambi has increased.

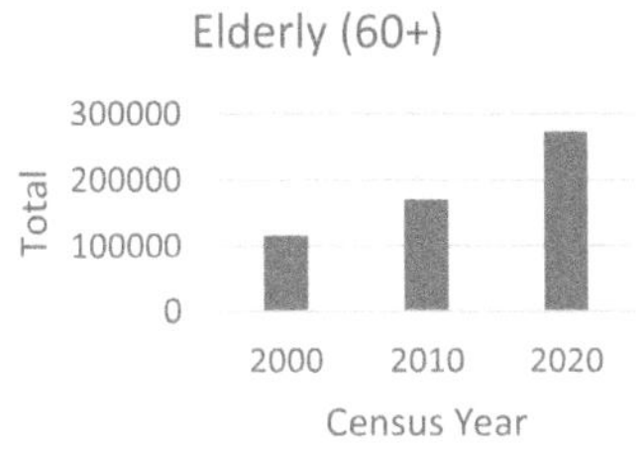

Figure 1. Elderly in Jambi province.
Source: Census 2000, 2010, and 2020.

*Corresponding Author: syukronprasaja@uinjambi.ac.id

DOI: 10.1201/9781003322054-32

The current Life Expectancy has increased over time (Medford 2017; Pascariu et al. 2018). Based on Life Expectancy data from the Indonesian Central Statistics Agency (BPS), The life expectancy of the population in Indonesia has experienced positive development, as shown in Figure 2. In 2011 Indonesian Life Expectancy was 68.09 for men and 72.02 for women, and it increased in 2020 to 69.59 for men and 73.46 for women. Similarly, Jambi Province has a positive Life Expectancy value development, even though the Life Expectancy value of Jambi Province is still below the national Life Expectancy value. In Jambi Province, the Life Expectancy score increased by 1.13 for males and 1.04 for females from 2011 to 2022. The increase in Life Expectancy occurs due to a decrease in the risk of death at young ages (Yang et al. 2022). However, since the 1950s, the increase in Life Expectancy has been driven more by survival at an old age than early in Life (Zuo et al. 2018) and also due to the advancement of the health field in an area (Lestari et al. 2020). The high Life Expectancy indicates that the number of elderly populations is also high, called population ageing, and tends to increase rapidly (Šídlo et al. 2020).

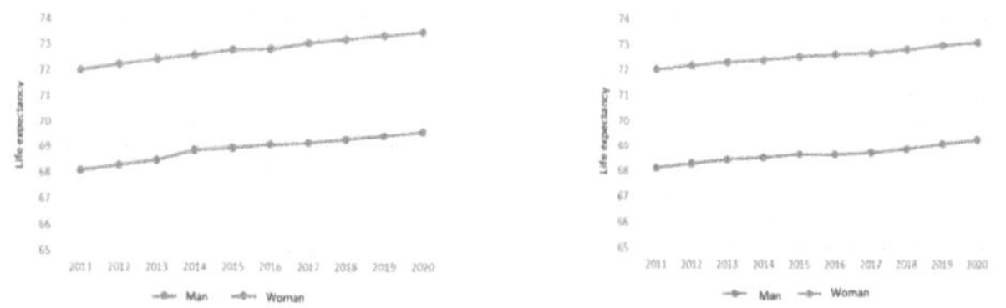

Figure 2. Life expectancy in (a) Indonesia and (b) Jambi 2011–2020.

Population ageing is a phenomenon that can cause a change in the structure of the population from young to old. This phenomenon will provide new challenges in the regional development process because it directly affects economic growth in the present and future (Heffner et al. 2019). The increase in Life Expectancy indicates it is one of the indicators of the success of management development. However, if the increase in Life Expectancy is not in line with the improvement of population quality, it can delay the development process or cause a burden on development (Wolff et al. 2018). It is exacerbated by the increase in Life Expectancy, which is not evenly distributed due to the close relationship between health and death and the socioeconomic status of the community (Yu et al. 2022). From the economic side, population ageing will increase the dependency ratio between the productive and non-productive age populations.

Moreover, population ageing will also increase government spending to provide public facilities. So that the high level of participation of the elderly population in Jambi Province will help reduce the dependence rate on the productive age population and help to reduce the budget spent by the government on providing public facilities used by the elderly. The elderly population still actively participating in productive activities can be influenced by several factors. However, the elderly with limited skills will find it challenging to find a job (Nilmini & Samaraweera 2022).

SUSENAS Data in 2020 shows that more than half (50.69%) of the elderly in Jambi Province have the main working activity. The high percentage of the elderly still working can also indicate that there are still many productive elderly people. However, this may also indicate that the level of welfare of the elderly is still low. The elderly still have to work to fulfill the necessities of life (Pazos & Ferreira 2022). Based on the research background described above, it is necessary to analyze the work participation of the elderly population in Jambi Province related to factors that affect the work participation of the elderly in terms of socio-demographics.

2 METHODS

The study was conducted in Jambi Province. The selection of the research location was carried out by purposive sampling, where Jambi Province experienced an increase in the value of Life Expectancy, which indicates that the number of elderly people in Jambi Province has also increased. The data used in this study is secondary data sourced from SUSENAS (*Survei Sosial Ekonomi Nasional*/ National Socio-Economic Survey) (2021) surveys designed to collect social population data conducted by BPS annually. The data used in this study is 2020, with the elderly population being the sample of 2,105 residents and 1,067 elderly workers.

The data analysis technique in this study is multiple linear regression to determine the influence between independent variables, namely age, education, health, marital status, and gender, on the dependent variable, namely the work participation of the elderly population in Jambi Province.

3 RESULTS AND DISCUSSION

The analysis results on the influence of age, education, health, marital status, and gender on the simultaneous work participation of the elderly can be seen in Table 1. Where a calculated F value of 21,556 was obtained with a significance value of 0.000 which means that age, education, health, marital status, and gender have a simultaneous effect on the work participation of the elderly population in Jambi Province. The F value of the independent variable > F of the table is 2.218, and the significance value is below 0.05.

Table 1. ANOVA.

Model	Df	F	Sig
Regression	5	21,556	3. 0,000
Residual	2099		
Total	2104		

Source: SUSENAS data processing, 2021.

The scale of the determinant coefficient R^2 can be seen in Table 2. The determinant coefficient in this study is 0.491, which means that independent variables can explain 49.1 percent of changes in work participation of the elderly population, and factors outside the study explain another 50.9 percent.

Table 2. Summary model.

Model	R	R Square
1	0,701	0,491

Source: SUSENAS data processing, 2021.

The analysis results on the influence of age, education, health, marital status, and gender on the work participation of the elderly can be partially seen in Table 3. Table 3 shows that three of the five variables have a significance level of < 0.05. It shows that the variables of age, education, and marital status are significant to the dependent variables of the elderly population in Jambi Province to choose to work or not work.

Table 3. Regression results.

Variable	Coefficient	Std. Error	t	Sig
Age	−0,005	0,001	3,683	0,000
Education	0,012	0,002	5,505	0,000
Health	0,006	0,005	2,137	0,050
Marital Status	0,078	0,010	7,569	0,000
Gender	−0,006	0,020	−0,303	0,762

Source: Susenas data processing 2020, 2021.

The population will experience an increase in age every year, and the increase in age will affect the ability of the population to carry out activities in their daily lives. Age will naturally decrease the ability to take care of oneself, interact with the surrounding community, and become increasingly dependent on others (Lestari et al. 2020). The result showed a calculated t-value of 3.683 with a significance of 0.000. It is shown that the age of the elderly has a partially negative influence on the work participation of the elderly in Jambi Province. The regression coefficient from age is −0.005, which means that the increase in age of 1 year in the elderly will reduce work participation by 0.005 hours per week. So, it can be interpreted that the higher the age of the elderly population, the greater the probability of the population not working. The older a person is, the more productivity will decrease. It is in line with research conducted by Mckee (2006), which states that the work participation of the elderly will decrease in line with the increase in age. Work participation decrease of the elderly due to the age factor is also mentioned in the research by Junaidi *et al.* (2017) where age will significantly affect the probability of the elderly working. The education variable has a coefficient of 0.012 with a significance of 0.000 (<0.05), meaning that the higher the level of education of the elderly population, the higher the probability of the elderly population going to work. It is caused by the elderly population with higher education having high creativity as well as being able to use time productively.

The elderly population whose work generally has good health conditions that allow the elderly to work. The calculated t value based on the regression analysis results between the elderly work participation variable and the health variable is 2.137, with a signification value of 0.050. So, the results indicate that a good level of health will positively affect the work participation of the elderly population in Jambi Province. The calculated t value is greater than the table t value of 1.961, with a significance value of more than 0.05, Where the elderly population with good health will have high productivity or willingness to work. Research conducted by (Doan et al. 2022) under the title "Health and occupation: the limits to older adults work hours" states that health affects the work participation of the elderly population.

Elderly residents who have marital status have a high probability of working. The analysis showed that the calculation of the regression results between the dependent variables of elderly work participation and the independent variables of marital status was 7.569 with a significance of 0.000. It means that there is a difference between the elderly who are married and unmarried status/widowed/widower because the calculated t value > of the Table t is 1.961, and the significance is below 0.05. These results align with the study Retkno *et al.* (2020), where the marital status of the elderly negatively influences the decision to stop working. This is possible due to marital status, which will indirectly increase material responsibility. Marital status becomes one of the variables that affect the elderly desire to continue work after retirement (Kantachote & Wiroonsri 2022). The gender variable has a coefficient of −0.006 with a significance value of 0.762 (<0.05) which can be interpreted to mean that the job market in Jambi Province does not depend on the gender of the population.

Some of the variables that have been analyzed show that the marital status variable is the most dominant factor affecting the work participation of the elderly population. The value of the existing coefficient of 0.078 indicates this. The analysis of the work participation of the elderly can be used as reference material by local governments to prepare programs that are population responsive to support population conditions when population aging occurs. So that the region will benefit because this elderly population is still productive, healthy, and able to contribute to economic growth and not become a dependent material for families and the government.

4 CONCLUSION

Population-minded development to improve the welfare of the population, especially the elderly population, can be done by knowing the factors that affect the work participation of the elderly. Some variables that affect the work participation of the elderly in Jambi Province simultaneously and significantly are age, education, health, marital status, and gender. The age of the elderly population has a negative and partially significant influence on work participation. The work participation of the elderly population partially and significantly influences the variables of education, health, and marital status. As for the gender variable, it does not have a partial or significant influence on work participation. Marital status is the most dominant variable of the five independent variables. To ensure that the high life expectancy will not burden regional development but instead can be used as a potential in regional development. The local governments can use this analysis to make population-responsive policies or programs that can increase the potential and quality of existing resources.

ACKNOWLEDGMENT

Thanks to *Badan Pusat Statistik* (BPS) through *Sistem Layanan Data Statistik* (SILASTIK) for providing SUSENAS data for free for this research. In addition, thank the Center Studies of Demography, Ethnography, and Social Transform UIN Sutlhan Thaha Saifuddin Jambi for helping to fund this research.

REFERENCES

Badan Pusat Statistik. (2021). *Survei Sosial Ekonomi Nasional (Susenas) Tahun 2020.*

Doan, T., Labond, C., Yazidjoglou, A., Timmins, P., Yu, P., & Strazdins, L. (2022). Health and occupation: the limits to older adults' work hours. *Ageing and Society*, 1–29. https://doi.org/10.1017/S0144686X22000411

Heffner, K., Klemens, B., & Solga, B. (2019). Challenges of regional development in the context of population ageing. Analysis based on the example of opolskie voivodeship. *Sustainability (Switzerland)*, *11*(19). https://doi.org/10.3390/su11195207

Junaidi, E., & Purwaka. (2017). Faktor sosial ekonomi yang mempengaruhi keterlibatan penduduk lanjut usia dalam pasar kerja di provinsi jambi. *E-Journal Masyarakat,Kebudayaan Dan Politik*, *30*(2), 197–205.

Kantachote, K., & Wiroonsri, N. (2022). Do elderly want to work? Modeling elderly's decision to fight aging Thailand. *Quality & Quantity*. https://doi.org/10.1007/s11135-022-01366-0

Lestari, M. D., Stephens, C., & Morison, T. (2020). Constructions of older people's identities in Indonesian regional ageing policies: The impacts on micro and macro experiences of ageing. *Ageing and Society*, *2050* (May), 1–21. https://doi.org/10.1017/S0144686X20001907

Mckee, D. (2006). A dynamic model of retirement in Indonesia. *Journal International: California Center For Popoulation Research.*

Medford, A. (2017). Best-practice life expectancy: An extreme value appr. *Demographic Research*, *36*, 989–1014.

Nilmini, P. G. N., & Samaraweera, G. R. S. R. C. (2022). Beyond the working age: Labour supply of elderly men and women in Sri Lanka. *Journal of Social Sciences and Humanities Review*, *7*(1), 29. https://doi.org/10.4038/jsshr.v7i1.105

Pascariu, M. D., Canudas-Romo, V., & Vaupel, J. W. (2018). The double-gap life expectancy forecasting model. *Insurance: Mathematics and Economics*, *78*, 339–350.

Pazos, P. de F. B., & Ferreira, A. P. (2022). Aspects of occupational aging according to elderly workers: old age, work, and worker health. *Research, Society and Development*, *11*(10), e507111032960. https://doi.org/10.33448/rsd-v11i10.32960

Šídlo, L., Šprocha, B., & Ďurček, P. (2020). A retrospective and prospective view of current and future population ageing in the European Union 28 countries. *Moravian Geographical Reports*, *28*(3), 187–207. https://doi.org/10.2478/mgr-2020-0014

Turner, A. (2021). Community development: A critical and radical approach. *Community Development Journal*, *56*(3), 547–550. https://doi.org/10.1093/cdj/bsaa023

Undang-undang (UU) *Tentang Kesejahteraan Lanjut Usia*, Pub. L. No. 13 Tahun 1998 (1998).

Widiastuty, I. L. (2019). Pengaruh kualitas hidup perempuan terhadap dinamika angka harapan hidup di jawa barat. *Jurnal Kependudukan Indonesia*, *14*(2), 105–118.

Wolff, J. K., Beyer, A. K., Wurm, S., Nowossadeck, S., & Wiest, M. (2018). Regional impact of population aging on changes in individual self-perceptions of aging: Findings from the german ageing survey. *Gerontologist*, *58*(1), 47–56. https://doi.org/10.1093/geront/gnx127

Yang, Y., Yang, M., & Zhang, X. (2022). An end-to-end perceptual enhancement method for UHD portrait images. *IET Image Processing*, *16*(7), 1988–2000. https://doi.org/10.1049/ipr2.12464

Yu, K., Chen, L., Fu, Z., Wang, Y., & Lu, T. (2022). A coding layer robust reversible watermarking algorithm for digital image in multi-antenna system. *Signal Processing*, *199*. https://doi.org/10.1016/j.sigpro.2022.108630

Zuo, W., Jiang, S., Guo, Z., Feldman, M. W., & Tuljapurkar, S. (2018). Advancing front of old-age human survival. *Proceedings of the National Academy of Sciences*, *115*(44), 11209–11214.

Religion, Education, Science and Technology towards a More Inclusive and Sustainable Future – Rahiem (Ed.)
© 2024 the Author(s), ISBN: 978-1-032-56461-6
Open Access: www.taylorfrancis.com, CC BY-NC-ND 4.0 license

The challenges faced by modern Muslim Indonesian society in the Industrial Revolution era

S. Ramadhan*
Institut Agama Islam Nasional Laa Roiba Bogor, Indonesia

A. Zamhari, M.I. Helmi, A.M. Albantani & I. Subchi
UIN Syarif Hidayatullah, Jakarta, Indonesia

ABSTRACT: Religion is vulnerable to being distracted by modernity due to classical traditions and irrational paradigms that are difficult for religious communities to abandon. However, the era of the Industrial Revolution 4.0 has made it difficult for Muslim communities to be resistant so they need to respond to the emerging wave of modernity. This paper aims to analyze the response of the Muslim community to the currents of modernity that occurred in the era of the Industrial Revolution 4.0. This research is qualitative research that uses a normative approach. The data are sourced from books, journal articles, and encyclopedias. This study concludes that essentially, the response of the Muslim community in the Industrial Revolution era tended to be open. The fundamental values of modernity did not necessarily distance religion from technological progress because of the nature of Islam, which is open to changing times (adaptive-inclusive). Openness to religious values and modernity has paved the way for modern Muslim society to move together and collaborate to build an integrated system. In the economic aspect, for example, the halal industry is becoming more advanced and flexible. Openness to the education aspect has built a technology-based Islamic education ecosystem that is easily accessible to people in various regions.

Keywords: Religion, modernity, modernization, industry 4.0, Muslim society

1 INTRODUCTION

Modernization reaches human life in various ways, which are generally characterized by the presence of advanced science and technology from the 18th century until today. The convenience provided by advances in science and technology has generated a lot of critical responses from the public. This phenomenon occurs in almost all societies in the world, including the Muslim community. The understanding of the Muslim community toward modernity is certainly influential in responding to advances in science and technology. On the other hand, Muslim understanding of religious attitudes in responding to modernity also has a strong influence in responding to the inevitability of modernity that occurs. A good understanding of religion is an important basis for Muslims in implementing scientific and technological advances, in order to avoid the degradation of values and morals that modernity does not bring (Bellah 2011). This way, the Muslim community will know the effects of advances in science and technology, whether good or damaging to the system of human life.

*Corresponding Author: suciramadhan95@gmail.com

 DOI: 10.1201/9781003322054-33

The relationship between Muslims and modernity is not monolithic, but there are various responses from particular Muslim persons or groups (Mundzir 2013). The Muslim community has two different views. Some people think that Islamic values are guidance and obedience for humans as the caliph of Allah SWT. Another view believes that all the basic assumptions about modern Western civilization are the antithesis of the noble Islamic principles (Haryati 2012). In fact, modernization has influenced the lifestyle of the Muslim community in a positive way, such as open-mindedness, being dynamic, not being confined by classical and traditional things, anticipatory and selective attitude in accepting new things. In addition, modernity also negatively affects a person's attitude toward the social behavior of his community, such as tending to be more closed, apathetic, indifferent, and accepting everything without first selecting (Suradi 2018).

In the current era of the Industrial Revolution 4.0, the value of pragmatism that is channeled through modernization is increasingly shaping the life of the Muslim community and resulting in many changes (Nur 2019). Science and technology are establishing symbols of progress and benchmarks of success in the current industrial era (Mukri *et al.* 2019). Muslim communities are required to adapt their daily lives by using conveniences offered by the sophistication of technology and information, but eliminating the essential and sacred values of religion. Basically, as a Muslim, of course, it is very profitable to be able to carry out daily activities easily and practically. However, seeing the developments in this era, modernity has various impacts in life, giving rise to various responses and reactions from Muslims (Bartodziej 2017). This study intends to investigate and analyze the response of the Muslim community in Indonesia to modernity in the industrial revolution 4.0 era.

2 METHODS

This research is a type of qualitative research with an integrative approach. The data used is secondary data sourced from various scientific literatures (books, journal articles, and research results) that discuss religion, modernity, and the industrial era 4.0. The limitation of discussion specifications regarding the halal industry and educational technology is due to the current development and progress of the halal industry world has seized a lot of attention from various religious figures and experts. Educational technology is also relevant to be studied because the development of the world of education in Indonesia is experiencing many problems, especially in the midst of the spread of the COVID-19. Data collection techniques used documentation techniques and then the data were analyzed using content analysis. The data search process was carried out through Google Scholar using the keywords "Religion", "Muslim Society," "Modernization," "Modernity," "Modern Era," and "Industrial Revolution 4.0" with a limit of 2012–2021; as many as 179 written results were found in the early stages. The articles are then grouped, analyzed, and sorted based on scientific articles from accredited journals. After going through a critical review and assessment, 13 scientific articles were selected that specifically related to the focus of this research, namely religion and modernity in the era of the Industrial Revolution 4.0. The articles are then analyzed descriptively and discussed in the research results and discussion section.

3 RESULTS AND FINDINGS

After analyzing the 13 selected articles, the authors found two important themes that became the locus of discussion of these articles, namely: 1) The existence of religion in the modern era, and 2) the challenges of the Muslim Community in Indonesia toward the Halal Industry and Educational Technology in the Era Industry 4.0.

3.1 *The existence of religion in the modern era*

The existence of religion in modernization has become a category that is reified through political, social, and economic means. Marx considered that the concept of reification has treated religion as a single phenomenon (Goldstein 2006). Indeed, modernity will always suppress traditions and cultures that are considered no longer relevant. Meanwhile, the technological movement brought about by modernization tends to be adaptive and fast in providing unlimited convenience. Religion as an element that is considered a part of tradition and culture is being eroded because it is no longer relevant in providing answers and convenience for the community. In various countries, there are always sections of religious groups that tend to reject modernity in order to refocus on the core teachings (sacred texts) which are understood as absolute truths. This group is usually called the fundamental-mental group (McDaniel 2015). In fundamental Islamic culture, modernity is seen as a change that deviates from the texts and tradition and also because it is disseminated and inherited from colonialism (Eyadat *et al.* 2018). Seeing this clash, it is certain that violence and chaos are part of the problems that will arise from a space where there is a meeting between modernity and tradition (Narayanan 2007).

Modern society in the industrial era has succeeded in developing sophisticated technology to overcome the problems of human life; however, this sophistication has an impact on moral degradation. In this regard, religion serves as a moral controller, such as by imposing moral prohibitions on sex and alcohol, making the oppressed feel safe, limiting tyrannical power, and limiting the tendency to commit suicide (Blasi & Pickering 1986). Durkheim, as quoted by Giddens, states that religion plays an important role in society, which he calls the original source of all developing moral, philosophical, scientific, and juridical ideas. He states that beliefs collectively tend to have a religious character, although this is only conjecture and needs further research. However, at least it seems that the possibility of religious significance related to collective influence in society is offset by the awareness and the fact that major changes have occurred with the emergence of modern social groups (Giddens 1971).

Muslims, as part of the world community, need to understand the tendency of modern society to solve problems by considering the causal factors. This humanistic decadence in modern times occurs due to the loss of self-worth that humans should have. Modern society creates a lot of advanced technology, not based on intellectual light, but rather on positivism. This causes a lot of damage in the world such as environmental pollution, and an imbalance of ecosystems and ecology, caused by the determination of modern humans to act like God by removing the transcendental and religious dimensions (Haryati 2012).

In today's modern world, many social and individual pathologies have occurred due to the absence of traditional boundaries between countries and cultures with the advancement of advanced information systems and technology. Religious spiritual showers that have a positive influence are no longer considered in modern society. The repressive nature of modern culture and values can only be resolved with an intensive appreciation of religion (Anwar 2010). Therefore, the rational use of science and technology in the Industry 4.0 era needs to be accompanied by strong and inclusive religious values. This model of religious understanding will be able to overcome moral decline and the decline of one's Islamic values, thus increasing the morals and self-worth of the religious community.

3.2 *The exposure of modernity to the halal industry and educational technology*

Whether we realize it or not, the entire world community today has undergone many changes, both quickly and slowly. These changes have led to the construction of a modern society with various dynamics, including the modern Muslim society today. Modernity has brought many positive and negative influences so that problems and challenges arise in the life of modern Muslim society at that time. In responding to the growth of modernity and the progress of science and technology that is taking place, the Muslim community responds to the progress of the 4.0 era with various actions, including in the economic and educational aspects.

In the current economic aspect, modern technology has become an important tool for developing the halal industry in a short time (Suhaimi 2020). We can see how the halal industry plays a role and competes in the global economy. In this era of Industry 4.0, food is no longer just an item for consumption, and more attention needs to be paid to the production mechanism (Arifin & Hatoli 2021). The food industry needs to maintain the quality of production by leveraging technological advancements. This is not only to see the nutritional value of a food but also, crucially for Muslims, to ensure the halal status of consumed food. Food crime in the modern era is real today, and it is necessary to increase the attention of the Muslim community to the pattern of food production and industrial technology offered in the modern era. Research by Ramli *et al.* (2021) stated that food crime has affected the halal industry ecosystem through the use of advanced technology. The role of halal authorities and the Muslim community progressively needs to adopt the latest technology to ensure the supply of halal, safe, and good food for consumption.

In this era, the halal industry is increasingly affordable with the development of science and technology, both based on industrial technology and information technology. The tools and machines for testing halal products and food provide many conveniences, such as simplifying the process of ordering halal products and food, simplifying price estimation and menu updates, simplifying the process of ordering halal food, increasing product competitiveness, and providing better customer experience. The sophistication of technology offered today has changed the pattern of life of traditional society to a modern religious society. Several applications that enhance halal status include halal tests, halal quests, zabihah, halal advisors, halal scans, halal trips, halal spots, halal trips, and halal spots. In addition, there is a halal laboratory to test the halalness of products and the Modern Halal Valley as an integrated halal industrial complex (Zahrah & Fawaid 2019). Through the economic aspect in relation to halal food, the Muslim community does not appear to be a consumptive human being who only thinks about how to eat but also considers the quality of the food consumed based on religious teachings (Nurrachmi & Setiawan 2020).

In the aspect of education, efforts to bring Islam and modern secular sciences closer are not a way to weaken Islam, but rather a necessary condition for survival in modern conditions (Zaman 2012). With this integrated and quality education, an individual or Muslim intellectual will be able to contribute to community development, because it is impossible to build a quality society without quality education (Shadiqin 2011). Education in modern society in the Industrial Revolution 4.0 era, basically functions as a liaison between students and the constantly changing social environment (Suryono 2014). This is in line with the objectives of the Indonesian national education law, which is to provide the widest possible learning opportunities for every citizen and to ensure equal distribution of educational opportunities to face challenges in accordance with the demands of changing local, national, and global life. Therefore, it is necessary to reform education in a planned, directed, and sustainable manner (Law of the Republic of Indonesia No. 20 of 2003 on National Education System, 2003).

The real challenges in the world of education felt by the people of Indonesia, particularly during the spread of the COVID-19 virus in 2020, have caused chaos in traditional education, which is now largely directed toward online-based distance education through information and communication technology. The application of technology in the education system is urgent, especially amid the current pandemic. However, it seems that the 4.0 education system cannot run without harmony between humans and technology (Afif 2019), and creative ideas are needed in order to solve the problems of human life. In this case, the teacher is the main guard to understand technology in the field of education so that they are able to transfer their knowledge to students. In addition, students are also required to be able to apply technology that supports the continuity of the teaching and learning process (Yunita 2021).

The transformation of the education system must be carried out in accordance with the demands of the times as well as a bulwark against moral decadence. In this case, it is necessary to have an integrative curriculum based on religious content and modern, practical content (Hajriyah 2020). The subjects included in the secular school curriculum are clearly important for the accumulation of economically useful knowledge. Economically useful knowledge can prepare students to learn more about the general knowledge needed for the profession and the world of work (Squicciarini 2020). Some of the compatibility that might be applied in the curriculum content are changes in character education content (*akhlâq*) with competitive advantage, and understanding cognitive factors, affective factors, psychomotor factors, and even spiritual ones. This way, students can become human beings who are innovative, creative, democratic, characterized, and religious. Character values originating from religion such as honesty, tolerance for differences, discipline and responsibility, hard work, love for peace, and having a national and cultural spirit are an important part of the construction of the character education curriculum (Priyanto 2020).

The development of modernization in the halal industry and education provides two important values: the benefits provided as well as concerns about the value of freedom brought by modernity. The most important thing that needs to be understood in carrying out modern life is the existence of human cooperation based on goodness and responsibility to God, not based on sin and enmity. This is an open view of life that needs to be held in the face of modernity because no matter how great modernity is, not all problems of human life can be understood by humans. Despite technological and scientific advancements, much of the world's knowledge and secrets remain beyond human reach, raising worries and fears. Modernity, with its destructive elements, contributes to the loss of peace, human rights violations, mental stress, and increasing hatred, strife, and hostility. Nurcholish Madjid views modernity as a form of limp due to excessive pressure on contemporary and worldly things without paying attention to deep and long-lasting things (Majid 2008).

4 CONCLUSION

Modernity brought major changes to the life of the Muslim community in the era of the Industrial Revolution 4.0 which is currently in progress. In this era, Muslims are faced with the problem of adjusting between religious and modern values. This needs to be resolved so that both of them move together in solving problems that arise in society such as in the fields of economy and education. The Muslim community in Indonesia seems to use a lot of applied technology in the economic field of the halal industry to ensure the halalness of a product or food to be consumed. Of course, this is a form of Muslim attention to religious and spiritual values that should not be eliminated in technology. This is not like modernity which only offers convenience and emphasizes materialistic aspects. Likewise, in the educational aspect, the application of educational technology is widely used to spread the values of Islamic education without eliminating the essence of religion in the application of the technology. Thus, the Muslim community seems aware that the use of science and technology in the Industrial Revolution 4.0 era needs to be accompanied by strong and inclusive religious values. It aims to ensure that the religious values and spiritual spirit of the Muslim community do not disappear with the emergence of the convenience provided by modernity. Furthermore, religious values will be able to overcome and reduce moral decline in self-worth, and, conversely, can increase the morals and self-worth of the Muslim community.

REFERENCES

Afif, M., 2019. Meningkatkan kreatifitas dan inovatif dalam lintasan pembelajaran hipotetis pendidikan Islam pada era industri 4.0. *Progressa: Journal of Islamic Religious Instruction*, 3 (2), 1–8.

Anwar, A., 2010. Al-Qur'an dan modernitas (Pergeseran paradigma pemahaman Al-Qur'an). *Al-Fikra: Jurnal Ilmiah Keislaman*, 9 (2).

Arifin, Z. and Hatoli, H., 2021. Application of halal certification by Indonesian ulema council on electronic and non-consumption products: Maslahah perspective. *Justicia Islamica*, 18 (1), 115–131.
Bartodziej, C.J., 2017. *The Concept Industry 4.0 – an Empirical Analysis of Technologies and Applications in Production Logistics*. Germany: Springer Gabler.
Bellah, R.N., 2011. *Religion in Human Evolution: From the Paleolithic to the Axial Age*. United States of America: Harvard University Press.
Blasi, A.J. and Pickering, W.S.F., 1986. *Durkheim's Sociology of Religion. Themes and Theories*. Sociological Analysis.
Eyadat, Z., Corrao, F.M., and Hashas, M., eds., 2018. *Islam, State and Modernity: Mohammed Abed al-Jabri and the Future of the Arab World*. New York: Palgrave Macmillan.
Giddens, A., 1971. *Capitalism and Modern Social Theory: An Analysis of the Writings of Marx, Durkheim, and Max Weber*. United Kingdom: Cambridge University Press.
Goldstein, W.S., ed., 2006. *Marx, Critical Theory, and Religion: A Critique of Rational Choice*. Leiden: Brill.
Hajriyah, H.B., 2020. Modernisasi pendidikan agama Islam di era revolusi industri 4.0. *Momentum: Jurnal Sosial dan Keagamaan*, 9 (1), 42–62.
Haryati, T.A., 2012. Modernitas dalam perspektif seyyed hossein nasr. *Jurnal Penelitian*, 8 (2).
Majid, N., 2008. *Islam, kemodernan, dan keindonesiaan*. Edisi Baru. Bandung: PT Mizan Pustaka.
McDaniel, J., 2015. Indonesia, modernity and some problems of religious adaptation. *Wacana*, 15 (2), 314.
Mukri, M., Faisal, F., Anwar, S., and Asriani, A., 2019. Quran-integrated science in the era of industrial revolution 4.0. *Journal of Physics: Conference Series*, 1155 (1), 0–5.
Mundzir, I., 2013. Sikap muslim terhadap modernitas: Kasus gerakan khilafatull muslimin di lampung. *Afkaruna*, 9 (1), 65–82.
Narayanan, V., 2007. Explaining the global religious revival: A response. In: G. ter Haar and Y. Tsuruoka, eds. *Religion and Society: An Agenda for the 21st Century*. Leiden: Brill.
Nur, K., 2019. Menggagas konsep teologi kekinian di era industri 4.0. *Jurnal Theosofi dan Peradaban Islam*, 1 (2).
Nurrachmi, I. and Setiawan, S., 2020. Pengaruh religiusitas, kepercayaan, dan kepuasan terhadap keputusan pembelian ulang produk halal. *Iqtishadia Jurnal Ekonomi & Perbankan Syariah*, 7 (2), 126–137.
Priyanto, A., 2020. Pendidikan Islam dalam era revolusi industri 4.0. *J-PAI: Jurnal Pendidikan Agama Islam*, 6 (2), 80–89.
Ramli, M.A., Afiq, M., and Razak, A., 2021. the Emergence of halal-related food crimes in the era of industry 4.0. *Al-Qanatir: International Journal of Islamic Studies*, 24 (2), 45–50.
Shadiqin, S.I., 2011. Islam dan modernitas dalam pandangan fethullah gulen. *Jurnal Subtantial*, 13 (2), 98–111.
Squicciarini, M.P., 2020. Devotion and development: Religiosity, education, and economic progress in nineteenth-century france. *American Economic Review*, 110 (11), 3454–3491.
Suhaimi, S., 2020. Sistem ekonomi syariah sebagai eebuah solusi dalam mengembangkan ekonomi ummat di era revolusi industri 4.0. *Ahsana Media: Jurnal Pemikiran, Pendidikan dan ...*, 6 (2).
Suradi, A., 2018. Konsepsi pendidikan agama islam dalam menyikapi modernitas. *Jurnal Manajemen dan Pendidikan Islam*, 4 (1), 50–70.
Suryono, 2014. Pendidikan islam dan modernitas. *Jurnal Pemikiran Keislaman*, 25 (1), 98–112.
Undang-Undang Republik Indonesia Nomor 20 Tahun 2003 Tentang Sistem Pendidikan Nasional, 2003.
Yunita, L.A., 2021. Tantangan pendidikan di era revolusi industri 4.0 di tengah pandemi COVID-19.
Zahrah, A. and Fawaid, A., 2019. Halal food di era revolusi industri 4.0: Prospek dan tantangan. *Hayula: Indonesian Journal of Multidisciplinary Islamic Studies*, 3 (2), 121–138.
Zaman, M.Q., 2012. *Modern Islamic Thought in a Radical Age: Religious Authority and Internal Criticism*. United States of America: Cambridge University Press.

Religion, Education, Science and Technology towards a More Inclusive and Sustainable Future – Rahiem (Ed.)

Digital participatory archive for justice: An inclusive future for vulnerable groups

T.Y. Sari*
UIN Syarif Hidayatullah, Jakarta, Indonesia

K.P. Silalahi
University College London, UK

ABSTRACT: This article discusses the participatory digital archive as a means to advocate justice for vulnerable groups. According to Brown (2004) and a report series issued by the Center for Innovation Policy and Governance (2013), groups such as religious minorities, ethnic minorities, women, children, people with different abilities (diffable), and sexual minorities around the globe, including in Indonesia, are considered vulnerable to victimization and violence. The combination of weak human rights enforcement and media oligopoly has resulted in severe conditions for creating an open and democratic society in Indonesia. Hence, by drawing on the contemporary archive studies approach, we recommend a participatory digital archive as an alternative advocacy model for vulnerable and minority groups in Indonesia. This model of participatory digital archives enables vulnerable groups to submit their own representation of their cultural identity as well as self-determination through a digital-based archival platform.

Keywords: digital archive, social justice, vulnerable groups, inclusive society

1 INTRODUCTION

Archives of activism have been existing for decades. Many communities have been canalizing their struggle against the power structure through the community media. In Indonesia, for instance, as a form of resistance against Dutch colonialism, Indonesian young scholars have produced a number of newspapers, such as *Medan-Moeslimin* (1915) dan *Islam Bergerak* (1917). However, *Poetri Hindia* (1909), *Soenting Melajoe* (1921), and *Jurnal Perempuan* (1996 till today) are a few examples of feminist media that uplift women's movements against patriarchy.

Archiving is more than documenting. According to Schwartz and Cook (2002: 13), archives are always about power, whether it be the power of the state, the church, the corporation, the family, the public, or the individual. They further asserted that a dichotomy exists between the power to privilege and the power to marginalize. In sum, archives can act both as a hegemonic and as a resistance tool.

By employing the notion of a participatory archive, this study highlighted the importance of archival autonomy for seeking social justice. Archival autonomy is a condition in which archival and recordkeeping systems are social constructions, artificial, and products of countless contextual contingencies (Evans *et al.* 2015: 347). This means that the archive can only be an effective means to resist the oppressive structure by empowering the weakened

*Corresponding Author: trie.yunita@uinjkt.ac.id

 DOI: 10.1201/9781003322054-34

groups to share, to record their story, and to exercise their human rights. This approach is in contrast with traditional archiving, which is solely centered upon exclusive notions of custodianship and ownership, behind the walls of the archival institution.

Along with the advancement of information and communications technologies (ICTs) in the internet era, participatory archives can take place in digital media settings. Digital media is an alternative media for weakened groups to represent multiple perspectives in more liberated ways compared with conventional media that have been mostly co-opted by the government and or oligarchs. The significance of new digital technologies for participatory archives has been noted by McKemmish (2011: 142). He asserts that new digital technologies enable shared control and the exercise of negotiated rights in records. Besides, it offers alternate and contested views in parallel or together in a shared archival space; these allow community organizations to integrate government records into their own knowledge and records systems, and individuals to interact with public and community archives.

Given the above explanation about the participatory digital archive framework, this study offers a beneficial yet novel approach to advocate for vulnerable groups. According to Brown (2004) and a report series issued by the Center for Innovation Policy and Governance (2013), vulnerable groups such as religious minorities, ethnic minorities, women, children, people with different abilities (diffable), and sexual minorities around the globe, including in Indonesia, are often vulnerable to victimization and violence. The combination of weak human rights enforcement and media oligopoly has resulted in severe conditions for making them oppressed and marginalized. Hence, by drawing on the contemporary archive studies approach, this study recommends a participatory digital archive as an alternative advocacy model for creating an inclusive future for vulnerable and minority groups in Indonesia.

2 METHODS

This study is a qualitative research, using social and archive studies as a study approach. We collected data by interviewing a few respondents who actively engage in participatory archive activism. In addition to that, we did a desk study to identify and evaluate the three kinds of digital archives that exist in the Indonesian context: (1) *E-Religious Literature Library* by the National Library of Indonesia and the Ministry of Religious Affairs; (2) I-KHub by the National Counter Terrorism Agency; and (3) *DEMAND* (community archive platform to record the voice of victims of street harassment). Thus, the method used in this research is descriptive-analytic.

This study employed the concept of a digital participatory archive by McKemmish (2011). McKemmish emphasizes on the progress of participatory archives within the era of the internet. He argues that due to the impact of ICTs, the information landscape in which archives and records management are situated is changing. The internet challenges many old assumptions about creation, curation, privileging, and access to information. In the digital environment, shared and networked archival spaces can be created by communities maintaining records in partnership and exercising mutual rights and responsibilities. Communities can archive their records voluntarily, and their participation can be linked with the records of other individuals, communities, and organizations. In addition, they are able to contribute annotations that elaborate, clarify, or contextualize information content sourced from other records.

3 RESULTS AND DISCUSSION

The results of this study showed that in the context of Indonesia, there are a few digital archives for social initiatives, yet among those, we found an impartial approach that makes the archives still insignificant in building a just society. First, is the electronic library (e-library) that builds on the spirit of mainstreaming religious moderation in Indonesia. The so-called *Kepustakaan Keagamaan* or the e-religious literature library is a joint project between the National Library of Indonesia and the Ministry of Religious Affairs that presents various interfaith information. This e-library is accessible at the following site: Kepustakaan-keagamaan.perpusnas.go.id/lintas-agama. In this e-library, the government facilitates religious communities to share their knowledge products, such as religious textbooks, religious sermon materials, and articles on community activities.

Digital archives on religious moderation aim at socializing social harmony among Indonesian citizens despite religious pluralism. Religious moderation itself is a creative effort by multistakeholders such as government, civil society organizations, educational institutions, religious institutions, and religious communities in response to various inter- and intra-religious conflicts in Indonesia. Religious moderation is understood as a moderate and tolerant type of religion that steers clear of absolute truth and subjectivity, literal interpretations, and arrogant rejection of religious precepts, as well as radicalism and secularism (Ditjen Bimas Kristen 2019).

Ironically, in this digital archive platform, we found some omissions against marginalized religious groups. Even though the religious moderation project attempts to create a platform for constructive interaction among religious communities, the archivists restrict the access to unrecognized communities. The *E-Religious Literature* contains various literatures of official religions or state-recognized religious libraries only, such as Islam, Christianity (Protestant and Catholic), Hinduism, Buddhism, and Confucianism. While there are about 245 unofficial religions in Indonesia (Aritonang 2014). In addition to that, the archiving process does not invite the community to represent themselves. As entitled in the section on frequently asked questions (FAQs)[1], it is stated that the sources of content on the Religious Library website are collections owned by the National Library of Indonesia and the Ministry of Religious Affairs. Hence, this digital archive is still biased since it does not allow individuals and communities to participate in societal memory, with their own voice, and to become participatory agents in recordkeeping and archiving for identity, memory, and accountability purposes.

Archival activism using digital technology can also be found in I-KHub. The I-KHub stands for Information and Knowledge Hub, which is initiated between the government, developmental organizations, and civil society organizations in compiling various policy papers, knowledge, and best practices to prevent and counter violent extremism (PCVE) in Indonesia. This joint hub is launched as a mechanism between ministries, civil society organizations, and development partners to coordinate, monitor, and report the efforts of preventing and countering violent extremism by utilizing a technology platform to ensure accountability, transparency, and participative action.[2]

Different from the abovementioned digital activism, I-KHub provides a directory of all organizations that are concerned with the issue of PCVE. Besides, those civil society organizations held a mandate to archive various knowledge, such as policies and lessons learned, or experiences in responding to terrorism or violent extremism. On the platform, I-KHub is also equipped with a digital map of events where violent extremism occurred. This map is a window for the public to be aware of the threat of terrorism and violent extremism in their surroundings. Nonetheless, we find that participatory archives on this platform are still partial. This digital archive does not feature the experiences of victims and witnesses, including vulnerable groups of terrorism propaganda, families of perpetrators, families of victims of terrorism, and law enforcement officers.

To sum up, we find both digital archives have not yet provided a place for survivors or vulnerable groups to speak up for a few reasons. First, both *Kepustakaan Keagamaan* and

---

[1]See https://kepustakaan-keagamaan.perpusnas.go.id/

[2]See https://ikhub.id/about-us

I-KHub are managed under government control. Instead of constructing a healthy public discourse on religious moderation, the government actually increases the stigmatization or labeling of communities that do not represent in archives, for example, *Ahmadiyya,* indigenous beliefs, and other denominations. Second, the I-KHub shows a lack of empathy toward the survivors, ex-combatants, or their families in the archiving process. Here it is argued that the stakeholder perceives them as insignificant in telling their own stories.

While the two above-mentioned digital archives seem to fail to accommodate the vulnerable groups in participatory groups, there is a grassroots movement initiated by Hollaback Jakarta! (now renamed DEMAND: Di Jalan Aman Tanpa Pelecehan). DEMAND is a movement that initiated a digital participatory archive for women to share their sexual harassment experiences. The movement creates a documenting space in which issues of the community archive, identity, feminist ethics, and trauma combine. The platform enables anyone who has experienced street harassment to share their story and feel supported. The community does not merely recount instances of harassment, sexism, or reasons why feminism is necessary, but also enables the sharing of a range of emotions, reactions, and effects that accompany the experiences. Clearly, rather than proposing a totalizing framework contrary to a myriad of myths around sexual violence and the ways a "typical" or "legitimate" victim should respond, the activism tends to encourage women's voices regarding everyday sexual assault experiences. The repository itself can build self-confidence, self-belief, the ability to shape and participate, ingest justice, and equal opportunities, and it also offers some limited protection from gender-based violence.

In contrast to women as a vulnerable group being treated as objects in archiving, DEMAND archival activism continues to promote women as actors. Based on our analysis of the DEMAND project, we found this movement provides a greater sense of justice for women, which is lacking in two previous archive infrastructures, by upholding what is called the ethics of care. The term "ethics of care" was coined by a feminist named Carol Gilligan (1993). According to her, morality is supposed to be centered on interpersonal relationships and moral judgments based on the context of an issue. In the study of archives, such moral values are marked as radical empathy by Casswell & Cifor (2016). It means that archivists should recognize their powers and positionalities. Furthermore, empathy in such practices is a means to challenge, subvert, undermine, make possible, and engender change. In other words, the invitation of "the other" into the archive (space for representation) and the promotion of hospitality toward them are then expected to provide a means to defend the rights of vulnerable groups.

Despite the accomplishments of the movement, DEMAND faced a challenge in terms of its sustainability in 2022. According to the founder of DEMAND project, Anindya Restuviani[3], in an interview, told us that ironically, while DEMAND used to provide meaningful data for stakeholders, they are insufficiently aware of the movement's sustainability. All these years, the people behind DEMAND relied on an external donor to run the project. This issue also reflects how grassroots initiatives often face serious challenges when it comes to scarce resources. From this case, we can see that the participatory digital archive would have been successful and sustained, as long as there is supportive and collaborative work among stakeholders, archivists, as well as the community or volunteers to support the operation.

4 CONCLUSION

Based on the three above cases we summarize that when an archivist with an ethic of care refuses to align themselves with lasting oppressive structures, they open up the right to a

[3]Interview was conducted on 20th of June 2022, 15.30 WIB via Zoom Platform.

story, self-determination, and representation to vulnerable groups. The model of digital participatory archives enables vulnerable groups to submit their own representation of their cultural identity as well as self-determination through a digital-based archival platform. This model is ideal to acknowledge that multiple parties, including the oppressed and marginalized groups, have rights, responsibilities, needs, and perspectives with regard to the record and digital platform archive. Nonetheless, the result of this study has shown that in the Indonesian setting, the use of digital participatory archives for justice is still impartial. We argue that nearly all digital archive infrastructure controlled by authority focuses solely on migrating records from conventional to digital media. However, it disregards archival autonomy belonging to individuals and/or communities affected by human rights abuse in the record-making and management process. Based on that context, first, we suggest that digital participatory archives should be adopted into the Indonesian archives' legal framework, for instance in archives of law and national action plans for implementing sustainable development goals. We believe that policy initiatives should stimulate various good practices at all levels to achieve a just and inclusive society. Second, we suggest more collaborative work among parties, such as affected societies, human rights activists, scholars, and relevant stakeholders, to create safe and secure archive platforms so that they can express their own agency, reality, or representation.

REFERENCES

Aritonang, Margareth S. 2014. "Government to recognize minority faiths". *The Jakarta Post*. Retrieved on 16 June 2022 16.38 WIB.

Brown, Hillary. 2004. *Violence Against Vulnerable Groups: Integrated Project "Responses to Violence in Everyday Life in a Democratic Society."* Strasbourg Cedex: Council of Europe.

Caswell, M. & Cifor, M. 2016. *From human rights to feminist ethics: radical empathy in the archives. Archivaria* 81: 23–43. https://www.muse.jhu.edu/article/687705.

Direktorat Bimas Kristen. (2019). *Mozaik Moderasi Beragama Dalam Perspektif Kristen oleh Thomas Pentury*. Jakarta: BPK Gunung Mulia & Ditjen Bimas Kristen.

Evans, J. *et al.* 2015. Self-determination and archival autonomy: Advocating activism. *Arch Sci* 15: 337–368. https://doi.org/10.1007/s10502-015-9244-6.

Evans, J., McKemmish, S., Daniels, E. *et al.* (2015). Self-determination and archival autonomy: Advocating activism. *Arch Sci* 15, 337–368 https://doi.org/10.1007/s10502-015-9244-6

Gilligan, Carol. (1993). *In a Different Voice: Psychological Theory and Women's Development*. Cambridge: Harvard University Press.

Gilliland, A.J., & McKemmish, S. (2014). *The Role of Participatory Archives in Furthering Human Rights, Reconciliation and Recovery*.

Kawangung, Y. (2019). Religious moderation discourse in plurality of social harmony in Indonesia. *International journal of social sciences and humanities*, 3(1), 160–170.

McKemmish, S. 2011, Evidence of me ... in a digital world. in CA Lee (ed.), I, Digital: Personal collections in the digital era. *Society of American Archivists*, Chicago IL USA: 115–148.

Nugroho, *et al.* 2012. *Media and the Vulnerable in Indonesia: Accounts from the Margins*. Report Series. Engaging Media, Empowering Society: Assessing media policy and governance in Indonesia through the lens of citizens'rights. Research collaboration of Centre for Innovation Policy and Governance and HIVOS Regional Office Southeast Asia, funded by Ford Foundation. Jakarta: CIPG and HIVOS.

Schwartz, J.M., Cook, T. 2002. Archives, records, and power: The making of modern memory. *Archival Science* 2: 1–19.

Tewksbury, D. & J. Rittenberg. 2012. *News on the Internet, Information and Citizenship in the 21st Century*, Oxford: Oxford University Press.

Religion, Education, Science and Technology towards a More Inclusive and Sustainable Future – Rahiem (Ed.)

Can wives deradicalize their husbands?

E. Kurniati* & A. Zamhari
UIN Syarif Hidayatullah Jakarta, Indonesia

ABSTRACT: This article investigates the role of wives of terror convicts in the deradicalization of their husbands. This article uses a case study that the results supported by the micro and macro-level analysis of the deradicalization process by Doosje *et al.* (2016) and the importance of the family's role in radicalization and deradicalization by Sikken *et al.* (2017), El-Amraoui & Ducol (2019), and Yayla (2020). Thus, I argue that there are two factors, external and internal to deradicalize terror convicts. The external factor is an intervention that presents many actors that the terror convicts do not know such as academics, religious leaders, and psychologists that influence terror convicts' ideology. Whereas, the internal factor comes from people who have emotional bonds such as mothers, wives, and children. This study is expected to become a vital addition to the literature on Islam and deradicalization study.

Keywords: deradicalization, family resilience, PCVE, women

1 INTRODUCTION

Demant *et al.* (2008) explained that deradicalization is the opposite word of radicalization. They stated that deradicalization is "the process of becoming less radical" (Demant *et al* 2008:13). Doosje *et al.* (2016) define radicalization as a process where people motivate to use violence toward other groups that are targeted to create a behavioral change and reach political goals. Many studies have found that family influences an individual to join radical groups, which can be divided into internal or external factors (Barricman 2019; Bloom 2010, 2011; Curtis 2020; Dass 2021; Doosje *et al.* 2016; Gitaningrum 2021; Jadoon 2020; Kasanah; Kurniati 2021; Musfia 2017; Noor 2019; Rasyid 2018; Rozika 2017; Sikkens *et al.* 2017; Wickham *et al.* 2019; Vergani *et al.* 2018; Zuhdi & Syauqillah 2019). However, there are few studies that analyze the role of the family in the deradicalization of terror convicts. Gazi (2016) has examined the reasons why former convicts and/or members of Jamaah Islamiyah (JI) abandoned terrorism. The results of his findings showed that former convicts and/or members of JI left the path of terror because of three factors: personal, organizational, and social. The form of regret of the formers can be a way to be an agent of change to prevent/counter violent extremism (PCVE). It can be seen in Schewe & Koehler's (2021) study that explained former extremists were motivated to PCVE as a redemption. Then, the other study by Elga Sikkens *et al.* (2017) focused on the role of the family in preventing and deradicalizing family members. The results of this study found that the knowledge about the different ideologies and tools on how to respond to their children's radicalization affects the process of deradicalization of family members. This study also was strengthened by El-Amraoui & Ducol (2019) and Yayla (2020). Nevertheless, those studies do not explore more the role of wives as a member of the family in the deradicalization process. Thus, this article focuses on the role of wives that can be a tool to deradicalize their husbands and/or member of the family.

Burgess & Locke (1976) define a family as a group of people united by ties of marriage, blood, or adoption which is one household that interacts with each other in their respective

*Corresponding Author: kurniati1402@gmail.com

DOI: 10.1201/9781003322054-35

social roles as husband and wife, mother and father, brother and sister, and women who create a shared culture. This is in line with the definition that Tallman & Babcock (2000) described, namely the family as any group of individuals bound together by generally recognized and socially recognized kinship ties. The key to this definition is in the concept of kinship. Kinship is a special type of social relationship determined by heredity, marriage, and adoption (Tallman & Babcock 2000). The family has the main task of meeting the physical, spiritual, and social needs of family members (Wirdhana *et al.* 2013).

The involvement of women in terrorist acts is a worrying phenomenon. This is because in most parts of Indonesia women have an important position as caregivers in a family and at the same time are responsible for the education of the children. Patriarchal hegemony greatly contributes to women in giving the meaning of jihad and its implementation (Asiyah *et al.* 2020). The involvement of women is a form of recognition by radical groups on the issue of inequality and injustice which always positions women as weak creatures (Qori'ah 2019). Sukabdi's (2021) study explained that in Indonesia women have a high risk to involve in terrorism. There are reasons for the emergence of the phenomenon of female terrorists in Indonesia. First, the decreasing number of male jihadist cadres or combatants is due to the large number of those arrested by law enforcement officers and undergoing the sentencing process. Hence, the choice to make female "martyrs" became urgent. Second, women are considered as not suspicious so do not make law enforcement officers as an alert. Third, with the increasingly sophisticated information technology, such as the emergence of social media networks, various types of recruitment patterns and jihadist propaganda are increasingly accessible even to women who can stimulate them to carry out acts of terrorism (Hartana 2017).

The involvement of women in terrorist networks illustrates how fragile a family is or vice versa. A family is like a coin that has two sides, besides being a place for radicalization, the family can also be an effective deradicalization place for family members. In this article, an ethnographic approach is used to gain a detailed understanding of the influence of the wives in the deradicalization of their husbands. I was involved in the Family Resilience Program and visited 17 wives of former convicts to observe the involvement of wives in the program and their roles to influence their husbands. This study also consists of an interview with a former terror convict to understand the factors behind the involvement of wives in the program and the role of wives in the deradicalization process of their husbands. I argue that there are two factors that are internal and external factors to deradicalize terror convicts. The external factor is an intervention that presents many actors that the terror convicts do not know before such as academics, religious leaders, and psychologists that influence terror convicts' worldviews. The internal factor comes from people with emotional bounding such as mothers, wives, and children.

2 METHODS

In this study, a case study approach was used to gain a detailed understanding of the role of wives in the deradicalization of husbands. This study consists of an observation during Family Resilience Program in 2019 that was held by Division for Applied Social Psychology Research. This article is also strengthened by interviewing the ex-terror convict and senior consultant of the Division for Applied Social Psychology Research, Nasir Abas.

3 FINDINGS AND DISCUSSION

3.1 *Involving the wife in the husband's deradicalization*

Doosje *et al.* (2016) describe how the process of individual deradicalization can be divided into three aspects or levels: micro, meso, and macro. The micro-level can be a loss of ideological appeal. It is an internal factor because it occurs in the deradicalization process itself. Experiencing other major life events (marriage, birth of a child) influences an individual to change his mind and even his behavior. Another micro (internal) factor is intellectual doubt ('Do I want to live my life like this forever?'), which is sometimes supported by exposure to alternative viewpoints, such as through relevant books and media. Self-debate is supported by Sirry's (2020) study on youth in Indonesian universities. The study explained that "among students who have been exposed to radical ideologies in Indonesia, it is evident that they can easily join a radical group; however, they may also easily decide to leave" (Sirry 2020: 11). The next factor is at the meso level. An important element at the meso level is a separation from the group and its activities, sometimes caused by intragroup conflict and disappointment in the group (leader). At the macro level, prisons can sometimes create contexts in which people want to start afresh and experience deradicalization. The other finding from Ayuningtyas (2022) found that deradicalization can occur because of the arrest which ultimately makes the person imprisoned and limit his movement space so that he questions his ideology. This internal factor usually arises from the individual himself who has experienced a critical thinking process about the events he has experienced, as experienced by Ika Puspitasari as an ex-woman terror convict (see Ayuningtyas 2022).

To understand more about involving personal issues as factors to deradicalize an individual, I found that the role of wives and giving an intervention to the wives of terror convicts can be a strategy for the government or non-government institutions to deradicalize a terror convict in a prison. Family Resilience Program is a social, psychological, economic, and religious intervention that is given to the women's families and/or wives of ex/terror convicts held by the Division for Applied Social Psychology Research (DASPR) since 2015 (interviewed with Abas 2022; see Magrie *et al.* 2022). I have been directly involved in observing the program from 2019 to 2021. The field data showed that the wives have the potential to become agents of change. They can approach their husbands and share their knowledge and feelings (sadness because of the stigma and bullying that their children got from society because their father was arrested) (see also Rufaedah & Putra 2018) to affect their husbands' emotionally to leave their radical path and get back to the family after they are freed. Here is a quote from the interview session with Nasi Abas, a consultant of DASPR, that wives could be a mouthpiece to influence husbands.

"So that this wife can be a mouthpiece and can remind her husband, namely internal factors that we expect from the wife to give advice, yes, open her husband's mind so that he does not commit acts of terrorism again." (Interviewed with Nasir Abas, former terror convict, May 11, 2022)

Nasir Abas (2022) reminded us that the role of wives is very important to be part of the deradicalization strategy. He explained that in fact, deradicalization should be constructed in two ways or fulfill two elements, internal and external elements. The first, external element, the actors of deradicalization (instead of the government or non-governmental actors) provide interventions such as presenting religious leaders, academics, and psychologists into the prison to reeducate the terror convicts (see Erikha & Rufaedah 2019; see Subagyo 2021). According to the second, internal element, the actors of the deradicalization should intervene with the wives of terror convicts such as giving them workshops or pieces of training to build their skills to be more resilient (see also Putra *et al.* 2018). The aim of giving intervention to the wives is to prevent the ideology of their husbands to be followed by their wives and children. The role of wives is to be a mouthpiece to deradicalize their husbands when they visit their husbands in prison.

While observing the wives in the Family Resilience Program in 2019, I found that bringing wives of terrorist convicts in a workshop series together could be a support system for the wives to cope with their doubt, sadness, and stress when their husbands were arrested. In the workshop series, the wives gained a psychological session to release their emotions and thoughts about their husbands' cases. The psychologist was presented to guide the wives in writing and talking about their unhappiest and happiest stories. I found that mostly the wives mentioned that their husband's case is the saddest moment that they felt. They lost the husband's role as father and spouse. It made them realize that what their husbands did was wrong and affected their family's happiness. I also saw that most of the children had negative impacts such as bullying and traumatic moment because they were there while their father was arrested. But I have to highlight that in this program, especially in 2019, the families that were involved in the program (1) mostly did not know their husband's case, and (2) could cooperate during the workshop series, which meant they were not extremists and did not agree with their husbands. In this program, I found that their husbands' cases were mostly as members of a radical group affiliated with ISIS, so their role as a follower rather than a leader. Otherwise, the Surabaya case in 2018 was a big blow for Indonesia because it showed the development of an extreme action involving all family members carrying out suicide bombings. In Dass's (2021) study, this is due to two reasons: first, it shows a new trend in terrorism where perpetrators are motivated mainly by a misinterpreted religious ideology as compared to most of the previous suicide terrorism cases where attackers were motivated by nationalistic and liberation ideologies from foreign occupying forces. Second, it is the first case in the region involving the use of the entire family including parents to arm their own young children. So, in this case, it is assumed that family involvement can also have the opposite effect, i.e. deradicalizing individuals to get out of terrorist networks. Family-based deradicalization is a preventive effort carried out by families because the family is the first and foremost institution for children to get protection and Islamic religious education with an integrated and balanced strategy and approach (Shofiyah 2018; see Wirdhana *et al.* 2013). Children who grow up in a tolerant family will be taught the importance of being tolerant and will be spared from radical thoughts (Botma 2020). This is supported by the study of Fikriyati (2017) showing that wives have an important role in the deradicalization process of their husbands. The Peace Pioneer Women's Group demonstrated the success of the involvement of families and women in dealing with the issue of radicalism (Syaifuddin & Belida 2019; Yayla 2020).

Another factor that can deradicalize individuals is to provide space for women or wives to have a voice. When women are given the same rights as men, this can reduce the incentive to participate in terrorism because women will have other opportunities to voice their opinions and gain power in a society that is not limited to participating in violence (Wickham *et al.* 2019). Cultivating an inclusive political culture, encouraging dialogue on issues, and expressing opinions should reduce the possibility of radicalization and violent tendencies exhibited by some countries (Wickham *et al.* 2019). Family situations (problems) can affect the process of radicalization and family support may play a role in deradicalization. It was also emphasized that parents have a need for knowledge of different ideologies and tools on how to respond to the radicalization of their children. Family support programs can focus on this void to help families fight radicalization (Sikkens *et al.* 2017; Sulistjaningsih & Mukhlasin 2019). Involving the family in deradicalization can be started by maximizing the function of the family in the household, so that family members are given space to be empowered and can embrace those who are exposed to radicalism back to mainstream teachings (see Wirdhana *et al.* 2013). In addition, children should be taught the values of religious and racial tolerance from a young age. The topic of religious extremism should be included in the curriculum to prevent young people from falling into perverted ideologies and enable them to distinguish between normal and extremist beliefs. Religious education and religious spaces such as mosques, madrasas, religious schools, lectures, and da'wah must be regulated so as not to spread radical and deviant ideas to the community (Dass 2021).

Hartana (2017) also mentioned three recommendations that need to be carried out in deradicalization: first, increasing counter-radicalization efforts on a massive and sustainable basis, both conventionally and using information technology; second, increasing deradicalization efforts carried out by related parties (Polri, BNPT, Correctional Institutions, etc.) with methods that not only focus on suspects/defendants/convicts/former convicts of terrorism cases but also on their families; third, blocking social media accounts or websites that spread teachings of radicalism and terrorism as a form of government firmness against terrorism movements in the cyberspace area (Hartana 2017). In addition, the government should consider the fate of the wives and children of terrorists. They should not only be supported financially but also by programs so that the families of terrorists will less likely be radicalized. This is because there is a potential that a child will make his father, who committed a terrorist act an idol. The family, more precisely the mother, plays important role in the education process. For the social context of countering violent extremism in the family, the authorities need to explain to the public the important difference between terrorism and jihad, at least so that society can prevent the children of terrorists from following their father's footsteps (Saputro 2010).

4 CONCLUSION

This article emphasizes the importance of two elements in the deradicalization process that are internal and external factors. The internal factor involves the wives of terror convicts convincing their husbands to leave radical groups. Involving wives can be involved by giving them workshops such as the Division for Applied Social Psychology Research's intervention in the Family Resilience Program. The use of wives as agents of change can be part of a tertiary intervention that government needs to deradicalize the terror convicts besides they can be part of a secondary intervention to prevent the radicalization of a family member. The external factor refers to actions that involve outsiders to influence the terror convicts in prison such as presenting academics, religious leaders, and/or psychologists. Family is a crucial aspect of a human being. It can be a way to radicalize or deradicalize depending on the direction of the family's ideology. Hence, this article suggests to the actors of deradicalization, the government or non-governmental actors, to counter violent extremism from the family of terrorists.

REFERENCES

Asiyah, U., Prasetyo, R.A., & Sudjak. 2020. Jihad perempuan dan terorisme. *Jurnal Sosiologi Agama* 14(1): 125–140.

Ayuningtyas, K. 2022. Galang dana dan siap jadi pengantin bom, ika kini pilih hidup damai. *BeritaBenar*. Diakses dari https://www.benarnews.org/indonesian/berita/eks-perempuan-militan-kini-pilih-hidup-damai-05022022165535.html

Barricman, B. 2019. *Deradicalizing and Disengaging the Children of the Islamic State*. Naval Postgraduate School.

Bloom, M. 2010. Death becomes her: The changing nature of women's role in terror. *Georgetown Journal of International Affairs* 11(1), 91–98.

Bloom, M. 2011. *Women and Terrorism: Bombshell*. University of Pennsylvania Press.

Botma, A. 2020. Deradikalisasi paham keagamaan melalui pendekatan pendidikan agama islam dalam keluarga. *Jurnal Ilmiah Iqra'* 14(2): 171–185.

Burgess, E. & Locke, J. 1976. *The Family*.

Curtis, G. 2020. Apa yang menjadi kesalahan indonesia tentang perempuan dan ekstremisme berkekerasan. *THC Insights* 18.

Dass, R. A. S. 2021. The use of family networks in suicide terrorism: A case study of the 2018 surabaya attacks. *Journal of Policing, Intelligence and Counter Terrorism*.

Demant, F, *et al.*, 2008. Decline and disengagement: An analysis of processes of deradicalisation. *Institute for Migration and Ethnic Studies*.

Doosje, B., Moghaddam, F. M., Kruglanski, A. W., Wolf, A. D., Mann, L., & Feddes, A. R. 2016. Terrorism, radicalization and de-radicalization. *Current Opinion in Psychology* 11: 79–84.

El-Amraoui, A. F. & Ducol, B. 2019. Family-Oriented P/CVE programs: Overview, challenges and future directions. *Journal for Deradicalization*.

Erikha, F. & Rufaedah, A. 2019. Dealing with terrorism in Indonesia: An attempt to deradicalize, disengage and reintegrate terror inmates with a social psychology approach. *Terrorist Rehabilitation and Community Engagement in Malaysia and Southeast Asia*: 145–173. Routledge.

Fikriyati, U. N. 2017. Perempuan dan deradikalisasi: Peran para istri mantan terpidana terorisme dalam proses deradikalisasi. *Jurnal Sosiologi Reflektif* 12(1): 1–16.

Gazi. 2016. *Dinamika Relasi Sosial dalam Proses Meninggalkan Jalan Teror, a Dissertation*. Jakarta: Universitas Indonesia.

Gitaningrum, I. 2021. Children and terrorism: Human rights for Indonesian cubs of caliphate. *Jurnal Penelitian* 18(2): 171–180.
Hartana, I. M. R. 2017. Teroris perempuan; ancaman faktual di Indonesia. *Jurnal Ilmu Kepolisian*: 45–50.
Jadoon, A. 2020. Gendering recruitment into violent organizations: Lesson for counter-terrorism operations. *Texas National Security Review* 4(1): 168–173.
Kasanah, N. Perempuan dalam jerat terorisme: Analisis motivasi pelaku bom bunuh diri di Indonesia. *IJouGS: Indonesian Journal of Gender Studies* 2(2): 34–43.
Kurniati, E. 2021. The recruitment of parking attendants as members of radical religious study group in Indonesia. *Indo-Islamika: Jurnal Kajian Interdisipliner Islam Indonesia* 11(2): 123–140.
Magrie, M. F., Putra, I. E., Rufaedah, A., & Putera, V. S. 2022. An empowerment program for spouses of convicted terrorists in Indonesia. In G. Barton, M. Vergani, and Y. Wahid (eds), *Countering Violent and Hateful Extremism in Indonesia*: 145–173. Palgrave Macmillan.
Musfia, N. W. 2017. Peran perempuan dalam jaringan terorisme ISIS di Indonesia. *Journal of International Relations* 3(4): 174–180.
Noor, H. 2019. *Memperkuat Hubungan Keluarga untuk Mencegah Kerentanan Pekerja Migran terhadap Radikalisasi*. Buruhmigran.or.id. Diakses dari https://buruhmigran.or.id/2019/03/27/memperkuat-hubungan-keluarga-untuk-mencegah-kerentanan-pekerja-migran-terhadap-radikalisasi/
Putra, I. E., Danamasi, D. O., Rufaedah, A., Arimbi, R. S., & Priyanto, S. 2018. Tackling islamic terrorism and radicalism in Indonesia by increasing the sense of humanity and friendship. *Handbook of Research on Examining Global Peacemaking in the Digital Age*: 94–114. IGI Global.
Rasyid, M. 2018. Perempuan dalam jaringan radikalisme vis a vis terorisme global. *Muwazah: Jurnal Kajian Gender* 10(2): 162–182.
Rozika, W. 2017. Propaganda dan penyebaran ideologi terorisme melalui media internet (Studi kasus pelaku cyber terorisme oleh bahrun naim). *Jurnal Ilmu Kepolisian*: 122–134.
Rufaedah, A. & Putra, I. E. 2018. Coping with stigma and social exclusion of terror-convicts' wives in Indonesia: An interpretative phenomenological analysis. *The Qualitative Report* 23(6): 1334–1346.
Saputro, M.E. 2010. Probabilitas teroris perempuan di Indonesia. *Jurnal Ilmu Sosial dan Ilmu Politik* 14(2): 211–228.
Schewe, J. & Koehler, D. 2021. When healing turns to activism: Formers and family members' motivation to engage in P/CVE. *Journal for Deradicalization.*
Shofiyah. 2018. Deradikalisasi berbasis keluarga. *Madinah: Jurnal Studi Islam* 5(1): 80–95.
Sikkens, E., San, M. V., Sieckelinck, S. & Winter, M. D. 2017. Parental influence on radicalization and de-radicalization according to the lived experiences of former extremists and their families. *Journal for Deradicalization.*
Sirry, M. 2020. Muslim student radicalism and self-deradicalization in Indonesia. *Islam and Christian–Muslim Relations.*
Subagyo, A. 2021. The implementation of the pentahelix model for the terrorism deradicalization program in Indonesia. *Cogent Social Sciences* 7(1): 1–21.
Sukabdi, Z. A. 2021. Risk assessment of women involved in terrorism: Indonesian cases. *International Journal of Social Science and Human Research* 4(9): 2495–2511.
Sulistjaningsih, S. & Mukhlasin, L. 2019. Re-education: A treatment to revise the misunderstanding of terrorist religion, a study case on first female terrorist in Indonesia. *Empati: Jurnal Ilmu Kesejahteraan Sosial* 8(2): 92–108.
Syaifuddin & Belida, O.O. 2019. Strategi komunikasi kelompok perempuan pelopor perdamaian dalam menghadapi isu radikalisme. *Kalbisocio: Jurnal Bisnis dan Komunikasi* 6(2): 167–173.
Tallman, I. & Babcock, G. M. 2000. Family policy in western society. In Edgar F. Borgatta & Rhonda J. V. Montgomery (eds), *Encyclopedia of Sociology* Second Edition, Volume 2: 962–969. New York: Macmillan Reference USA.
Vergani, M., Iqbal, M., Ilbahar, E., & Barton, G. 2018. The three Ps of radicalization: Push, Pull and Personal. A systematic scoping review of the scientific evidence about radicalization into violent extremism. *Studies in Conflict & Terrorism.*
Wickham, B.M., Capezza, N.M., & Stephenson, V.L. 2019. Misperceptions and motivations of the female terrorist: A psychological perspective. *Journal of Aggression, Maltreatment & Trauma*: 1–16.
Wirdhana, I., dkk. 2013. *Buku pegangan kader BKR tentang delapan fungsi keluarga*. Jakarta: Badan Kependudukan dan Keluarga Berencana Nasional Direktorat Bina Ketahanan Remaja.
Yayla, A. 2020. Preventing terrorist recruitment through early intervention by involving families. *Journal for Deradicalization.*
Zuhdi, M. L. & Syauqillah, M. 2019. Analysis identity fusion and psychosocial development: How the role of Father, Mother and son on radicalization within family. *Journal of Strategic and Global Studies* 2(2):64–84.

Religion, Education, Science and Technology towards a More Inclusive and Sustainable Future – Rahiem (Ed.)

Role of SUPKEM leadership in fighting against extremism and terrorism in Kenya

H.Y. Akasi*
Umma University, School of Sharia and Islamic Studies, Kenya

ABSTRACT: The Supreme Council of Kenyan Muslims (SUPKEM) was established in 1973 as an umbrella body to bring together all Muslims in Kenya with the intention of addressing the needs of the Muslim community in Kenya. It also acts as a link between the Kenyan government and its Muslim citizens. Extremism is one of the major threats to the East African region and Kenya in particular. The ideology of violent extremism is spread through different media and radicalizes individuals into terrorists who kill and cause devastation. The Muslim community in Kenya has a role to play in combating violent extremism and terrorism in the country because conversion to Islam has been linked to radicalization and violent extremism, particularly along Kenya's coast. The aim of this study is to demonstrate the role that SUPKEM leadership can play in the counterterrorism measures implemented in Kenya. Qualitative research was used to study the role that Muslim leaders in SUPKEM play in countering violent extremism and terrorism. Qualitative interviews with selected Islamic leaders provided rich data that helped to understand terrorism and violent extremism in Kenya. The results showed that SUPKEM leadership is well-informed about the problem of terrorism and violent extremism in Kenya. Islamic leadership can play a role through alternative narratives, religious dialogue, and more scholarly research to guide their believers against dangerous religious ideology.

Keywords: SUPKEM, extremism, terrorism

1 INTRODUCTION

Religious leaders have the strategic advantage of religious authority, as they are accorded respect by their followers and society in general. In both Islam and Christianity, religious leaders have legitimate authority and the right to direct others. They are therefore able to maintain sobriety among the believers, especially in the face of problems such as extremism and terrorism (Harris-Hogan *et al.* 2016). There is a complex relationship between religion and violent extremism. It is difficult to determine the direct causation of this relationship. In the past, violent extremism has happened without being directly linked to religion as the primary driver, while in other instances, religion has played a big role (Botha 2013). The Supreme Council of Kenyan Muslims (SUPKEM) is a religious organization that was founded in 1973 with the goal of uniting all Muslims and meeting their needs. It also enables the communication between the government and the Muslim community. (Omari 2014). Terrorism is a threat to public safety as it is meant to cause death and immense bodily harm to civilians and noncombatants. It is also aimed at intimidating the public and coercing the government or international entities to desist from championing or performing their duties (Karlsrud 2017). Terrorism threatens societal existence because it is an illegal act against the

*Corresponding Author: hakasi@umma.ac.ke

DOI: 10.1201/9781003322054-36

laws of the land (Liht & Savage 2013). Since the 1998 terrorist attacks on the US Embassy, the East African region has been on the list of countries affected by terrorism. After the 1998 incident, in 2002, another attack targeting Paradise Hotel in the coastal town of Mombasa claimed the lives of 12 people and injured 80. The majority of the terrorist attacks in the region are perpetrated by al-Shabaab, which is part of the al-Qaeda network (Ali 2016). Globally, jihadist groups have continued to advance their agendas and insurgencies in conflict zones such as Afghanistan, Iraq, Somalia, Yemen, Syria, and other North African and Middle Eastern countries (Lemu 2016). Occasionally, they have successfully launched attacks against the United States, such as during the September 11, 2001, attacks. There have been attacks in Europe, with countries such as the UK and France being affected. The attacks are viewed in terms of global terrorism campaigns that are planned and implemented by terrorists and extremist groups. (Abdel-Fattah 2020).

There is a sustained effort by Kenya to counterterrorism from Somalia and within its borders. The government has a responsibility to offer national security, which is of interest to its citizens. Strict measures have been put in place to achieve the counterterrorism goals. This is because if the problem is left unattended, more lives are likely to be lost in future attacks. At the institutional level, the involvement of SUPKEM has been one of the ways to ensure that the war against terrorism and violent extremism is well-balanced (Badurdeen & Goldsmith 2018). A number of studies have been conducted to help counterterrorism in the Horn of Africa and Kenya. The implementation of counterterrorism actions in Kenya has been done in the past without proper involvement or legislation.

1.1 *Statement of problem*

The presence of violent extremism and terrorist activities can result in the loss of lives if something is not done. Peace and tranquility may become elusive within the Muslim community and the entire country. Kenya and other countries affected by terrorism have implemented counterterrorism measures aimed at preventing future attacks. For instance, the deployment of the military in Somalia was part of Kenya's strategy to combat terrorism and violent extremism (Demuynck & Julie Coleman 2022). The country has also used other strategies such as active community engagement, multidisciplinary case management, and other measures that help to keep vulnerable groups or individuals on the right path. A myriad of approaches are therefore necessary to deal with the problem of terrorism. The research thus looks into how Islamic organizations and institutions in general, and SUPKEM in particular, have dealt with situations such as the one described above in the past and asks what they have done so far in response to what is happening now with the killings of innocent people, and why they haven't, given that they are Kenya's highest Muslim leadership organization.

2 METHODS

Qualitative methods were used to study the role of SUPKEM in countering terrorism and violent extremism in Kenya. Qualitative approaches allowed the researcher to get attitudes and insights from the participants. The study identified SUPKEM leaders as an informed group that has a lot of knowledge about Islam. This group has an influence on young people who are the targets of radicalization into violent extremism and terrorism recruitment. Purposive sampling was used to identify leaders who possess relevant knowledge about terrorism and violent extremism. This ensured that the data collected from such participants was in-depth and captured the insights of the leaders. The approach also ensured that high-quality participants were included in the study. The researcher used publicly available directories to identify people regarded as community leaders and invited them to participate. Within the SUPKEM, there are scholars, youth leaders, imams, and other people who perform the role of leaders. The selection of scholars was done in consultation with the SUPKEM leadership. The researcher was directed to find the most qualified person to participate in the study (Wilson 2017).

2.1 *Secondary data*

Secondary data in this study was obtained through a systematic search and review of the literature. To achieve this type of data, online databases were used. A combination of keywords was used to conduct searches on the major databases, and this led to the identification of articles that have been published about terrorism and extremism in Kenya and other parts of the world. This step led to the identification of important resources that were included in the literature review. This review of literature was also used to identify future resources, such as books, articles, and videos that can be used to train more on the project.

2.2 *Observation*

Observations were used to collect data from the selected Muslim clerics. The specific observation was made on sermons delivered by the Muslim leaders, who are also members of SUPKEM.

2.3 *Interviews*

Qualitative interviews were useful in collecting the opinions of the participants. This allowed the participants to go in-depth into discussions about the role of Muslim leaders in combating terrorism and violent extremism in Kenya. The participants provided explanations regarding what people expect from the leaders and what they are likely to gain from the knowledge and experience of such leaders. Relevant interviews were crafted with the aim of collecting valid and reliable data from the participants. The researcher guided the interview to ensure that the respondents did not veer too far from the main topic.

The collected data from the interview was subjected to thematic analysis. There are five major themes and sub-themes about the role of leaders in fighting against terrorism and extremism that were identified during the study. This also made it possible to organize the final results properly based on the identified themes.

3 RESULTS AND DISCUSSION

Terrorism is defined as an act intended to cause death or serious bodily harm to civilians or noncombatants to intimidate a population or coerce the state government or an international entity to cease or refrain from performing any acts or duties (Neria *et al.* 2006). In contemporary times, terrorism is a major risk to societal existence and, hence, illegal based on the laws of the United States. Terrorism is also considered a war crime under the laws of war when applied to target non-combatants, such as unprejudiced military personnel or civilians (United Nations 2008). The symbolism of terrorism can harness human fear to help achieve certain goals (Momanyi 2015).

The East African region has also had its share of terrorist attacks. The 1998 attack on US embassies in Kenya and Tanzania and the foiled attack in Uganda confirmed the presence of international terrorism (in this case, Al-Qaeda) in the region (Woldemichael 2006). Accordingly, in 2002, Kenya witnessed a major attack on the Paradise Hotel along the Kenyan coast in the town of Mombasa. The attack killed 12 people and injured 80 people (Botha 2013). It is worth noting that the first successful terrorist attack in Kenyan history dates back to the Norfolk Bombing of 1980 (Kiruga 2013). The region's current terrorist threats emanate from Al-Shabaab, an al-Qaeda affiliate that has local support from Al-Hijra

(Torbjörnsson & Jonsson 2016). Al-Hijra is a local terrorist group that mainly operates in Nairobi and Mombasa. The alleged first al-Shabaab attack occurred in Kampala in 2010 (Shuriye 2012). The group also attacked the Westgate shopping mall in Nairobi in 2013; nonetheless, there was also a spate of attacks using hand grenades (Blanchard 2013) across the country. Some of these attacks have been a counterattack against Kenya's military intervention in Somalia in an attempt to restore peace.

Based on the interview with a SUPKEM officer, the first interviewer mentioned that Muslim leaders can play a role in promoting counter-narratives among their community members. "Alternative" or "counter narrative" refers to online or offline communication activities that directly or indirectly challenge extremist propaganda. This can be achieved in different ways, including face-to-face sermons, testimonials, blogs, chat rooms, and social media profiles (Avis 2016).

> *"We as Muslim clerics believe we can play a role in engaging the young people about the misinformation and propaganda that is conveyed to them by the extremists. We know the truth about Islam, and we can change our society through what we tell people about our faith."*

Also, SUPKEM played a vital role in fighting against in the form of Religious dialogue, the second interviewer mentioned that Muslim leaders have a role to play in enhancing unity and trust among religious leaders. One of the strategies for countering terrorism and violent extremism is through engaging in religious dialogue. Muslim leaders can also establish good relations with other faith leaders, like Christian faith leaders, to help counter the negative perceptions and narratives about Muslim youths and terrorism.

> *"I believe that Muslim leaders should create awareness about terrorism in mosques because they have some authority and can speak directly to the youths who are at risk of radicalization".*

Correspondingly, SUPKEM plays a vital role in fighting against extremism in the form of public meetings. In the third interview, we found that counterterrorism can also be achieved through public meetings. Religious leaders have a role to play in organizing and engaging community members through such gatherings (Abdel-Fattah 2020). Through public meetings, Muslim leaders can join other leaders in discussing the problems affecting their communities and try to engage other stakeholders, such as political leaders and security agencies. Challenges were also identified with regard to public meetings as avenues of countering terrorism and violent extremism. It emerged that there was low participation of young people in such meetings, though it was considered to be a good avenue for sharing ideology about Islam and creating awareness about radicalization among the youth.

Last, by giving the formation of inter-religious committees a vital role in fighting against terrorism, from the fourth interview, it emerged that forming inter-religious committees is one of the strategies that can help to counter terrorism and violent extremism. Religious leaders should take on leadership roles at local and national levels. Inter-religious committees should comprise Christians, Muslims, and other faith leaders. "*To address the problem of terrorism and violent extremism, we should love one another. Both Christian and Muslim leaders should hold regular meetings that can help to find solutions to the problem. Leaders should show a good example by attending committees and making the right decisions for the communities."*

4 CONCLUSIONS

SUPKEM's leadership is aware of the threat of terrorism and violent extremism in Kenya. The Muslim leaders at SUPKEM have a big role to play in counterterrorism measures in the country because of the position they hold and the authority they have over their believers. With the knowledge they have about Islam and ideology, SUPKEM leaders should engage in alternative narratives that should counter the extremist ideology that is advanced by terrorist groups. Online spaces, where youths are likely to be exposed to propaganda about Islam, are great avenues through which religious leaders can enhance alternative narratives. Publishing literature is also a great move that can be used by SUPKEM leadership for counterterrorism. There is a need for more scholarly literature that young people should access to influence them in the right way instead of being left to read propaganda about Islam on the internet. SUPKEM leadership also has a role to play in creating programs where they can engage the community to create awareness about terrorism and violent extremism. Community programs should include victims of terrorism and vulnerable groups that are at risk of being radicalized by violent extremists. Islamic leaders should also be encouraged to promote interfaith dialogues in Kenya. Creating forums where Muslim and Christian leaders come together is a way of promoting unity within diverse communities. Through such dialogues, misconceptions about Islam and terrorism can be clarified, and believers can gain access to the right information to counterterrorism and violent extremism.

REFERENCES

Abdel-Fattah, R., 2020. Countering violent extremism, governmentality and Australian Muslim youth as 'becoming terrorist.' *Journal of Sociology*, 56 (3), 372–387.

Ali, A.M., 2016. *Islamist Extremism in East Africa.* National Defense University Fort McNair DC Washington.

Avis, W., 2016. *The Role of Online/Social Media in Countering Violent Extremism in East Africa.* Helpdesk Report.

Badurdeen, F.A. and Goldsmith, P., 2018. *Initiatives and Perceptions to Counter Violent Extremism in the Coastal Region of kenya.*

Botha, A., 2013. Assessing the vulnerability of Kenyan youths to radicalization and extremism. *Institute for Security Studies Papers*, 2013 (245), 28.

Demuynck, M. and Julie Coleman, J.D., 2022. *Customary Leaders and Terrorism in the Sahel: Co-opted, Coerced, or Killed?*

Harris-Hogan, S., Barrelle, K., and Zammit, A., 2016. What is countering violent extremism? Exploring CVE policy and practice in Australia. *Behavioral Sciences of Terrorism and Political Aggression*, 8 (1), 6–24.

Karlsrud, J., 2017. Towards UN counter-terrorism operations? *Third World Quarterly*, 38 (6), 1215–1231.

Kiruga, M., 2013. *20 Killed in Bomb Attack on Norfolk* [online].

Lemu, M.N., 2016. Notes on religion and countering violent extremism. *Countering Daesh Propaganda: Action-Oriented Research for Practical Policy Outcomes*, 43–50.

Liht, J. and Savage, S., 2013. Preventing violent extremism through value complexity: Being Muslim being British. *Journal of Strategic Security*, 6 (4), 44–66.

Momanyi, S.M., 2015. *The Impact of Al-Shabab Terrorist Attacks in Kenya.* Master's Thesis. UiT Norges Arktiske Universitet.

Neria, Y., Gross, R., and Marshall, R., 2006. *9/11: Mental Health in the Wake of Terrorist Attacks [online].* Cambridge University Press. Available from: http://libgen.rs/book/index.php?md5=818A96B946D9B6147F034FDB17F87478.

Omari, H.K., 2014. *Islamic Leadership in Kenya: A Case Study of The Supreme Council of Kenya muslims (Supkem).* Doctoral Dissertation. University of Nairobi.

Shuriye, A.O., 2012. Al-Shabaab's leadership hierarchy and its Ideology. *Academic Research International*, 2 (1), 274–285.

Torbjörnsson, D. and Jonsson, M., 2016. *Containment or Contagion? Countering Al Shabaab's Efforts to Sow Discord in Kenya.* Retrieved September 15, 2017.

United Nations, 2008. *Office of the United Nations High Commissioner for Human Rights United Nations Office at Geneva.* United Nations.

Wilson, T.P., 2017. Normative and interpretive paradigms in sociology. In: *Everyday Life*. Routledge, 57–79.

Woldemichael, W., 2006. International terrorism in East Africa: The case of Kenya1. *Ethiopian Journal of the Social Sciences and Humanities*, 4 (1), 33–53.

Religion, Education, Science and Technology towards a More Inclusive and Sustainable Future – Rahiem (Ed.)

Consequeces of panai money in Siri and value of Maqasid Al-Syariah in tribe marriage Bugis in Indragiri Hilir Riau

B. Maani*, Syukri & Aliyas
UIN Sulthan Thaha Saifuddin, Jambi, Indonesia

ABSTRACT: This article aims to examine the effects and essence of panai money (spending money given to a wife) in adat perkawinan Bugis (marriage customs) in Indragiri Hilir Regency. This area has a plural society, but there are many locations of the Bugis tribe who settled and colonized and then had their own customs for holding traditional ceremonies, especially panai money as a marriage custom. Is the panai money high only because it is based on an attitude of siri' (self-respect), or does this Bugis community have a feodal-mindedness with only wealth (money oriented)? The high panai money is a necessity. The majority of the Bugis are reluctant to marry; even if they want to marry, they tend to take shortcuts, such as elopement, or seek from non-Bugis tribes. As a result, most of the girls of the Bugis tribe became spinsters; crimes occurred by taking other people's girls away, and committing unlawful acts by impregnating them. This happens because it maintains *siri*, and seeks maqa>s}id al-syari>ah but appears in another series with greater consequences. The research is carried out through the sociology of law, which consists of primary and secondary data using book references, direct interviews with the Bugis community, observation, and documentation methods in the triangulation corridor.

Keywords: Panai Money, Si>ri, Maqa>s}id al-Syari>ah

1 INTRODUCTION

During the end of the 17th century, the Bugis community migrated from South Sulawesi to the Malay village area, Indragiri Hilir Regency Riau Province, such as Kuala Enok Luar and Kuala Enok Dalam, Pulau Kijang, Benteng, Pulau Kecil, Pebenaan, Sanglar, Kota Baru Seberida, and Teluk Kelasa. The Kotabaru Sebrida area has an 80% and Benteng 100% Bugis population, with a total population of 652,342 in Indragiri Hilir, Riau Province (Fattah 2020).

In these areas, the Bugis people have wedding customs, which grow and develop, but still adhere to traditional and religious traditions through local wisdom that has been developed from generation to generation and is believed to be accurate (Nurnaningsih 2015). Panai money is one of the social and cultural arrangements in Bugis marriage customs that must be fulfilled by giving a high amount of money from the groom to the bride. Whether the panai money is high or not depends on the social status, such as education level, physical status, widowhood, or virginity, of the bride and groom (Ikbal 2016).

The agreed amount of panai money always varies, for fellow aristocratic 150 million rupies, aristocratic women and non-aristocratic men 100 million rupies, aristocratic men and nonaristocratic women 80 million rupies; for men and women who are not aristocratic, it is

*Corresponding Author: bahrulmaani2018@gmail.com

DOI: 10.1201/9781003322054-37

60 million rupies; and for men and women who are not from the Bugis ethnic group, it is 50–40 million rupies. The method of organizing panai money is called the *Mapettu Ade'* (decided and determined in a traditional event) (Interview 2022).

1.1 *Panai money in the application of legal rules and siri's attitude*

1.1.1 *Panai money in the rule of law*

A positive tradition is a legacy of the past. In Indonesia it is seen as something that must be preserved and does not even contradict what is expressed by a group as a recognized unit (Muh. Sudirman Sesse 2017).

According to qaidah fiqh}iyah, customs can be used as a legal basis (Rachmat Syafi'i 2010: 274). With these rules, Islamic law can be formulated and applied in accordance with the existing customary traditions. The essence of the Qur'an and Hadith explains the basic principles and general characteristics of Islamic law, which can be explained by looking at local wisdom of each region.

In addition, through these rules, in the field of trade (economics), fiqh}iyah rules provide the flexibility to regulate various forms of trade or cooperation: If the text of the legal provisions is not explained, this rule allows us to formulate legal provisions. Research on adat ('urf) is needed as a basis for defining legal terms, but not all human habits can be used as a legal basis and what can be used as a legal basis is adat, which does not conflict with the basic principles and basic objectives of Islamic law itself (Sesse 2017: 78).

The Ulama have developed various rules related to customs to clarify their status and role in implementing Islamic law. According to these rules, adat can become law. The customs of these rules include 'urf qau>li>y and amali>y, this rule means the habit of syara, i.e., punishing humans when making laws, is both general and specific. After all, as long as there is no such thing as human behavior is evidence that must be done, it can be valid evidence. When it is generally accepted or adopted, these habits are taken into account. This rule explains that one of the conditions for calculating 'urf is that it must be universally accepted, this is the previous rule (al-'a>dah al-muhakkamah).

So what is generally accepted is the concept and substance that cannot be abandoned based on habit. This rule only applies to 'urf lafaz} writes the same as words. This rule states that one of the reasons for 'urf lafz}i is that the text appears on the word stupid sign of understanding such as verbal explanation. This rule also shows that the intelligible signs of a mute person can take the place of a normal speaking person.

So what is generally accepted is the concept and substance that cannot be abandoned based on habit. This rule only applies to 'urf lafaz} writes the same as words. This rule states that one of the reasons for 'urf lafz}i is that the text appears on the word stupid sign of understanding such as verbal explanation. This rule also shows that the intelligible signs of a mute person can take the place of a normal speaking person (Azis (2021).

Customs or traditions can change over time and place. Therefore, the previous law can be changed after changing the shariah law, ('urf as-s}ahi>h. Salafi>) the previous scholars are of the opinion that teachers who teach prayer, read the Quran, fast, and perform hajj should not be given money and should not perform the honorary duties of Imams and mosque muazzins, because their welfare is borne by bai>t al-ma>l. Due to time changes, bai>t al-ma>l will not be able to perform this function. Therefore, al-'urf as-s}ah}ih can conclude that 'urf can be used as a method of Islamic law to replace the views of previous scholars, as long as it does not conflict with previous laws, previous laws can be used as long as they do not conflict with previous laws. Nor does it conflict with the text (Mahmud Huda *et al.* 2018: 148).

From the various forms of 'urf mentioned above, several conditions determine the use of the word 'urf as a source of Islam: 'urf must be applied continuously or primarily and must become a habit used as a particular source of Islam. There is no statement (nas}) that contradicts ‘urf when appealed in court, and the use of ‘urf does not disqualify the explicit text of Islamic law (Faiz Zainuddin 2015: 403–405).

1.1.2 *Siri's attitude*

In the Bugis community, there is a uniqueness also owned by other tribes, but in the Bugis community, it is very prominent in daily behavior. *Siri* means self-respect that should be maintained and upheld as much as possible and not to be involved in any disgraceful acts. *Siri* also means pacce, namely a sense of empathy that can be felt in the form of emotion or pain as felt by other people, relatives or family so that there is a sense of always helping each other in any circumstances (Pelras C. 1996).

It is *sunnatullah* (natural law) that everyone has shame, therefore faith is part of shame or vice versa shame is part of faith. In the *Bugis* community, *siri* (shame) is categorized into two types: first, positive, which is used as a tool to compete and to get academic, social, political, and scientific achievements (Errington S. 1977).

Every time there is a form of application, there is always deliberation and consensus to propose to someone's child. This is done due to society's very high form of appreciation and respect. Second, negative *siri* includes anger, jealousy, uncontrolled emotions, humiliation as a result of fights and quarrels, which may end in murder (Idrus N.I. & L.R. Bennett 2003).

2 PANAI MONEY FROM THE MAQA>S}ID AL-SYARI>AH PERSPECTIVE

According to Ash-Syatibi Maqasid al-Syariah, in general it is divided into two, referring to shariah's objectives. Then the second, refers to the purpose of the mukallaf (person who can act legally). This can be seen from two points of view: Maqasid al-Syariah means the purpose of the revelation of shariah, namely realizing the benefit of the people. These benefits can be realized if there are efforts to protect in terms of religion, soul, lineage, reason, and property (Ahsin 2013) and (Auda 2010).

So panai money based on maqasid al-shariah aims to achieve Islamic law by opening the way of good or by closing the way of evil with Islamic law. It is also a set of divine intentions and moral concepts of Islamic law (Asrizal 2021:33–40).

Maqasid al-Shariah views marriage as a form of worship, with the essence of the meaning of mi>s\a>qan ghali>z}a (a strong bond). Islam provides full legal certainty based on the legal rules of maqa>s}id al-shari>ah, namely hifz}uddi>n (religious maintenance), hifz}u an-nasab. Marriage is also one of the basic human needs and instincts. According to Islamic historians, Adam had other needs besides food and drink. After getting introduced to Eve, it became clear that Adam was very happy and content, and his longing for the feelings that surrounded him disappeared (Muhlis Hadawi 2009; Al-Raysuni 2006).

The rituals that accompany panai money are the tradition of giving pattenre āda (promise), massarāpo (installing a tent), cemme passīling (special bath), tudang penni (sitting at night), sowing rice in the maddupa botting (welcoming the bride and groom) tradition, welcoming the presence, and the tradition of mappacci (berinai), as a gift from the prospective groom to the prospective bride as a historical, sociological, and philosophical basis, as a cause of high profits (Yayan Sopyan 2018).

Bugis customs in marriage can be divided into three groups, forbidden for reasons, containing elements of polytheism, bid'ah, and extravagance, making it difficult for oneself. Second, maybe because the traditional ceremonies performed in community marriages are adapted to the concepts and values contained in Islamic teachings. Third, al-‘urf al-Sahīh, namely habits that apply in society and do not conflict with the text, do not eliminate the benefit.

The word maqa>s}id, the plural of maqs}id which means a strong desire, holding fast and deliberately (Ahsan Lihasanah 2008). Shari>'ah literally means religion (addi>n), manhaj (al-minha>j), way, way of life (at-thari>qa>t) (Ahmad Sarwat 20190).

3 METHODOLOGY

This research adopts a qualitative method by examining the phenomenon in the Bugis community regarding expensive marriages. Data was obtained through interviews with religious leaders, community leaders, and traditional leaders, including stakeholders who are directly involved in the traditional marriages with panai money as the object of research in the Indragiri Hilir Regency, Riau Province. Thus, this research was carried out through the documentation of the KUA office and observations and the consequences arising from panai money. The sociology of law, which consists of primary and secondary data by using references from books and the field by direct interviews from the Bugis community.

4 ANALYSIS

The marriage customs of the Bugis are almost the same as those of other ethnic groups, only that the panai money is very high and this is a burden that must be borne by the men in Indragiri Hilir Regency. On the one hand, if the money is very high, there is a certain pride, respect, and glory in the family. But when the money is paid low, the family feels insulted. The logical consequences of this attitude can have negative consequences, such as, teenagers getting reluctant to get married, taking shortcuts to avoid high costs, getting pregnant out of wedlock, remaining as spinsters, or taken away. In the maqasid al-syariah aspect, the public benefit takes precedence over personal or group interests. There is a consensus to set a standard for pennies so that you can get married easily and avoid negative actions.

5 CONCLUSION

The nominal value of the panai varies according to the status of the family, whether of nobles or of ordinary people. It can even be judged that marriage between a noble's child and a commoner's child or servant is an offense called Nasoppa' Tekkenna, which means being stabbed with one's own stick. But on the other hand, a man is considered successful and has succeeded in choosing women among the nobility if the man is warani (brave), rich (tau sugi), and intellectual in the field of religion.

From the aspect of maqasid al-Shariah, he wants his child to be preserved from nasabiyah (descendants) and the religion that will protect him. So a very high panai money can guarantee his future. Men can also be creative and have a high work ethic, but the middle-income community bears a heavy burden. This tradition (adat) prioritizes the marriage process to seek big profits rather than seeking Allah's pleasure with the Shari'ah. Based on this phenomenon, the panai money is quite large, many young people refuse to get married immediately, even though it is not uncommon to do something contrary to the Shari'ah that affects their psychics.

Expensive panai money has significant consequences, such as cases of silariang (eloment), nalariang (kidnapping), and erangkale (women carrying themselves). This will also create a mate *siri* burden, reduce the dignity of family and relatives, so that they are not recognized as children and are expelled from the village because of *siri* (shame), creating psychological pressure on both parents and the girl who was taken away.

REFERENCES

Ahsan Lihasanah, (2008) *al-Fiqh al-Maqashid Inda al-Imami al-Syatibi*, Mesir: Dar al-Salam.
Ahmad Sarwat, (2019) *Maqasid Syari'ah*, Jakarta: Rumah Fiqh Publishing.
Ahsin W. Al-Hafidz, (2013) *Kamus Fiqh*, Jakarta: AMZAH.

Al-Raysuni, A. (2006). *Imam Al-Shatibi's Theory of the Higher Objectives and Intents of Islamic Law* (1st ed.). The International Institute of Islamic Thought.

Alimuddin, Interview, Tokoh Masyarakat Kuala Enok Indragiri Hilir Riau, tgl. 20 Februari, 2022).

Auda, J. (2010). *Maqasid Al-Shariah as Philosophy of Islamic Law* (1st ed.). The International Institute of Islamic Thought.

Azis, A. D. (2021). Symbolic meanings of equipments used in mappacci buginese traditional ceremony. *Jurnal Ilmiah Profesi Pendidikan, 6*(1).

Fattah, R.A.H.A., 2020. *Sejarah Suku Bugis Hijrah ke Tanah Melayu Indragiri Hilir (Part II)* [online]. Available from: https://indovizka.com/news/detail/750/sejarah-suku-bugis-hijrah-ke-tanah-melayu-indragiri-hilir-part-ii. Accessed on. 9-2-2022 [Accessed 22 Jun 2022].

Idrus, N.I. and L.R. Bennett (2003) "Presumed consent: Marital violence in bugis society", in L. Manderson and L.R. Bennett (eds) *Violence against Women in Asian Societies*. London and New York: Routledge Curzon. Pp. 41–46.

Ikbal, (2016) "Uang panaik dalam Perkawinan Adat Suku Bugis Makassar", *The Indonesian Journal of Islamic Family Law* 6, no.1 (Juni 2016): 202, https://doi.org/10.15642/al-hukama.

Lahsasna, A. (2013). *Maqasid Al-Shariah in Islamic Finance* (1st ed.). Islamic Banking & Finance Institute Malaysia.

Laldin, M. A., & Furqani, H. (2014). Maqasid Al-Shariah and stipulation of conditions (shurut) in contracts. *ISRA International Journal of Islamic Finance, 6*(1), 173–182.

Lamido, A. A. (2016). Maqasid al-Shari'ah as a framework for economic development theorization. *International Journal of Islamic Economics and Finance Studies, 2*(1), 27–49.

Muh, (2017) Sudirman Sesse, *"Eksistensi Adat Perkawinan Masyarakat Bugis Parepare Dalam Perspektif Hukum Islam,"* Disertasi UIN Alauddin, Makassar.

Muhlis Hadrawi, *Assikalabineng: Kitab Persetubuhan Bugis*, Makassar: Ininnawa, 2009.

Pelras, C. (1996). *The Bugis* (1st ed.). Cambridge: Blackwell Publishers Ltd.

Pelras, C. (1975). Introduction à la littérature Bugis. *Archipel, 10*(1), 239–267. https://doi.org/10.3406/arch.1975.1252

Yayan Sopyan and Andi Asyraf, Mahar and Paenre'; Regardless of Social Strata Bugis Women in Anthropological Studies of Islamic Law, *Universitas Islam Negeri Syarif Hidayatullah Jakarta 110 – Jurnal Cita Hukum (Indonesian Law Journal)*. Vol. 6 No. 1 (2018). P-ISSN: 2356-1440. E-ISSN: 2502-230X.

Religion, Education, Science and Technology towards a More Inclusive and Sustainable Future – Rahiem (Ed.)

Knowledge sharing on local wisdom in the workplace: A systematic literature review

R.J. Fakhlina*
UIN Imam Bonjol Padang, Indonesia

Zulkifli, S. Basuki & A. Rifai
UIN Syarif Hidayatullah, Jakarta, Indonesia

ABSTRACT: Knowledge is the intellectual capital of an organization. The important aspect of knowledge management (KM) is how people in an organization share their knowledge. Knowledge sharing (KS) and different local wisdom (LW) in the workplace will provide different patterns. To know the tendency of researchers about KS on LW in the workplace, it is necessary to conduct a systematic literature review using publish or perish (PoP) and preferred reporting items for systematic reviews and meta-analyses (PRISMA 2020) applications, which collected 37 articles published from 2017 to 2021. The results show that there are several tendencies of researchers in examining KS on LW, but few that relate to the workplace. The research of KS on LS in the workplace, such as libraries, is still very interesting to study.

Keywords: Knowledge management, knowledge sharing, and local wisdom

1 INTRODUCTION

One of the most basic activities in an organization's operational activities in the workplace is knowledge sharing (KS); these activities are known as knowledge management (KM) (Allameh 2018; Hislop 2013; Singh *et al.* 2021). KM is an organizational learning management tool (Hovland 2020). As a management tool, Martensson (2000) explores the origins of KM theory and sees KM as an integral part of the broader concept of "intellectual capital" (Abualoush *et al.* 2018; Jordão & Novas 2017). The strategy of managing knowledge resources is an important aspect that must be done by the organization/workplace (Valmohammadi 2015). However, just having a source of knowledge does not ensure an organization's success (Hislop *et al.* 2018). To develop a sustainable competitive advantage, the staff of an organization/workplace must share and apply knowledge practices. (Dalkir 2017).

KS activities require employees' willingness to share their valuable knowledge and behavior, which facilitates the exchange of relevant information among organizational staff (Anser *et al.* 2020). KS can be conceptualized as staff engagement in exchanging valuable knowledge, experience, and skills with other members of the organization (Mohajan 2019). KS occurs when individuals in the organization are willing to share and acquire knowledge from others so that competency can be created (Law *et al.* 2017). KS will be successful if the organization builds strategic logic, and this is dependent on the characteristics that are built because of local wisdom (LW) in the organization. In this case, an organization must make its staff carry out knowledge sharing and use activities automatically to achieve the goals of the organization (Setini *et al.* 2020).

*Corresponding Author: restyjf@uinib.ac.id

DOI: 10.1201/9781003322054-38

Several SLR studies that discuss KS have not focused on discussing KS tendencies toward LW in the workplace. The SLR articles include KS mechanisms and techniques in the project team (Navimipour dan Charband 2016); online KS mechanisms (Charband dan Navimipour 2016); KS practice in global software development (Anwar *et al.* 2019); KS and social media (Ahmed *et al.* 2019; Sarka dan Ipsen 2017) and KS with IT in multinational companies (Ira *et al.* 2020).

The characteristics of people in different LWs will lead to different ways of sharing knowledge, especially in the workplace. LW has also been widely studied by researchers over the last five years, and some are LW toward KM and KS. Therefore, this article focuses on how research trends over the last five years (2017–2021) relate to the KS process on LW in the workplace.

2 METHODS

The systematic literature review (SLR) approach was chosen to review and analyze KS on LW in the workplace. SLR research is carried out to identify, review, evaluate, and interpret all previous research on interesting and relevant topics so that novelty is obtained from the topic that will be discussed in the next research in the future (Wienen *et al.* 2017). Research questions for articles include: (1) the trend of KS research per year; (2) co-authorship network analysis that examines KS; (3) clusters of KS study; and (4) the tendency of KS toward LW.

3 RESEARCH FINDINGS

3.1 *Framework of KS on LW in the workplace*

There are three stages of the framework of KS on LW in the workplace, as shown in the figure below:

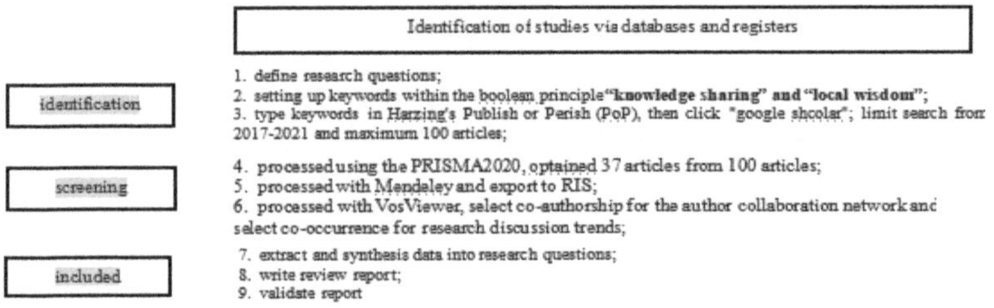

Figure 1. Framework of KS on LW in the workplace with PRISMA2020 diagram.

3.2 *Discovery of articles about KS on LW in the workplace*

The findings for this research is 37 articles. The distribution of articles per years is presented in Table 1.

Table 1. Articles about KS on LW in the workplace.

Year of articles	Total articles
2017	5
2018	8
2019	9
2020	7
2021	8
Total articles	**37**

4 DISCUSSION

4.1 *The trend of KS research per year*

Researchers began to be interested in KS tendencies related to organizational culture, information and communication technology, social capital, and LW. In Figure 2, it can be seen that in the last 5 years, the tendency of researchers toward KS has been almost evenly distributed every year. In 2017, there were 5 articles/13.52% related to KS; in 2018, it increased to 8 articles/21.62%; in 2019, there were 9 articles/24.32%; in 2020 and 2021, it decreased to 7 articles/18.91% and 8 articles/21,62%, repectively.

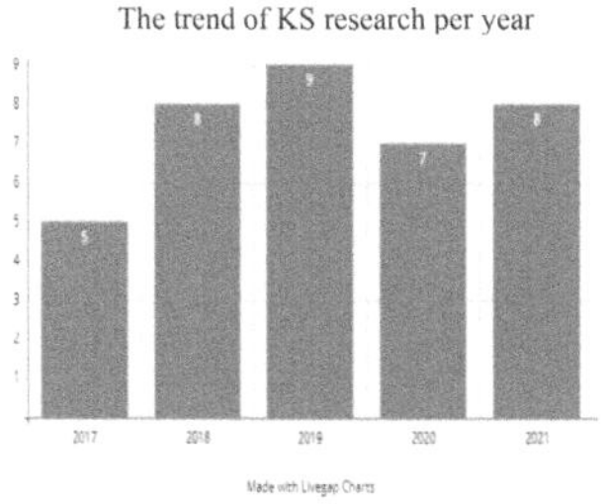

Figure 2. The trend of KS research per year.

Figure 3. The trend of KS research per year based on overlay visualizations.

In 2017, the trend of KS research on ICT, organizational police, learning culture, leadership, psychological environment, social exchange, and others. Although researchers discuss KS on LS, not yet the research main focus. Then in 2018, researchers became interested in developing KS research toward organizational culture. In this to the relationship between KS and LW. Research of KS on LW increased in 2019. In 2020 and 2021, KS research is favored by researchers, but we argue the situation and conditions of the COVID-19 pandemic have caused the quantity of decrease, especially for social humanities research.

4.2 *Co-authorship network analysis that examines KS*

Co-authorship network is a social network that allows authors to participate in one or more articles through indirect but connected channels (Dang-Pham 2018). Co-authorship network analysis is the best bibliometric indicator to illustrate different patterns of co-authorship in academic disciplines (Newman 2001). Figure 4 shows the indirect path mapping that connects them in the network because they are connected to one academic discipline. Each of the authors – Mustof, Agistiawati, Asbari, Basuki, Chidir, Novitasari, Silitonga, Sutardi, and Yuwono – have 8 total link strength for co-authorship network analysis.

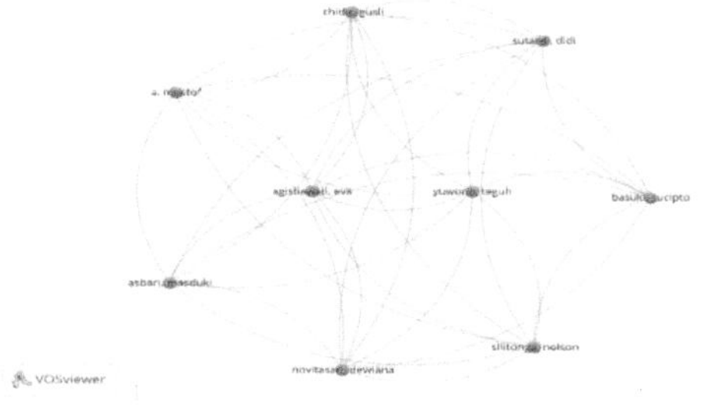

Figure 4. Mapping of Co-authorship network.

Authors included in the co-authorship network analysis were dominated by those who published an article about KS on local wisdom. For the author who produces the most articles (3 articles), there is not even a co-authorship network with other authors. But for the authors with 2 articles each, 4 authors had 7 total link strength and 3 authors had 4 total link strength. KS research on local wisdom is mostly done by researchers from Southeast Asia, such as Thailand and Indonesia. This trend occurs in connection with the culture that affects the characteristics of society from all sides, even in office organizations in the workplace.

Table 2. Total link strength for co-authorship network analysis.

No.	Author	Total Link Strength	Document
1	A, Mustof	8	1
2	Agistiawati, Eva	8	1
3	Asbari, Masduki	8	1
4	Basuki, Sucipto	8	1
5	Chidir, Gusli	8	1
6	Novitasari Dewiana	8	1
7	Silitonga, Nelson	8	1
8	Sutardi, Didi	8	1
9	Yuwono, Teguh	8	1
10	Giantari, I Gusti Ayu Ketut	7	2
11	Setini, Made	7	2
12	Supartha, I Wayan Gede	7	2
13	Yasa, Ni Nyoman Kerti	7	2
14	Farsizadeh, Hossein	4	2
15	Feiz, Davood	4	2
16	etc.		

Table 3. Trends of authors on KS research.

No	Author	Document	Total Link Strength
1	Mulyaningsih	3	0
2	Giantari, I Gusti Ayu Ketut	2	7
3	Setini, Made	2	7
4	Supartha, I Wayan Gede	2	7
5	Yasa, Ni Nyoman Kerti	2	7
6	Farsizadeh, Hossein	2	4
7	Feiz, Davood	2	4
8	Soltani, Mahdi Dehghani	2	4
9	A, Mustaf	1	8
10	Agistiawati, Eva	1	8
11	Asbari, Masduki	1	8
12	Basuki, Sucipto	1	8
13	Chidir, Gusli	1	8
14	Novitasari, Dewiana	1	8
15	Silitonga, Nelson	1	8
16	etc.		

4.3 *Cluster of KS study*

KS research on local wisdom in the workplace in 37 articles over the last 5 years, was analyzed through KS discussion clusters using the VOSviewer application. It is known that these articles consist of 75 keywords. These keywords are dominated by knowledge sharing as much as 86.48% (32 articles with 94 total link strength). Then those who directly discuss KS with local wisdom are 24.32% (9 articles with 24 total link strength). Keyword **knowledge management** as much as 21.63% (8 articles with 25 total link strength). These keywords are grouped into 18 clusters by analyzing them using the VOSviewer application. This can be seen from the cluster mapping of the KS discussion in Figure 5. The keyword distribution in the clusters is as follows:

Figure 5. Cluster pembahasan KS.

First cluster 9: business performance; information and communication technology; leadership; management support; organizational rewards; reward system; social exchange; and structural equation model. Second cluster 8: innovation; innovation capability; knowledge quality; knowledge reciprocity; marketing performance; social capital; tacit knowledge; and women entrepreneurs. Third cluster 7: explicit knowledge; organizational; organizational commitment; organizational culture; organizational structure; teacher innovation capability; and job performance. Fourth cluster 6: knowledge in works; local wisdom; problem solving; religious moderation; sufficiency economy philosophy; and thematic learning. Fifth cluster 5: knowledge sharing; sense of virtual community; shared emotional connection; shared language; and symbolic convergence theory. Sixth cluster 5: free-zone authority; intellectual structure; knowledge management; muslim academic; and theory of planned behavior. Seventh cluster 4: informal knowledge sharing; information; trust; and workplace gossip. Eighth cluster 4: knowledge sharing benefits; knowledge sharing effect; knowledge sharing outcomes; and knowledge sharing review. Ninth cluster 4: local communities; organizational justice; power; and stakeholders. Tenth cluster 3: collaborative cultures; non-profit organization; and organizational learning community. Eleventh to fourteenth cluster each consisting of 3 keywords. While the fifteenth to eighteenth clusters each consist of 2 keywords.

4.4 *The tendency of KS toward LW*

Figure 6 presents previous research regarding the tendency of KS toward LW, which is also influenced by several scientific fields, such as problem solving; community of inquiry; women's community; intercultural communication; intercultural competence; knowledge in

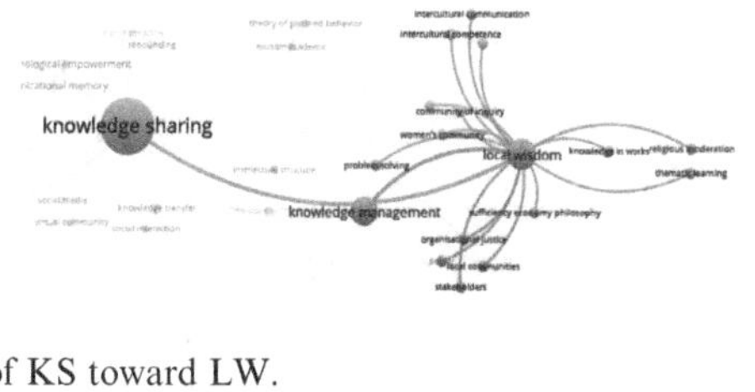

Figure 6. The tendency of KS toward LW.

work; religious moderation; thematic learning; sufficiency economy philosophy; organizational justice; local communities; and stakeholders.

5 CONCLUSION

There are many things that can be explored for research, such as the factors that influence KS on LW in the workplace. KS activities in the workplace are generally dominated by research focused on ICT. Matters related to the characteristics of human resources who carry out KS, such as employee motivation, employee psychology, organizational culture, and LW, have only been studied in the last few years. Even previous research topics related to KS can be researched by studying LW in the workplace.

REFERENCES

Ahmed, Y.A., Ahmad, M.N., Ahmad, N., Zakaria, N.H., 2019. Social media for knowledge-sharing: A systematic literature review. *Telematics and Informatics* 37, 72–112. https://doi.org/10.1016/j.tele.2018.01.015

Allameh, S.M., 2018. Antecedents and consequences of intellectual capital: The role of social capital, knowledge sharing and innovation. *Journal of Intellectual Capital* 19, 858–874. https://doi.org/10.1108/JIC-05-2017-0068

Anser, M.K., Yousaf, Z., Khan, A., Usman, M., 2021. Towards innovative work behavior through knowledge management infrastructure capabilities. *European Journal of Innovation Management* 24, 461–480. https://doi.org/10.1108/EJIM-09-2019-0250

Anwar, R., Rehman, M., Wang, K.S., Hashmani, M.A., 2019. Systematic literature review of knowledge sharing barriers and facilitators in global software development organizations using concept maps. *IEEE Access* 7, 24231–24247. https://doi.org/10.1109/ACCESS.2019.2895690

Dalkir, K., 2017. *Knowledge Management in Theory and Practice*. MIT Press, Britania Raya.

Donald Hislop; Rachelle Bosua; Remko Helms, 2018. *Knowledge Management in Organizations: A Critical Introduction*, 4th ed. Oxford University Press, Oxford.

Duy Dang-Pham; Karlheinz Kautz, 2018. *A Social Network Analysis of Co-Authorship at the Australasian Conference on Information System (ACIS)*.

H.C.A. Wienen; F.A. Bukhsh; E. Vriezekolk; R.J. Wieringa, 2017. *Accident Analysis Methods and Models – A Systematic Literature Review*, Centre Telematics Inf Technology.

Hislop, D., 2013. *Knowledge Management in Organizations: A Critical Introduction*, Third Edition. ed. Oxford University Press.

Hovland, I., 2020. *Knowledge Management and Organizational Learning: An International Development Perspective, in NGO Management*. Routledge, pp. 353–368.

Ira Putri Hutasoit, D., Elisabeth, D., Indra Sensuse, D., 2021. Knowledge sharing for multinational corporations using technology: A systematic literature review, in: *2020 3rd International Conference on Algorithms, Computing and Artificial Intelligence, ACAI 2020*. Association for Computing Machinery, New York, NY, USA. https://doi.org/10.1145/3446132.3446418

Jordão, R.V.D., Novas, J.C., 2017. Knowledge management and intellectual capital in networks of small- and medium-sized enterprises. *Journal of Intellectual Capital* 18, 667–692. https://doi.org/10.1108/JIC-11-2016-0120

Law, K.K., Chan, A., Ozer, M., 2017. Towards an integrated framework of intrinsic motivators, extrinsic motivators and knowledge sharing. *Journal of Knowledge Management* 21, 1486–1502. https://doi.org/10.1108/JKM-03-2016-0119

Mårtensson, M., 2000. A critical review of knowledge management as a management tool. *Journal of Knowledge Management* 4, 204–216. https://doi.org/10.1108/13673270010350002

Mohajan, H.K., 2019. Knowledge sharing among employees in organizations. *Journal of Economic Development, Environment and People* 8, 52–61.

Mulyaningsih, M., 2018. The transformation of sharing culture organization characteristics as a rebounding result of local wisdom value in improving indonesia's community competence in the 21st Century. *JBFEM* 1, 79–89. https://doi.org/10.32770/jbfem.vol179–89

Newman, M.E.J., 2001. The structure of scientific collaboration networks. *Proceedings of the National Academy of Sciences* 98, 404–409. https://doi.org/10.1073/pnas.98.2.404
Nima Jafari Navimipour; Yeganeh Charband, 2016. Knowledge sharing mechanisms and techniques in project teams: Literature review, classification, and current trends. *Computers in Human Behavior* 62, 730–742. https://doi.org/10.1016/j.chb.2016.05.003
Sarka, P., Ipsen, C., 2017. Knowledge sharing via social media in software development: a systematic literature review. *Knowledge Management Research & Practice* 15, 594–609. https://doi.org/10.1057/s41275-017-0075-5
Setini, M., Yasa, N.N.K., Supartha, I.W.G., Giantari, I.G.A.K., Rajiani, I., 2020. The passway of women entrepreneurship: Starting from social capital with open innovation, through to knowledge sharing and innovative performance. *Journal of Open Innovation: Technology, Market, and Complexity 6*. https://doi.org/10.3390/joitmc6020025
Shadi Abualoush; Ra'ed Masa'deh; Khaled Bataineh; Ala'aldin Alrowwad, 2018. The role of knowledge management process and intellectual capital as intermediary variables between knowledge management infrastructure and organization performance. *Interdisciplinary Journal of Information, Knowledge, and Management* 13, 279–309. https://doi.org/10.28945/4088
Singh, S.K., Gupta, S., Busso, D., Kamboj, S., 2021. Top management knowledge value, knowledge sharing practices, open innovation and organizational performance. *Journal of Business Research* 128, 788–798. https://doi.org/10.1016/j.jbusres.2019.04.040
Valmohammadi, C., Ahmadi, M., 2015. The impact of knowledge management practices on organizational performance. *Journal of Enterprise Information Management* 28, 131–159. https://doi.org/10.1108/JEIM-09-2013-0066

Religion, Education, Science and Technology towards a More Inclusive and Sustainable Future – Rahiem (Ed.)
© 2024 the Author(s), ISBN: 978-1-032-56461-6

Grave pilgrimage as a transcendental communication media (Ethnographic study of Diamlewa and Oimbani grave pilgrimage)

Jamaludin* & S.N. Yaqinah
UIN Mataram, Indonesia

ABSTRACT: The phenomenon of rituals and religious traditions still exists in the midst of today's modern society, namely tomb pilgrimage. The tradition of tomb pilgrimage contains many important meanings and is practiced by the Muslim community in the archipelago. This paper intends to examine the tradition of tomb pilgrimage in Bumi Pajo Village, Donggo District. Using ethnographic research methods and collecting data through interviews and observations, this study explores the form of transcendental communication carried out by tomb pilgrims, as well as the symbolic and historical meaning of two tombs, namely Diamlewa and Oimbani, in Bumi Pajo Village. The results of the study indicate that the two tombs are sites that are considered sacred by pilgrims, and they believe that the Diamlewa tomb still resides with the ancestral spirits (parafus) who are in that place to be asked for safety, while at the Oimbani site, tomb pilgrims ask for strength. Tourists who come to visit various places in Bima, in addition to visiting historical sites, also carry out rituals and prayers as a form of transcendental communication with their ancestors. In view of the pilgrims, communication can not only be done between people who are still alive but also with those who have died through rituals as cosmological objects.

Keywords: culture, sacred, communication, perception

1 INTRODUCTION

1.1 *Background*

Auguste Comte's theological view of religious phenomena in society states that all phenomena that occur in the world are caused by supernatural things, so society or individuals in that phase manifest more theological awareness in the form of fetishism or belief in things (Wahyuni 2018). The concept described by Comte can be simply seen in the primitive community belief system, such as myths, sacred objects, dogmas, or legends, which are considered sacred by the community (Dvamonny 2015). The sacred concept or religious concept contains various elements of belief in which there is a dimension of experience, including all feelings, perceptions, and sensations experienced when communicating with supernatural reality, as well as the dimensions of rituals, ceremonies, behavior, and mindset (Sumerta 2021). These religious dimensions are performed in the form of activities such as praying, sacrificing, imprisoning, meditating, providing offerings, and eating together. In today's society, belief systems still maintain some archaic traditions, despite the fact that society is currently at a positive level of monotheism in Comte's view (Wahyuni 2018). This is the focal point of this research. This is also what is still preserved by some modern societies today, where the beliefs of the ancestors are still being practiced in life. As is done by some

*Corresponding Author: 2010405025.@uinmataram.ac.id

 DOI: 10.1201/9781003322054-39

people in Bumi Pajo Village who still believe in the world of animism and dynamism. Some people in the village still visit the sacred spring (Oimbani) and the graves of the ancestors (Diamlewa tomb) to pay respect, and ask for safety, and other rituals.

Pilgrimage activities to tombs and other sacred sites, such as those carried out by the people in Bumi Pajo Village above, if viewed from the perspective of communication science, is a form of transcendental communication constructed by the community or the pilgrims. Transcendental communication is a form of communication between God and humans using metaphysical media (Janu 2018). This describes the basic concept of humans as social beings who are identical with their limitations in carrying out life; therefore, building communication between fellow humans and God, or vertical and horizontal communication, is something that must be carried out by every individual and society while at the same time depicting humanist attitudes (Ibrahim 2016). The pattern of interaction between humans and God always leads humans or individuals to represent being fully human beings (Supriadi 2021). In this study it was found that the old belief in the form of *imbi parafu* (believing in ancestral spirits) is still preserved and believed by some people in Bumi Pajo Village. This explains that a belief for a group is difficult to forget, even though there is already a religion that has become the identity of a community, and in other aspects, the religious practice of the community is still elaborated with the practice of religious beliefs of their ancestors, such as the belief in *parafu*.

The primitive beliefs of the people in the village and the Bima community in general is known as *ma kakamba* (dynamism) (Waris 2021) and *ma kakimbi* (animism). Beliefs related to animism include large trees, springs, graves, and large stones, which are believed to provide help and have a soul that must be respected and believed to be a helper (Szufur 2017). This belief tradition is still preserved by people in one area in Bima, precisely in Ncuhi Hamlet, Bumi Pajo Village, Donggo District, as a form of transcendental interaction with the sacred. The people of Bima in general, before entering monotheistic religions in the form of the teachings of Islam and Christianity, had already built transcendental communication in the form of their beliefs in supernatural powers, which were manifested in the form of fetishism or polytheism known as animism and dynamism. Long before the religions of Christianity or Islam, they had their belief in the existence of supernatural powers or God (Mazni *et al.* 2020).

The research is very different from the previous research conducted by Ginting (2017), Rosmana (2009), Alfian (2014), and Tutiana (2017). Previous researchers were only able to examine the problems in terms of the motivational level of the pilgrims. However, research studies are now much more in-depth and comprehensive in relation to the problem of how to build transcendental communication with the development of Laswell's theory of communication patterns. Besides that, previous researchers only revealed transcendental communication that had to do with religion, while current researchers are much deeper into this old belief that is continuously maintained by the local community to form a belief system other than the religion they believe in. The importance of this research is to inform people that in a community belief system, it is difficult to forget old beliefs even though they are now living in modern times, such as people still believing in sacred sites.

2 METHODS

This qualitative research with an ethnographic approach to understanding the conditions of reality and natural conditions that occur in the field, also includes the constructivist paradigm in communication science (Rachmat Kriyatono 2019). The researcher is the key instrument in descriptive and ethnographic approaches (Maleong 2011). The data collection techniques include observation, interviews, and documentation. Sources of research data include primary data sources (such as the wisdom of the local community), and secondary data in the form of journals and books related to the object that the researcher wants to study (Sugiono 2016). Data is analyzed using the data analysis techniques of Huberman and Miles, which include data reduction, data display, and data verification (Raco 2010).

3 RESULTS AND DISCUSSION

3.1 *Process of transdental communication at the Oimbani sacred site and Diamlewa grave by Pilgrims*

3.1.1 *Diamlewa grave*

It is no longer a secret for the people in Bumi Pajo Village that some people still preserve old traditions in life as a form of manifestation of classical beliefs describing the relationship between God and humans. One of them is the Diamlewa tomb, which is one of the objects or media for the community to carry out transcendental communication with their God. People who visit Diamlewa's tomb still believe that a visit to the place would provide strength and safety to anyone who asks for it.

On the other hand, the perception and experience of the pilgrims toward this sacred site has become part of some people's self. The experience of the people, in Rudolf Otto's view, is included in the religious experience in the aspect of mysticism (Kahmad 2000) i.e., practicing spiritual experiences. The question therefore is: Why do people believe in such mysticism even today when they already have religion as their identity and belief in One God:? However, for some people, the old tradition, especially visiting the tombs of their ancestors as a form of transcendental communication, cannot be forgotten. It is possible that many people have started to forget and want to destroy this old culture.

This illustrates that each individual must have different experiences and perceptions in terms of beliefs. Of course, the pilgrims' perception of the Diamlewa tomb site is classified as a sacred and sacred place for certain individuals so that the tradition of visiting this sacred site is still carried out. With the belief that there is a sacred element to this sacred object, people continue to come or make pilgrimages. Behind the sacredness of this place there are figures who are mythical as *parafus* or ancestral spirits who were previously considered to have power and merit in the journey of mankind or the surrounding community.

At this point the community believed that their ancestors had merged with God. So that by making a pilgrimage or visiting the tomb or sacred place of Oimbani, it can be believed that all celebrations in this place will be quickly granted. But what we need to realize is that each pilgrim has a different goal and perception of the sacredness of the object, because this depends on the goals and needs of the individual. What we need to know is that the traditions of primitive society in worshiping sacred places are identical with several rituals and holding religious symbols by pilgrims. Before people make a pilgrimage to the two sacred sites in Oimbani and the Diamlewa (Radekarama) tomb, several offerings have to be prepared, including black or white free-range chicken, bananas, karodo, kalempe (traditional food for the people of Bumi Pajo Village made of glutinous rice, which is mashed with a mortar and mixed with sugar and young coconut), karaba (rice roasted to form flowers), betel leaf, areca nut, ro'o wau (wood leaves). These materials are often brought down by every pilgrim who visit Oimbani and to the Diamlewa tomb as tohora dore (offerings) that are offered to the ancestors.

If we investigate related to the transcendental communication pattern built between the above instruments, starting from the source and encoder or message sender, marked by the presence of pilgrims as sources or senders, the concrete form is where the prayer holder or guide will call several ancestors, such as "eee ..warombuku, oimbani, gaja mada, la hila", mai tiopo anara waro ro suri. Ma mai raho ake da wa'u kaina, bantu wea po" (O Warombuku, Oimbani, Gajamada, La Hila, let's see your children and grandchildren, who come to visit your place, in asking for help and supplication to all of you), these words are spoken by the prayer leader in calling the ancestral spirits. According to the messages conveyed by the pilgrims, there are those who ask for children, or those who ask to be kept away from all calamities, to be blessed in life, and others. And equipped with these messages with several offerings, for example, bananas, karodo, karaba, chicken, and other things, as a form of symbolic submission to ancestral spirits with the aim of being an intermediary media so that all wishes are quickly granted. While the destination is where the pilgrims ask the ancestors or *parafus* to be granted for all their intentions and intentions. The last is feedback, or reciprocity,

In the Diamlewa tomb people make pilgrimages usually to ask for safety and blessings of life. On the other hand, the perception and experience of the pilgrims toward this sacred site has become part of some people's self. The experience of the people in Rudolf Otto's view is included in the religious experience in the mystical aspect.

3.1.2 *Transcendental communication process in Oimbani sacred site*

In this section we will describe the cycles and activities of transcendental communication by pilgrims at the Oimbani system. The public perception of the Oimbani site is almost the same as the public perception of Radekarama or the Diamlewa tomb. They consider Oimbani and Diamlewa's tomb are the sites of *parafus* or the spirits of previous ancestors who are considered to still have *mbisara guna* (or supernatural powers) for anyone who believes in and visits them. Historically, Oimbani was known as the place of the grave of the wife of Diamlewa who the cannibals buried him alive. Essentially the difference between the transcendental communication process at the Oimbani site and the Diamlewa tomb lies in the tohora dore, whereas the chicken in Oimbani is released, in the Diamlewa tomb, the chicken is kept in the grave. Here we will see how the communication process is built.

The sacred site of Oimbani in the public's perception illustrates the existence of supernatural powers behind the object itself. According to the previous explanation, Oimbani is a spring that is believed by the community to have power and blessing for anyone who asks for and performs rituals in that place. Oimbani is often visited by people who want strength, according to the information obtained by researchers. Usually people who visit Oimbani, for example, ask for their children to pass in one of the higher education institutions, pass to the POLRI, TNI, and others related to strength. Sometimes people also feel that there is a lineage from the descendants of Oimbani and Radekarama came to him as a place to heal from illness.

The transcendental communication process is built starting from the source and encoder or message sender, marked by the presence of pilgrims as source or sender(2019), the concrete form where the holder or prayer guide will call some ancestors. While the massage or messages conveyed among the pilgrims there are those who ask for their children to quickly pass the selection of the TNI, POLRI, etc., there are those who ask to be kept away from all calamities, blessed in life and so on. And equipped with these messages with several offerings, for example bananas, karodo, karaba, chicken and others to the release of chickens, as a form of symbolic submission to ancestral spirits. While the destination is where the pilgrims ask the ancestors or parafus to be granted for all their intentions and intentions. The last is feedback, or reciprocity,

The procession of religious rituals in the form of a request to the ancestors who can use or have ancestral spirits who have power must be carried out starting from the preparation of tohora dore or the preparation of offerings in the form of chicken, bananas, karaba, karodo, and several other offerings. After all the materials to be will be used as tohora dore are ready, they are placed on the stone, then the procession of petition to God through the intermediary of the spirits of the ancestors is immediately carried out. After the celebration or request is made, the procession of releasing the chicken is carried out. The release of the chicken is the starting point for the difference between the rituals of the people in Oimbani and the Diamlewa tomb. If in Oimbani the chickens that were brought by the pilgrims are to be released, in Diamlewa's grave the chicken that is brought must be eaten together.

4 CONCLUSION

From the process of transcendental communication built by the pilgrims as well as the spiritual experience of the two sacred sites of Oimbani and the Diamlewa tomb, it can be concluded that the first transcendental communication process is preparation. At this stage, it is the person who guides the pilgrims that there must be a lineage from Oimbani and Diamlewa's tomb, because the elements of prayer and worship are believed to be far more efficacious or acceptable. The second is the approach process, at this stage, before placing the offerings, one must clean the place, then prepare all the food that is brought, starting from chicken, bananas and other types of offerings to be included as a medium of submission to the ancestors or *parafus*. The third is message delivery, in this phase where a celebration or message is conveyed in the form of a complaint, supplications to Allah through the intermediary of melting between prayers or messages that are said include asking for the safety of life. Fourth, is feedback or acceptance of effects and feedback, the effects felt by the pilgrims can be seen in the peace of mind of the pilgrims, their prayers are answered from those who do not have children so that children, who are sick recover from their illness..

Transcendental communication is still not discussed in the science of communication both conceptually and theoretically, even though this transcendental communication has a very urgent role for each individual, especially in carrying out his relationship with God. Therefore, this research is expected to help in adding insight and knowledge as well as findings related to transcendental communication in communication science. Both theoretically, in the methods and framework of thinking.

REFERENCES

Alfian, M., 2014. *Tradisi Ziarah Kubur ke Makam Keramat Raden Ayu Siti Khotijah di Desa Pemecutan, Kecamatan Denpasar Barat, Kota Denpasar bagi Umat Hindu dan Islam.* Denpasar: Universitas Udayana.

Dvamonny, M., 2015. *Fenemonologi Agama.* Penerbita Kansius, Bandung.

Ginting, K.A., 2017. Kepercayaan masyarakat karo terhadap makam keramat sibayak lingga di bukit ndaholi desa perbesi kecamatan tigabinanga kabupaten karo. *Buddayah: Jurnal Pendidikan Antropologi* 1, 186–190.

Ibrahim, M.R., 2016. Persepsi masyarakat tentang makam Raja dan Wali Gorontalo. *El-harakah (Terakreditasi)* 18, 76–93.

Janu, L., 2018. Persepsi masyarakat terhadap mitos air matakidi. *Etnoreflika: Jurnal Sosial dan Budaya* 7, 41–48.

Maleong, L., 2011. *Metode Penelitian Kualitatif.* PT. Raja Posdakarya, Bandung.

Mazni, N., Arifuddin, A., Hasana, N., irfan, I., 2020. Reveal the mystery of ritual sesajen (Toho Dore) on mbojo tribe in Bima. *Jurnal Sosiologi Reflektif.*

Morissan, 2019. *Teori Komunikasi.* PT. Raja Posdakarya, Bandung.

Raco, J.J., 2010. *Metode Penelitian Kualitatif. Jenis, Keunggulannya.* PT Gremadia Wisdiasma Indonesia, Jakarta.

Rosmana, T., 2009. Budaya Spiritual: Persepsi Peziarah pada Makam Keramat Lelulur Sumedang. *Patanjala* 1, 243–257.

Sugiono, M., 2016. Peacbuilding dan Resolusi Konflik dalam Prespektif PBB. *Academia accelarating the world reasearch* 5. No,5.

Supriadi, I.B.P., 2021. Pola komunikasi transendental dalam konteks proses ngereh bagi tapakan bhatara di bali. *Jurnal Ilmiah Ilmu Agama dan Ilmu Sosial Budaya* 116–125.

Szutfur, 2017. *Fenomenologi Agama.* PT Gremadia Wisdiasma Indonesia, Bandung.

Tutiana, M., 2017. *Fenomena Ziarah Makam Keramat Mbah Nurpiah dan Pengaruh Terhadap Aqidah Islam.* Skripsi. Universitas Islam Negeri Raden Intan Lampung.

Wahyuni, W., 2018. *Agama & Pembentukan Struktur Sosial, Pertautan Agama, Budaya, Dan Tradisi.* Prenada Media Group, Jakarta.

Waris, M., 2021. *Spiritual Mappelelo Cakkuriri: Komuniakasi Transdental Masyarakat Mandar Sendana.* Nas Media Pustaka.

Yuniati, K., Sumerta, 2021. Komunakasi transdental dalam pelaksanaan upacara Tumpek Pengatag di desa banyyuatis kabupaten buleleng. PRESSARE; *Jurnal Ilmu Komunikasi.*

Religion, Education, Science and Technology towards a More Inclusive and Sustainable Future – Rahiem (Ed.)

The social movement of women in pesantren (Islamic boarding schools): From empowerment to resistance against patriarchal culture

K.N. Afiah*
UIN Sunan Kalijaga, Yogyakarta, Indonesia

Kusmana, A.N.S. Rizal, D.A. Ningrum & A. Khoiri
UIN Syarif Hidayatullah, Jakarta, Indonesia

A.J. Salsabila
Gadjah Mada University, Indonesia

ABSTRACT: This study will examine women's social movements in pesantren (Islamic boarding school) that provided various forms of empowerment to women in pesantren and their resistance to patriarchal culture. This is because women in pesantren have limitations due to the strong patriarchal culture. This study used descriptive qualitative methods and data collection techniques using interviews, observation, and documentation. The object of this research study is a community called FASANTRI (Female Islamic Boarding School Educator Gathering Forum or *Forum Silaturahmi Pengasuh Pondok Pesantren Putri*). The results of this study indicate that women's social movement in pesantren is manifested in two forms, namely empowering women through providing services for both prevention and handling of victims of sexual violence, and some forms of cultural resistance, namely by forming the forum itself, thus providing a medium for female educators to voice various issues that they want to be voiced. and providing a medium to develop the potential of female educators and students.

Keywords: social movement, women, and Islamic boarding school

1 INTRODUCTION

Patriarchal culture is one of the biggest challenges for activists who want to speak out for women's rights (Albelda n.d.). Women often get discriminated against or get non-strategic opportunities just because of their status as women, even though women have the potential to develop (Arif *et al.* n.d.) and take part in making various changes that are being pursued. Roviana (n.d.) explains that women activists who have power over women's issues make various efforts to voice women's rights in various mediums. Rosmer (2017) explains that the various efforts made to provide welfare for women also provide opportunities for women to get the same rights and opportunities to work in the public sphere (Bordat *et al.* n.d.). In this case, one community was found that has a role in providing welfare for women, especially in the realm of pesantren (Islamic boarding schools), namely the FASANTRI community.

The FASANTRI community was a community initiated by a female pesantren activist, namely Nyai Hindun Annisah. The FASANTRI secretariat office is located in Jepara, at the Wahid Hasyim Islamic Boarding School. This community consists of female educators from various female Islamic boarding schools in Indonesia. Nyai Hindun saw that there was a

*Corresponding Author: khoniqnurafiah@gmail.com

need for dialogue and a forum to achieve the potential of women in pesantren. Women in pesantren through FASANTRI can express and develop their potential. FASANTRI is also a great force that can then be used to make many changes for the better within the pesantren environment, especially for women in pesantren. Therefore, this study examines in detail FASANTRI as a social movement founded by women in Islamic boarding schools in an effort to make changes through empowerment efforts carried out by its members.

Social movements are carried out by people who share the same vision, mission, and goals on a particular issue (Abdellatif *et al.* 2007). The movement also has certain goals to be achieved, and the movement is organized through a certain social group that has both written and unwritten rules in this regard (Haris et al. 2019). There are many theories of social movements that can describe various kinds of actions, including resource mobilization theory, value-added theory, emergent-norm perspective, assembling perspective, and new social movement theory. This research perceives the movement using the new social movement theory. The new social movement theory developed in Europe in the 1950s and 1960s in the post-industrial era. This theory was born when issues regarding humanist, cultural, and non-material aspects were developing in Europe. The new social movement theory focuses on human rights issues. In the FASANTRI case, the social movement that is carried out leads to the feminist movement by representing the different voices of women to get justice.

According to Rajendra Singh, the new social movement (GSB) expresses the community's joint efforts to demand equality and social justice, reflecting the struggles the community has made to defend their identity and cultural heritage (Afdhalia *et al.* n.d.). Collective action is an essential reality and has always existed in social movements. Collective action is a joint effort of a group of people to achieve short and medium-term goals and values held by society (Hamidah 2016), even though they are faced with resistance and conflict. The GSB also mobilizes community members to voice complaints against the enemy, be it the state, institutions, or other parts of society. The GSB must be characterized by the existence of an ideology that is shared by its members. It's the same with the FASANTRI community, in this case, the movement that is being carried out is classified as a new social movement, which will then be explained in the discussion section.

Many similar studies have been carried out. Research on Islamic boarding schools and women has been studied in various aspects, such as empowering women in Islamic boarding schools (Prasetiyawan et al. 2019), women as leaders in Islamic boarding schools (Istiqlaliyani & Fikriyah 2022), studies on the authority of women's fatwas in Islamic boarding schools (Muhtador 2020), female agency in Islamic boarding schools (Dyah & Milla 2019), and research on pesantren doctrine related to women (Gazali & Akib n.d.). However, so far researchers have not specifically examined the social movement within the pesantren environment, specifically examining the women's community born from the pesantren environment to provide welfare for students who have various kinds of problems.

2 METHODS

This article was the result of a descriptive-qualitative type of research (Sugiono 2017). The data source of this research was obtained by interview technique (Sugiono 2019) with the general chairman of FASANTRI, members of FASANTRI, and the organizing team of the first FASANTRI congress. Another data collection technique is the documentation technique (Sugiono 2019). The documents used are owned by FASANTRI, in particular the Standard Operating Procedures for Handling Sexual Violence in FASANTRI Islamic Boarding Schools. This study uses the concept of social movement as an analytical tool, which is then expected to clearly see the movements carried out by FASANTRI as a women's social movement.

3 RESULTS

The social movement carried out by FASANTRI is in the form of an empowerment movement for victims who experience sexual violence in Islamic boarding schools (Hindun Anisah 2022a). Based on the results of interviews with victims of sexual violence handled by Nyai Hindun as the head of FASANTRI, Nyai took action by taking the victim to a safehouse at her Islamic boarding school in the Jepara area. The victim is still a child because, at the time of the incident, the victim was still in the sixth grade in elementary school. The violence received is in the form of forcing sexual intercourse (Hindun Anisah 2022b). Nyai did special treatment by separating the victim and other students in a safehouse, and then the victim lived with the students who were much older than her. Nyai also provided facilities to victims by providing guarantees for their needs. Nyai always took care of the victim's feelings by not bringing up the problems she was facing (Haydar 2022). Broadly speaking, the movements carried out by FASANTRI include: (1) women-friendly learning; (2) identification of the conditions and the services needed; (3) emergency assistance for victims; (4) shelters for victims; and (5) referrals and/or recommendations

3.1 *Women friendly learning*

3.1.1 *Learning by providing material on women's reproductive knowledge*

The learning indicators must include gender justice (Hamidah 2016). Learning began with the planning stage, selecting a facilitator under the criteria that they are already responsive to gender.

3.1.2 *The integration of commitment and consistency between actors*

After the planning stage, the second stage is implementation. This stage is the realization of the planning stage, which is realizing child-friendly learning. The method used must make female students explore more and think critically, one of which is the peer education method. Then, the final stage of this learning is monitoring and evaluation. At this stage, there are several indicators that are owned as a benchmark for the success of women-friendly learning.

3.1.3 *Identification*

Identification is the initial stage in dealing with cases of sexual violence in female Islamic boarding schools. The steps taken in identifying are (1) conveying the purpose of identification and its benefits to female victims; (2) ensuring the victim's consent for identification; (3) ensuring the presence of a facilitator if the victim is still a minor (under 18 years old); (4) obtaining a consent of the parents or guardians of the students who are victims; and (5) requesting assistance from the relevant institutions to provide translators who understand sign language in the event that the victim is a person with disability. The team members identify the steps in the form of interviews and observations.

3.1.4 *Emergency relief for victims*

This process is carried out if the victim suffers minor injuries. The victim is then directed to the community health center (*puskesmas*) for a post-mortem, and taken to the nearest hospital for further treatment. In this case, the members of the handling team will direct the victim to choose a friend from the pesantren administrator who can provide assistance when there is repeated violence. Victims are trained to fight or flee from the crime scene while using certain codes or signs when they need emergency assistance.

3.1.5 *Safehouse (Shelter)*

A safehouse is one of the facilities used as a place to save victims of sexual violence in the event of threats and intimidation that are very dangerous for the victim. Team members in this case coordinate with the Integrated Service Center for the Empowerment of Women and Children. If the victim has stayed in the shelter, the team will monitor the progress of the victim by visiting or through existing communication facilities to inquire about the condition of the victim to the shelter manager. After monitoring, the team will evaluate the next handling step.

3.1.6 *Referrals and recommendations*

Referrals and recommendations are provided as follow-up services to victims with severe categories because they require expert handling, services to victims can only be provided by technical service units, critical services are needed, and victims are related to law enforcement. This referral and recommendation step is in collaboration with P2TP2A, as an institution that can handle victims of sexual violence.

4 DISCUSSIONS

The FASANTRI is a forum for female educators to participate in supporting the welfare of women. FASANTRI was initiated by Nyai Hindun Anisah. Now, it is time for women to have the opportunity to develop themselves in accordance with the corridors that have been agreed upon by Islam. This means that FASANTRI can be an alternative for women who have problems developing their capacities. If there are problems that hinder women, such as sexual violence in Islamic boarding schools, then FASANTRI will carry out a resistance movement in defending women's rights. It is called a resistance movement because there has been no movement carried out by female boarding schools in handling various cases, one of which is sexual violence. The resistance became a form of strength for its members to carry out humanitarian action.

Based on the results of the research above, the victim who has been handled by Nyai as the head of FASANTRI, is now a woman who is empowered and has a high education. This is also an evidence of the concrete actions taken to provide safety and welfare for women. In line with the GSB theory, FASANTRI expresses an effort to demand social justice and defend identity as an independent woman. The action taken is a joint effort to achieve the goal despite facing resistance and conflict, in this case, Nyai faces a conflict with the perpetrator who continues to look for victims. The efforts that have been made have finally yielded maximum results through the empowerment of victims.

The resistance movement can be seen as a form of empowerment (Bordat *et al.* n.d.) carried out by FASANTRI in dealing with victims of sexual violence in Islamic boarding schools. Women-friendly learning is one form of activity that includes material on women's reproduction. Its intention is for women to have sufficient readiness in dealing with various kinds of situations. Then, FASANTRI conducts an identification stage in the form of interviews and observations carried out in a way that makes the victim feel calm so that the victim is willing to express the problems that are being faced. If the victim has an injury, the team will provide emergency assistance as the next solution. A safehouse also becomes a forum for victims with serious impacts so that they can be handled by parties who have more capacity in dealing with victims of sexual violence. After the series of steps are carried out, the final stage in dealing with victims of sexual violence is to make referrals and recommendations, with the aim that the handling process goes well until the victim's condition is better.

The resistance was carried out to show the concerns of the Nyais in providing welfare to women, especially within the Islamic boarding schools. The steps above are clear evidence that there is now a forum that wants to uphold the role of women so that there is no longer a perpetuated patriarchal culture. In cases of sexual violence, FASANTRI has made vigorous efforts to obtain justice for victims and uphold human rights. So far, the patriarchal culture is firmly embedded in various circles, such as in Islamic boarding schools. Patriarchal culture has also been resisted in various ways. Previous research has provided various ways to fight patriarchal culture, now our research intends to complement existing research. In other words, the researcher elaborated on previous research. So, the results will be more diverse, as described above.

5 CONCLUSION

FASANTRI is a forum for Islamic boarding schools to discuss and accommodate various problems that occur in female Islamic boarding schools. The real actions taken include the empowerment of women in pesantren. The empowerment of women in pesantren is realized by providing a medium for women to increase their potential and by providing a medium for women to formulate a mission, which in turn is also for the welfare of the women in pesantren. One of its practical products was the existence of a massive movement to prevent sexual violence in pesantren by compiling SOPs and various discussions that led to the issue of sexual violence in pesantren. This research is limited to studies on concrete movements, although some studies of the dynamics of the movements carried out conflicts that occurred in FASANTRI. Historical studies of the birth of FASANTRI as a women's movement in Islamic boarding schools can also be done. However, this study has several limitations, such as the lack of in-depth data acquisition, which causes researchers to not be able to see the full patriarchal culture that exists in Islamic boarding schools. This research would be even better if the researcher could conduct interviews with various actors in the pesantren. Thus, the data obtained will be richer, and the empowerment measures offered will be more diverse. Seeing these limitations, the researcher offers recommendations for further research by first dissecting the patriarchal model in Islamic boarding schools. If the researcher has obtained the data, then problem solving will be easier. Thus, the empowerment carried out will be in accordance with the problems that occur.

REFERENCES

Abdellatif, Omayma, and Ottaway, M., 2007. *Women in Islamist Movements: Toward an Islamist Model of Women's Activism.*

Afdhalia, Nabila, A., and Jannah, S.R., n.d. Educational social movement of indonesian women from pre-independence to reform era. *Innovatio: Journal for Religious Innovation Studies*, 21, 40–49.

Albelda, R., n.d. Women and poverty: Beyond earnings and welfare. *The Quarterly Review of Economics and Finance.*

Arif, Zainal, Bahri, E.S., Zulfitria, and Sibghotullah, M., n.d. Peran pesantren entrepreneur dalam pengembangan masyarakat. *Al Maal: Journal of Islamic Economics and Banking*, 1 (2), 207–217.

Bordat, Willman, S., Davis, S.S., and Kouzzi, S., n.d. Women as agents of grassroots change: illustrating micro-empowerment in Morocco. *Journal of Middle East Women's Studies*, 7 (1), 90–199.

Dyah and Milla, W., 2019. *Agensi Perempuan Dalam Manajemen Pendidikan Madrasah: Belajar Dari Pondok Pesantren Nurul Jadid Probolinggo.*

Gazali and Akib, M.N., n.d. Doktrin pesantren terhadap perempuan (Kajian terhadap Kitab-kitab dan realitas perempuan di dalam pesantren. *IQRA Jurnal Ilmu Kependidikan Dan Keislaman*, 14 (2), 71–77.

Hamidah, 2016. Indonesian Islamic movement of women: A study of fatayat muslimat NU (1938–2013). *MIQOT-Jurnal Ilmu Ilmu Keislaman*, 1 (IX), 162–174.

Haris, A., AB Rahman, A.B., and Wan Ahmad, W.I., 2019. Mengenal gerakan sosial dalam perspektif ilmu sosial. *Hasanuddin Journal of Sociology*, 15–24.

Haydar, 2022. *Wawancara Bersama Penderek tentang Kiprah Nyai Hindun Annisah.*

Hindun Anisah, 2022a. *Wawancara tentang Kiprah Nyai Hindun Annisah Sesi I.*

HIndun Anisah, 2022b. *Wawancara tentang Kiprah Nyai Hindun Annisah Sesi II.*

Istiqlaliyani and Fikriyah, 2022. Ulama perempuan di Pesantren: Studi tentang kepemimpinan Nyai Hj. Masriyah Amva. *Jurnal Educatio Fkip UNMA*, 8 (1), 104–109.

Muhtador, M., 2020. Otoritas keagamaan perempuan (Studi atas fatwa-fatwa perempuan di pesantren kauman jekulo kudus). *Kafaah: Journal of Gender Studies*, 10 (1), 39–50.

Prasetiyawan, Arian Agung, and Rohimat, A.M., 2019. Pemberdayaan perempuan berbasis pesantren dan social entrepreneurship. *Muwazah: Jurnal Kajian Gender*, 11 (2), 163–180.

Rosmer, T., 2017. Agents of change: How islamist women activict in Israel are challenging the status quo. *Die Welt Des Islams*, 57 (3–4), 360–385.

Roviana, S., n.d. Gerakan perempuan nahdlatul ulama dalam transformasi pendidikan politik. *Jurnal Pendidikan Islam*, 3 (2), 403–24.

Sugiono, 2017. *Memahami Penelitian Kualitatif.* Bandung: Alfabeta.

Sugiono, 2019. *Metode Penelitian Pendidikan.* Bandung: Alfabeta.

Religion, Education, Science and Technology towards a More Inclusive and Sustainable Future – Rahiem (Ed.)

The political Islam perspective of social movement theory: A case study of FPI in Indonesia

A. Khoiri*, Kusmana, H. Hasan, Nukhbatunisa, D.A. Ningrum & J. Azizy
UIN Syarif Hidayatullah Jakarta, Indonesia

ABSTRACT: Political Islam in Indonesia has always been a worrying twist. During the two decades of *Reformasi*, the emergence of a massive Islamic movement played a significant role in the national political climate on the one hand and the religious climate on the other. Terrorism itself, in its actions, was born from political Islam that wanted to destroy the state and the government system. This study seeks to examine the FPI from the perspective of social movement theory. Since its establishment as the Islamic Defenders Front, being involved in political dynamics for twenty years, getting banned, and then turning into the United Islamic Front, FPI is an interesting phenomenon that indicates its complexity. Using social movement theory, this study finds that FPI takes advantage of political opportunities, mobilization structures, and framing for each of their movements. FPI's big agenda, namely upholding Indonesia under the Sharia (*NKRI Bersyariah*), is a crucial war against the continued existence of FPI as an Islamic orthodox and opportunistic political group and the solidarity of Muslims toward it. Suggestions for further research on this topic include the addition of variables that allow in-depth exploration of the main ideology of the FPI itself and its long-term prospects according to its ideological demands.

Keywords: FPI, Political Islam, Social Movement Theory

1 INTRODUCTION

Discussing the Islamic Defenders Front (FPI) genealogically requires a discussion on Indonesian politics at the end of the second millennium. The massive protests that resulted in the collapse of the New Order regime not only had an impact on the national political stream but also on the religious climate (Pribadi 2018). After Suharto stepped down, the New Order cronies turned their backs on their political interests. The fall of Suharto greatly disrupted the pro-Islamic political configuration; Habibie was its main symbol and FPI became its stronghold (Hasan 2008a) FPI was declared on Rabiul Tsani 25, 1419 H, or August 17, 1998, by a number of habibs and scholars at the Al-Umm Islamic Boarding School, South Tangerang, as a cooperative forum for scholars and people to eradicate evil. The word 'front' was oriented toward concrete FPI actions that never compromise with all falsehoods (Shihab 2013). Established at critical times, namely three months after the fall of the New Order and amid national political instability, FPI is not a mass organization that is free from political prospects (Kuru 2019).

There are five principles of the FPI movement. First, God as God. All of its programs and activities are purely because of Him. Second, the Prophet Muhammad as an example. Third, the Qur'an as a guideline that must be followed and defended. Fourth, jihad as a way, which

*Corresponding Author: ahmad.khoiri20@mhs.uinjkt.ac.id

 DOI: 10.1201/9781003322054-41

is to mobilize all capabilities to enforce the Shari'a. Fifth, martyrdom as a principle of constancy (As'ad 2016). Of the five principles, it is clear that FPI opposes liberalism, secularism, and especially communism—which were strongly opposed by the New Order. FPI's conflict with the Liberal Islam Network (JIL), as well as the echo of establishing the *NKRI Bersyariah*, were the embodiment of the five principles of the movement (Hicks 2012). There are countless actions that FPI assumed were part of the practice of preventing evil (*nahi munkar*). They raided nightclubs and discotheques and once rioted against the National Alliance for Freedom of Religion and Belief (AKKBB) during the Monas Incident. And the most famous and, at the same time, drastically changed the public opinion about FPI until it was banned was Islamic Defense Action (*Aksi Bela Islam*).

FPI's track record is a kind of marriage between Islamic orthodoxy and political pragmatism. The emergence of Rizieq Shihab was a result of the political product itself and his elevated status as the Great Imam (*Imam Besar*) after the Islamic defense action (Hadiz 2019). FPI has always been a pragmatic political vehicle that has made its paramilitary units walk in the middle of the organization's orientation. To this day, the Soeharto family is known to accommodate the FPI movement to attack their political opponents. This study attempts to portray the FPI from the perspective of social movement theory. By looking at the Islamic principles of the FPI on the one hand and their national commitment on the other hand, the two problems will be answered here, namely the existential reflection of FPI as a political Islamic group and the analysis of social movement theory on FPI's political actions itself.

A social movement can be defined as a collective effort to advance and secure common interests (Karagiannis 2005) and goals through collective action outside the scope of established institutions (Giddens 1997). A revolutionary social movement has the goal of fundamental change, but not always with a confrontational method (Kammeyer et al. 1990). Therefore, social movements distinctly differ from other collective behavior such as crowds or masses by three characteristics: a higher level of internal organization, a longer existence of many years, and a massive effort to reorganize the society itself (Macionis 2001). Based on this definition, the FPI can be seen as a revolutionary social movement in terms of political Islam and not a mere mass group. After being able to survive for up to two decades and getting banned at the height of its confrontation, FPI has not completely disbanded as a political Islamic movement that wishes to declare *NKRI Bersyariah*. FPI has been disbanded, but the political Islamic movement is still strong.

2 METHODS

This research is qualitative, while the sampling technique is done through interviews, mass media, and literature review. Data collection from the mass media is categorized by period to form the construction of related events. To anticipate source bias, the construction used the principle of covering both sides. The method used in this research is a descriptive-analytical method with an available database. The analysis of this research used social movement theory. In Tilly's historical study, social movements are defined as political opposition, meaning that a movement emerges collectively to oppose the status quo or to become opposition to achieve certain goals, and the movement's activities extend mainly to the political realm (Tilly 2004). Tilly views emerging social movements as a synthesis of three important elements. First, it is a continuous and organized public effort that makes a collective claim against the target authority, or, in other words, a campaign (Tilly 2003). Second, the use of various forms of political action: the formation of associations, public meetings, demonstrations, petitions, and statements to and in the public media. Third, it is the public representation of worthiness, unity, numbers, and commitment (WUNC).

Social movement theory has several areas of analysis, and what social theorists often argue about is mobilization. Mobilization is the process by which a group acquires collective control over the resources needed for action. Those resources may be labor power, goods, weapons, votes, and many other things, just so long as they are usable in acting on shared interests (Tilly 1977). Mobilization is also associated with a contentious political trajectory, which runs through systematic structures and gets community support through framing. The structure of political opportunity can also explain the rise of a social movement triggered by major changes in the political structure (Hasan 2008b). The analysis of social movement theory in FPI's research as a political Islamic group helps to reveal the manipulation of politics and orthodoxization of the social organizations themselves. This method even analyzes the metamorphosis of FPI after it was banned by the government and transformed into a diaspora of Islamic political voices in Indonesia. For this purpose, the section on framing, which is the most important in social movement theory, is emphasized over political opportunity structures because framing can give rise to favorable political opportunity structures. In this context, how framing helped FPI become influential in its existence and movement became clear.

3 RESULTS

3.1 *FPI in social movement theory*

FPI's political pragmatism has always been linked to the New Order and its cronies (Hefner 2000). However, FPI's orthodoxy is rooted in the ideals of the *NKRI Bersyariah*. Rizieq Shihab inherited the thoughts of Abdul Qahhar Mudzakkar, who saw Indonesia as a legal state that tends to be secular and must be Islamized (Abady 2011). By looking at the political dynamics on the one hand and the actions of the FPI on the other, the FPI can be divided into three periods. First, the constructive period. This occurred from its inception, in 1998, until the end of the SBY administration. During this period, FPI moved freely and dominated the law. Second, the agitative period. This occurred when Ahok was promoted to governor of DKI Jakarta, culminating in the Islamic defense action, and subsided when Rizieq Shihab fled to Saudi Arabia. Third, the deconstructive period. It started with the return of Rizieq from Saudi Arabia. FPI tried to reconstruct the power of the masses again but failed; instead, it was designated as a banned mass organization. The restructuring was immediately carried out and gave birth to a new FPI, the United Islamic Front.

Three factors that serve as the starting point for the analysis of social movements include: (1) the structure of political opportunities and constraints faced by the movement; (2) the form of organization that accommodates the movement; (3) and the collective process of interpretation, attribution, and social construction that mediate between opportunities and challenges to action (McAdam *et al.* 2008). Briefly, McAdam termed it political opportunities, mobilizing structures, and framing processes.

3.1.1 *Political opportunities*

The principle of the political opportunity factor is that social movements are constructed by a series of constraints and/or national political opportunities (McAdam *et al.* 2008) This factor can be traced in FPI through its dynamic track record since its establishment. At the beginning of the Reformation, FPI pointed out that secular-communism threatened Muslims in Indonesia. When Rizieq Shihab intensively carried out raids, as well as other anarchists who were not touched by the law, FPI's motive was not only to carry out *nahi munkar* but also to have political opportunities, namely government accommodation and the assumption that Islam was cornered. Such political opportunities allow FPI to maintain its existence, in addition to guaranteeing its ideology, because it feels the need to supervise the government and Muslims. Due to this factor, FPI often carries out brutal and intolerant actions, but political opportunities support their movement. There is a political opportunity that allows FPI to act safely.

3.1.2 *Mobilizing structures*

It turns out that the strength behind FPI's raids on nightclubs and gamble, protesting or even persecuting anyone deemed deviant and dangerous to Islam, does not lie in Rizieq Shihab's charisma, but because of a common agreement or understanding that serves as background of their actions (Burhani & Saat 2020). In this perspective, in social movement theory, FPI has what is called a mobilizing structure, or "the collective vehicle, informal or formal, through which people mobilize and engage in collective action" (McAdam *et al.* 2008).

Mobilizing the structure in FPI is very complex. The Islamic Defenders Army (LPI) is a paramilitary wing that is present in every action. Meanwhile, the Sharia Markaz in Megamendung, Bogor, and the FPI headquarters on Jalan Petamburan III, Central Jakarta, are strategic locations that become the basis for their mobilizing structures. The paramilitary units (*laskar*) were educated, directed, and mobilized. FPI's indoctrination was successful because of its mobilizing structures. The FPI's mobilization structure also exists in each of their regional administrations. The masses at the grassroots, ordinary Muslims who do not know the real issue, are very easily mobilized through the local authorities affiliated with the FPI. In Madura, for example, there is the Madura Ulama Alliance (AUMA) and the Madura Young Kiai Forum (FKM), which are regional FPI communities. Every time there is a demonstration in Jakarta or local issues that are considered disturbing to Islam, each religious leader (*kiai*) in AUMA instructs the community, as well as facilitates them, to participate in demonstrations with the alibi of fighting for the religion of Allah (Beyer 1994). In this context, the mobilization structure is closely related to the social movement factor described by McAdam *et al.* (2008) as the framing process.

3.1.3 *Framing processes*

These two factors, i.e., political opportunities and mobilizing structures, were not enough to support the FPI as a social movement. Through the framing process, the people need to be made to feel disadvantaged about aspects of their lives and then be optimistic that, by moving collectively, they can overcome these problems (McAdam *et al.* 2008). Framing is a strategy for constructing a shared understanding of the world about themselves that legitimizes and motivates a social movement. In the case of the Islamic defense action, which politically took place because the Ahok case was deemed to have insulted the Qur'an (Burhani & Saat 2020), the framing occurred in two aspects: First, Narrative. The circulating narratives are that Jokowi is the PKI, that this regime is a communist regime, and that Ahok is part of a plan to destroy Islam in Indonesia. Second media (Lerner 2010) Social media, such as *Facebook*, *Twitter*, and *Instagram*, as well as websites, have become a means of FPI's propaganda for the moral degradation framing. Framing becomes a powerful strategy for any movement or organization to influence policies. FPI framed their activism as combating national moral degradation (Burhani & Saat 2020) with *Revolusi Akhlak* and *NKRI Bersyariah*.

However, social movement theorists use various terms in their analysis. McAdam himself, as quoted from Noorhaidi Hasan, also uses the term contentious politics, similar to the terminology of Sidney Tarrow. Mancur Olson, Mayer Zald, and Anthony Oberschall are well known for their resource mobilization theory, which defines collective movements as rational, prospective, and organized actions, while continental politics formulates collective challenges based on common goals and social solidarity rather than as expressions of extremity, violence, and violence disappointment (Hasan 2008b). What is clear is that the framing process has given birth to fragmentation, pluralization, and contestation among religious actors and has also changed the religious market in Indonesia (Isbah 2021). FPI's framing is related to this objective, namely that the mobilization is only one of their various religious-political goals as a political Islamic group.

4 DISCUSSION

I found that FPI, based on social movement theory, is a religious-political contention movement (McAdam *et al.* 2008). Referring to the social movement theories above, FPI exists, moves, and survives because of two factors. First, it is the political tactics. This relates to the FPI's tactics in each of their movements, which are not motivated by a political opportunity structure but rather the political tactic itself. This is different from framing. Outside of framing, FPI exists because of its political manipulation. *Second,* it is the orthodoxization. This is most existentially synonymous with FPI. FPI represents the tough character of Muslims in Indonesia. Moderation, which in many studies has been agreed to be the original identity of Indonesian Islam, is considered a liberal agenda that must be opposed. Islamic moderation, in the FPI worldview, has become foreign.

So it is not surprising that Rizieq Shihab is willing to label FPI as fundamentalist as long as the meaning is "the group that is most steadfast in defending Islam." FPI does not condone terrorism, and Rizieq Shihab opposes Salafi-Wahhabi *da'wa* (DeLong-Bas 2004). The orthodoxy pursued by FPI does not aspire to establish a caliphate like Hizb ut-Tahrir, but instead aims to cleanse Islam of all evils (*nahi munkar*). This doctrine is firmly engraved in every member of its paramilitary unit through the adage of Living Noble or Martyrdom (*Hidup Mulia Atau Mati Syahid/'Isy Kariman Aw Mut Syahidan*). This, then, no matter how dirty FPI's hands are from politics, does not dampen the sincerity of their *da'wa* to continue to be tough and uncompromising, and in general, become the foundation of all FPI social movements. FPI confirmed that it would not establish a caliphate and replace Pancasila; they would still adhere to Pancasila, with a record that Indonesian law is based on Islamic law.

Finally, from the perspective of social movement theory, FPI will be political Islam and ideological Islam, with two orientations. First, manipulation and orthodoxy. Islam in Indonesia for FPI has always collided with heterogeneity. They will label every step of renewal toward a progressive direction as an effort to weaken Islam by liberalism-secularism. At that point, FPI will always garner public participation because it is the personification of the hard-line spirit of Indonesian Islam. Second, it is opportunistic politics. FPI has become a populist political actor, using the voice of Muslims against the regime (Hara 2018). Since its establishment, FPI has been inconsistent and opportunistic, and it is this inconsistency that has made it survive the repression of every regime. Currently, the Cendana clan is using the FPI against the PDIP's repression by voicing the PKI-communist narrative.

5 CONCLUSION

FPI is an Islamic mass organization with the *Aswaja* school of thought that accepts the *NKRI* and *Pancasila* but wants Islamic law above the constitution. Therefore, the main agenda of the FPI is to establish the *NKRI Bersyariah*, which demands a moral revolution (*Revolusi Akhlak*) through jihad *nahi munkar*. From the perspective of social movement theory, FPI-style political Islam takes advantage of political opportunities, mobilization structures, and framing for their two main movements: manipulation of politics and orthodoxization. FPI uses democratic structures to injure democracy itself. Through political tactics, FPI garnered sympathy from Muslims. However, through orthodoxization, FPI maintains its existence as an Islamic social-*cum*-political movement. Therefore, unlike confrontational movements that quickly disband and die, FPI will always exist within the democratic space itself, regardless of the regime in power. While Islam is the weapon of their existence and, at the same time, their resistance.

REFERENCES

Abady, M.Y., 2011. *Konsepsi Dan Praksis Politik Islam Abdul Qahhar Mudzakkar.*

As'ad, 2016. *Politik Identitas Dan Gerakan Sosial Islam (Studi Atas Frot Pembela Islam).*

Burhani, A.N. and Saat, N., 2020. *The New Santri: Challenges to Traditional Religious Authority in Indonesia.* Singapore: ISEAS Publishing.

DeLong-Bas, N.J., 2004. *Wahhabi Islam: From Revival and Reform to Global Jihad.* New York: Oxford University Press.

Giddens, A., 1997. *Sociology.* Cambridge: Polity Press.

Hadiz, V.R., 2019. *Populisme Islam di Indonesia dan Timur Tengah.* Penerbit LP3ES.

Hara, A.E., 2018. *Populism in Indonesia and its Threats to Democracy.*

Hasan, N., 2008a. *Reformasi, Religious Diversity , and Islamic Radicalism after Suharto.*

Hasan, N., 2008b. *Laskar Jihad: Islam, Militansi, dan Pencarian Identitas di Indonesia Pasca-Orde Baru.*

Hefner, R., 2000. *Civil Islam: Muslims and Democratization in Indonesia.* Princeton University Press.

Hicks, J., 2012. *The Missing Link: Explaining the Political Mobilisation of Islam in Indonesia.*

Isbah, M.F., 2021. *Perspektif Ilmu-ilmu Sosial di Era Digital: Disrupsi, Emansipasi, dan Rekognisi.* Gadjah Mada University Press.

Kammeyer, K., Ritzer, G., and Yetman, N., 1990. *Sociology: Experiencing Changing Societies.* Allyn and Bakon.

Karagiannis, E., 2005. *Political Islam and Social Movement Theory: The Case of Hizb ut-Tahrir in Kyrgyzstan.*

Kuru, A.T., 2019. *Islam, Authoritarianism, and Underdevelopment: A Global and Historical Comparison.*

Lerner, M.Y., 2010. *Connecting the Actual with the Virtual: The Internet and Social Movement Theory in the Muslim World — The Cases of Iran and Egypt.*

Macionis, J., 2001. *Sociology.*

McAdam, D., Tarrow, S., and Tilly, C., 2003. Dynamics of contention. *Social Movement Studies.*

Pribadi, Y., 2018. *Islam, State and Society in Indonesia: Local Politics in Madura.* Routledge.

Shihab, M.R., 2013. *Dialog FPI Amar Ma'ruf Nahi Munkar: Menjawab Berbagai Tuduhan terhadap Gerakan Nasional Anti Ma'siat di Indonesia.*

Tilly, C., 1977. *From Mobilization to Revolution.*

Tilly, C., 2003. *The Politics of Collective Violence.*

Tilly, C., 2004. *Social Movements 1768–2004.*

Economy and science

Religion, Education, Science and Technology towards a More Inclusive and Sustainable Future – Rahiem (Ed.)

Sustainability challenges of community-based social security program in Indonesia: Effective collaboration and digital technology adoption

S. Hidayati*, N. Hidayah & Kamarusdiana
UIN Syarif Hidayatullah Jakarta, Indonesia

ABSTRACT: This paper discusses the governance of one model of community-based social security program, namely Bungkesmas, and analyses challenges to its sustainability. A qualitative research by conducting exploratory approach, this study looks inside into collaboration and digital technology adoption to the program. The data comprises semi-structured theme interviews with Bungkesmas management organizations, qualitative survey and passive participatory oversight by observing Bungkesmas information system (SIBungkesmas and Bungkesmas Android Mobile application). The results of this study reveal that appropriate business model and building commitment in collaboration becoming important for community-based social security program to provide optimal services, trust and sustainability. The study also raised the question whether technology digital really have an influence to the fundamental change in the nature of communication of community-social security program to the low-income group of people.

Keywords: Bungkesmas, social security, sustainability, collaboration, technology

1 INTRODUCTION

Health is imperative aspect and plays a key role in the economic development of a country. Therefore, improvement of quality and affordable health care for all is important to do. In order to achieve Universal Health Coverage (UHC) goals, Indonesian government have increased their efforts towards providing quality health care for poor people (Brooks et al. 2017). Initiatives to extend health coverage to the poor had been intertwined with social security programs since the late 1990s (Murphy 2019). The implementation of this initiatives is named with *Jamkesmas* (*Jaminan Kesehatan Masyarakat*) social health insurance, which formerly called *Askeskin* (*Asuransi Kesehatan Masyarakat Miskin*) (Wang et al 2017). The name of *Jamkesmas* is then merge to be BPJS (Health Care and Social Security Agency). Regardless to this insurance program for the poor, access becoming key issues for people live in rural and remotes area. Different standard on the household selection system to access *Jamkesmas* or BPJS facilities is another issue (Harimurti et al. 2013 & Wang et al. 2017) as well as poor referral systems, health facility, and non-facility-based expenditure (Brooks et al 2017). The implementation of the BPJS program also turned out to have problems in terms of funding. The social security fund has experienced a deficit even since the beginning of the program's implementation, especially during the pandemic (Poerwanto et al. 2021).

These problems have resulted civil society seek for effective alternatives health care, to provide poor people permanent solution to the problem of accessing health care (Donfouet & Mahieu, 2012). Development actors are increasingly considering community-based social security (CBSS) as an instrument that can enable not only easy access to quality health care, but also reduce absolute poverty among low-income populations. The practice of CBSS in Indonesia are vary. One study conducted by Samsudin and Prabowo (2022) highlighted the practice of Dana Sosial Muhammadiyah (DSM), an insurance product for students. Tambunan and Purwoko (2002) found

*Corresponding Author: s.hidayati@uinjkt.ac.id

traditional practices of CBSS as social protection practices exist in local level communities, such as the practice of joint money collected from community by head of RT or RW (head of community who represent local level government) to help people who affected by calamities. Another model of CBSS is Bungkesmas (Community Health Saving and Insurance) conducted by Social Trust Fund (STF) Syarif Hidayatullah State Islamic University (UIN) Jakarta. However, community-based social security program often do not run in the long term. Financial issues and low enrolment threaten the sustainability of the program (Adebayo et al. 2015). This paper focus to discuss about Bungkesmas program as a model of CBSS, and look into inside the organizer in developing collaboration system and applying digital technology to support sustainability program. Collaboration with trusted organization and digital technology adoption is belief to increase trust among consumers, including improving financial literacy, and reducing risk (Yang 2020).

2 METHODS

This research uses qualitative approach by doing exploratory research. This approach is used to investigate a social phenomenon and to gain a specific and in-depth understanding (Creswell JW & Poth CN. 2018) about Bungkesmas program as a model of Community Based Social Security. There are two types data collected, primary and secondary data. The primary data was collected through qualitative survey, in-depth interview, and observation. The researcher made qualitative survey to 16 Bungkesmas partnership, and in-depth interview to Bungkesmas program management. The second stage is observations by looking at two digital application SIBugkesmas and Bungkesmas Mobile Application. The qualitative survey and interview focus on finding out the impact of collaboration system and the digital technology applied in increasing enrolment and supporting sustainability. The secondary data from documents and Bungkesmas report, government regulation etc., was collected to support the primary data.

3 BUNGKESMAS AS A MODEL OF COMMUNITY BASED SOCIAL SECURITY

3.1 *A glimpse of Bungkesmas program*

Bungkesmas program is initiated by Social Trust Fund (STF) Syarif Hidayatullah State Islamic University (UIN) Jakarta in 2011. It is a kind of micro insurance intended for low-income market, to protect them from losing their income and assets when they are sick or get accidents. The program can also become the complementary to the government program (Jamkesmas/BPJS). The Bungkesmas scheme offers affordable, simple, easy claim and in accordance with the needs of low-income community. The scheme of Bungkesmas insurance applied to date is in health, accident and life insurance with premium price is IDR 150,000 per-year. and participants will get the insurance claim in the form of cash plan. The cash plan scheme enables participant to use the claim funds more flexible, such as to replace lost income during illness, replace loans, and pay expenses outside the hospital costs. Many Bungkesmas participants coming from micro and small entrepreneurs. They use insurance claim funds to rebuild their business after sick. The claim funds also can use to cover additional needs outside of hospital and medicine costs, such as for transportation to the hospital, food costs for families who look after or who are left at home. The Bungkesmas program has been operated in 10 provinces in Indonesia (Aceh, South Kalimantan, Banten, Jakarta, West Java, Center of Java, South Sulawesi, Center Sulawesi, South East Sulawesi, Maluku). To date number of Bungkesmas participants constitute 19.000 participants. The total insurance claim from 2015 to mid of 2022 amounted to IDR 1,248,527,000 (Bungkesmas report, mid 2022).

3.2 *Bungkesmas partnership collaborations*

The program of Bungkesmas operated through collaboration system. The collaboration system intended to create a workflow of information that allows individual or organization to share ideas and their talents with others so that the task can be finished both efficiently and effectively (Utama & Setyowati 2018). Collaboration model applied in Bungkesmas program is partner-agent model. The partner-agent model also has become the dominant approach to micro insurance in some country like India (Shetty & Veerashekarappa 2009). In partner agent model applied in Bungkesmas context, the collaboration conducted among three parties. *First*, STF UIN Jakarta as the initiator and fasilitator of the program. *Second,* is insurance company as the insurance product's provider. According to regulation no. 69/PJOK.05/2016, insurance products should be managed by insurance firm (https://www.ojk.go.id/). Since 2018, STF doing cooperation with Takaful Keluarga to provide the product of Bungeksmas insurance. Takaful Keluarga is Shari'a based insurance company. Prior to, the cooperation conducted with AIG and Zurich International insurance. Cooperation with the insurance conducted to serve more professional services.

Third, cooperation with local partner organizations, such as MFI (Micro Finance institutions, including cooperatives institution), local NGO, philanthropy organizations local communities, etc. The role of these partner organizations is vital to acknowledge and share comprehensive information about the Bungkesmas program to communities, particularly to the low income and poor communities. Information about the product should constantly provide to low-income people, because most of them has low level of education and poor literacy and knowledge of the insurance product (Gehrke 2014). Furthermore, the concept of microinsurance is quite new in emerging market. Therefore, there should be aggressive campaign to raise low people awareness about the importance of microinsurance and demonstrating the benefits of the service. By the involvement of local organizations can increase uptake through testimonials, referrals, and good-word-of mouth. Local organization can also function as channels that may also help to facilitate, collect premiums and handle claims (Mazambani & Mutambara 2018). In other name, these partner organizations become insurance agent. There are two types of relationships between insurance companies and their agents: agents that compete to represent several different insurers ("agents") and those that represent one company exclusively ("exclusive agents") (Safitri 2018). From 2011 until now, STF has had cooperation with more than 150 partner organizations, and large number cooperation conducted with Micro Finance Institutions. In addition, local partners such as local NGO and philanthropy organizations usually have program to improve quality of life of poor people. Inserting Bungkesmas program would strengthen their program for poor people.

Figure 1. Bungkesmas workflow.

3.3 *Digital adoption in Bungkesmas program*

Trust in CBSS products such as Bungkesmas is an important issue. Optimal services are part of increasing public trust. Particularly, the poor tend to think simply, they only need to get good, easy and quick services. Therefore, providing a fast, easy, simple service of Bungkesmas is imperative to do. Other issues are related to operational costs and the availability of human resources to operate the program. by the increasing number of people using internet and smartphone has prompted STF UIN Jakarta to applied digital application in Bungkesmas services. Digital technology applied is to improve Bungkesmas services such as to reduces the operational costs, ease and speed up the procedure of registration of insurance and claim, and also to organized database of participants. It is also to response trust issue exist among partner organization and community. Bungkesmas digital application will ease the work of Bungkesmas organizer either the STF or partner organizations, particularly on the issue of human resourcers. There are two digital applications built. First application is SIBungkesmas was built in 2017. SIBUNGKESMAS is a Web-based application designed to simplify and speed up the process of registration of new members and claim by partners' organizations. The application can store membership data as well as claim history. The usage of the application intended only for partners' organizations. *Second* application is mobile /Android base application. This application built in 2018, is provided for individual participants to facilitate people who do not attached to the partners organizations. The application can be accessed or downloaded from the play store in Android smartphone. The application provided information of Bungkesmas product, beside menu for registration and claim.

3.4 *Challenges for sustainability of Bungkesmas program*

This study found that Bungkesmas social insurance is quite complicated in management and face the issue of sustainability. *First*, Bungkesmas insurance as micro insurance product is created mainly to insure low-income market. In the low-income market, the insurance mostly is voluntarily. people can decide whether to insure or not. When the decision is not mandatory, people who has lower risk are unlikely to insure, while people who exposed to higher risk are likely to insure. This condition will put the insurance program at risk of adverse selection (Mazambani & Mutambara 2018). *Second,* people awareness on having insurance product is low. They do not think insurance as urgent. *Third,* people ability to buy insurance product remain low. In Indonesia, there are 27.54 million people who are below the poverty line, with an income limit of IDR 472,525 per capita per month (BPS 2021). These low-income people are mainly target of the Bungkesmas insurance. However, instead of buying the insurance product, these people remain struggle to fulfill their daily needs.

Fourth, financing operational of the program. Prior to, STF as the main organizer of the Bungkesmas program are highly dependent on donor assistants to support the operational cost such promotion activities. However, donor assistance cannot be permanent, because donor assistance will stop when they are no longer concerned with relevant issues. *Fifth*, cooperation with insurance company is fragile. The insurance company tend to take into account the benefits of the program to its organization. Therefore, they set a minimum number of targets for insurance enrolment that must be achieved. Inability the organizer to achieve the target will give consequences for the end of the collaboration. This is what causes the change of insurance companies that support Bungkesmas for several times.

Sixth, fragile condition of partner organizations. Partner organization involved in collaboration in average is coming from small size local organizations, such as MFI/cooperative, local NGO, etc. They often have fragile internal management. Many of these institutions eventually ceased to operate, which caused the Bungkesmas program also has eventually stop. Local NGO mostly works in ad hoc project following to project agenda and schedule. When the project stopped, the program also stopped. Another issues with partners local organization are they have less people who can organized and understand the insurance program, so they cannot promote the program effectively.

Seven, digital application has not been carried out optimally. The usage of web-based Bungkemas application to some extent is effective to ease and quick the process of registration and claim. But the web-based application can only operate and used by partner organizations. In term of supporting to increase number of enrolments, it is much depending on how actively partner organization promote the program. The Android based mobile application is highly expected to increase the enrolment indeed. However, the application is not yet suitable for targeting the poor. After two years applied and promoted, there is no significant number of people register through the application. Evaluation conducted to analyze the effective system of the mobile application shows some challenges such as: *1)* less people from low-income community who has gadget or handphone that support with the Bungkesmas android mobile application, so makes them difficult to access the application. The application should be downloaded through the play store, while download the application requires big data store because it take up big memory; *2)* it is not too effective to keep the application in the mobile phone, while it is used only for registration and claim, and only to be used to register one person. The other problem that is aligned with the previous research is about people low financial literacy including insurance literacy. Android mobile application might not really successful in improving Bungkesmas participants enrolment because mobile phones and internet for many people are still viewed as a tool for casual interaction. It is not yet fully reconceptualized as an effective tool to increase knowledge and access to insurance. In term of enrolment, the role of partner organization is still important as intermediary to facilitate low-income people to access insurance. Bungkesmas mobile application might be effective if it is also combined with social media. Social media such as Facebook, Instagram, Tik Tok, Youtube have becoming part of people daily life today, and also has becoming tools for business interaction and transaction.

4 CONCLUSION

Community based Social Security (CBSS) initiative such as Bungkesmas is a good alternative program to provide poor people access to health service, to protect them from calamities exposure and to avoid phenomenon of "poor after sick". At the same time, it also can reduce the level of inequality. In order to keep the program running and sustain, the program needs to overcome that various challenges as mentioned earlier. STF needs to find out effective strategies. *First*, it needs to find out appropriate business model to increase enrolment as well as to support its program financing. *Second,* Collaboration needs to be improved with high commitment, either with insurance companies or local partner. *Third*, digital adoption needs to be combined with social media in order to improve its efficiency to increase number of people in accessing Bungkesmas.

REFFERENCES

Adebayo *et al.*, 2015. A systematic review of factors that affect uptake of community-based health insurance in low-income and middle-income countries. *BMC Health Services Research* [online], 15 (543). Available from: https://bmchealthservres.biomedcentral.com/articles/10.1186/s12913-015-1179-3 [Accessed 11 April 2022]

Brooks, Mohammad I, *et al.*, 2017. Health facility and skilled birth deliveries among poor women with Jamkesmas health insurance in Indonesia: A mixed methods study, *BMC Health Service Research*, [online], 17 (105). Available from: https://bmchealthservres.biomedcentral.com/articles/10.1186/s12913-017-2028-3 [Accessed 11 April 2022]

Creswell JW & Poth CN, 2018. *Qualitative Inquiry Research Design: Choosing Among Five Approaches.* 4th Edition. Los Angeles: SAGE Publications.

Donfouet, Hermann Pierre Pythagore & Mahieu, Pierre-Alexandre, 2012. Community-health based insurance and social capital: A review. *BMC Health Economic Review* [online], 2 (5). Available from: https://healtheconomicsreview.biomedcentral.com/articles/10.1186/2191-1991-2-5 [Accessed 12 April 2022]

Gehrke, E., 2014. The insurability framework applied to agricultural microfinance: What do we know, what can we learn? *The Geneva Papers*, 39 (2), 264–279.

Harimurti P., *et al.*, 2013. *The Nuts & Bolts of Jamkesmas, Indonesia's Government-Financed Health Coverage Program for the Poor and Near-Poor* [online]. Washington DC, The World Bank. Available from: https://openknowledge.worldbank.org/handle/10986/13305 [Accessed 12 April 2022].

Mazambani, Last & Mutambara, Emmanuel, 2018. Sustainable performance of microinsurance in low-income markets. *Risk Governance and Control: Financial Markets & Institutions*, 8(2), 41–53.

Murphy, John, 2019. The Historical development of Indonesian social security. *Asian Journal of Social Science*, 47 (2), 255–279.

Naveen K. Shetty & Dr. Veerashekarappa, 2009. The microfinance promise in financial inclusion: evidence from India. *The IUP Journal of Applied Economics, IUP Publications*, 0 (5–6), 174–189.

Poerwanto, Eko Budi., *et al.*, 2021. *Kebijakan Jaminan Sosial di masa Pandemi*, 1th edition. Bogor: Pustaka Amma Amalia

Safitri, K. Amelia, 2018. Contribution of technology to insurance in Indonesia. In: *Advances in social science, education and humanities research, vol.426. 3rd International Conference on Vocational Higher Education (ICVHE)*, Atlantis Press: 78–83.

Samsudin, A. M & Prabowo, H., 2022. Community-based health coverage at the crossroad: The Muhammadiyah health fund in Indonesia. *Indonesian Journal of Islam and Muslim Societies*, 12 (1), 111–138.

Tambunan, TTH and B. Purwoko, 2002. Social protection in Indonesia. *In*: Erfried Adam, Michael von Hauff and John Marei (eds), *Social Protection in Southeast and East Asia*, Singapore: FES

Utama, A.A.G.S. & Setyowati,Y., 2018. Collaboration system and digital business efficiency in the accounting information system perspective (Case study Banyuwangimall.com). *In*: *Advances in economics, business and management research*, Vol. 117. *3rd Global Conference on Business, Management, and Entrepreneurship (GCBME)*, Atlantis Press: 129–134.

Wang, W., Temsah, G., & Mallick L., 2017. The impact of health insurance on maternal health care utilization: Evidence from Ghana, Indonesia and Rwanda. *Health Policy and Planning*, 32 (2), 366–375

Yang, J., Y. Wu & B. Huang, 2020. *Digital Finance and Financial Literacy: An Empirical Investigation of Chinese households.* ADBI Working Paper No. 1209. Tokyo: Asian Development Bank Institute. www.bps.go.id

Religion, Education, Science and Technology towards a More Inclusive and Sustainable Future – Rahiem (Ed.)

Governance and outreach of Islamic microfinance (BMT) toward cash waqf

A.S. Jahar*, R.A. Prasetyowati & I. Subchi
UIN Syarif Hidayatullah Jakarta, Indonesia

ABSTRACT: This paper examines the relationship between the determinants of corporate outreach and corporate governance in Islamic microfinance institutions (Baitul Maal wa Tanwil, BMT toward Cash Waqf management using a primary data set built from the concept of Islamic MFIs collected from internal and external respondents. Islamic Microfinance implements cash waqf management. Using the measurement model and the structural model in Partial least square (PLS), this paper examines the effects of the board and CEO characteristics, type of firm ownership, customer-firm relationship, and competition and regulation on the financial performance of Islamic MFIs and outreach to Islamic Microfinance customers and institutions Cash Waqf. This paper finds that Determinants of governance in microfinance institutions – BMT in cash waqf management is CEO/chairman duality, internal board auditor, board size, shareholder firm, individual loan, competition, regulated bank, urban market, Islamic microfinance Institutional age, portfolio at risk, firm size, and Human development index (HDI). Outreach governance of LKM-BMT on cash waqf management is divided into three dimensions, namely: first, the internal dimensions are; CEO/chairman duality, internal board auditor, the board size, and shareholder firm. Second, the external dimensions are; individual loans, competition, bank regulation, urban market, Islamic microfinance Institutional age. Third, the dimensions of the control factor are; portfolio at risk, firm size, and Human development index (HDI). LKM-BMT on cash waqf management in Indonesia discussed in this study proves that outreach governance of LKM-BMT occurs only in external dimensions, including; individual loan, competition, bank regulated, urban market, Islamic microfinance Institutional age. Our research concludes that MFIs in Indonesia focus on individual loans, competition between MFIs and BMTs, are subject to banking regulations, are more developed in rural markets than urban markets, and this dimension of outreach governance depends on the age of MFIs-BMTs.

Keywords: Islamic microfinance, Baitul Maal wa Tanwil (BMT), cash waqaf institution, outreach and governance, Partial Least Square (PLS)

1 INTRODUCTION

This study examines the impact of governance mechanisms on the dual mission of Islamic microfinance institutions (LKM Syariah) or Baitul Maal wa Tanwil (BMT) and Cash Waqaf Institutions as sustainability in providing financial and banking services for micro-enterprises and low-income families. This study identifies three dimensions of this problem: the vertical dimension between owners and management, the horizontal dimension between Islamic MFIs and cash waqf and their customers, and the external governance dimension. Recommendations for better governance are mainly on the first and third

*Corresponding Author: asepjahar@uinjkt.ac.id

 DOI: 10.1201/9781003322054-43

dimensions, according to Rock et al. (1998), Otero & Chu (2002), and Helms (2006) suggest importing best practices in governance from developed countries, such as board independence and shareholder ownership. Meanwhile, Van Greuning et al. (1999) and Hardy et al. (2003) argue for better regulation of MFIs.

BMT is a sharia financial institution that operates using a combination of the concepts of "Baitul Maal" and "Baitu Tamwil" with its operational target focusing on the Small and Medium Enterprises (SME) sector. The concept of Baitul Maal means that BMT acts as a socio-religious institution that has the function of receiving Zakat, Alms, Infaq, and Waqf funds and using them for the community. While in the Baitul Tamwil concept, BMT has a role as a business institution or financial institution that aims to seek profit (profit-oriented), such as opening a Toserba (convenience store) or offering savings and loan products to the public. However, if we look at the practice in the field, BMT is more likely to act as a sharia financial institution that offers savings and loan products to the public based on sharia principles.

1.1 *Governance and outreach of Islamic MFI*

Governance (Governance) is about achieving company goals. The first objective of the MFI is to reach more clients in the poorer sections of society, and the second goal is financial sustainability. This study analyzes the relationship between governance mechanisms and outreach.

The problem of incentives in Microfinance Institutions has at least two dimensions: one between owners and management (including the board), and the other between the MFI and its clients (Macey & O'Hara 2003). Overview by Becht et al. (2003) and Hermalin & Weisbach (2003) show that governance in the owner-board dimension is important in general, while Rock et al. (1998), Otero & Chu (2002) and Helms (2006) highlight its particular importance in microfinance. On the MFI-customer dimension, MFIs are subject to credit risk assessment and repayment problems because loan customers usually have little or no collateral (Armendariz de Aghion & Morduch 2005). Microfinance initiatives are finding new ways to address this problem through group lending, character lending, and the gradual development of credit histories. In group lending, using either a solidarity group or a village bank, the MFI delegates much of the screening and monitoring effort to the group.

In contrast, the customer's relationship with the MFI is more direct regarding individual lending. In addition, the unique nature of banks and financial institutions as providers of financial infrastructure often requires public regulation of bank/financial institution-customer relations. Therefore, studying MFI governance requires three considerations: the relationship between the owner and the board, the relationship between the MFI and its customers, and the external conditions of competition and regulation. Table 1 summarizes the independent variables, governance factors, and hypotheses related to outreach.

In the three dimensions of governance, namely: First, the board-owner dimension concerns the composition of the board and the type of ownership. The board composition variables are CEO/chairman duality, international director, internal auditor, and board size. CEO/chairman duality can signify CEO power (Hermalin & Weisbach 1991, 1998), whereby the CEO can pursue policies that result in personal gain. The Cadbury Committee (1992) advised against duplication of this role. However, duality can increase the effectiveness of decision-making. This ambiguity may explain why Brickley *et al.* (1997) do not find that firms with CEO-chair separation outperform those with CEO/chair duality. Oxelheim & Randøy (2003) find that firm performance increases with the presence of international directors.

Steinwand (2000) recommends that internal auditors in MFIs report directly to the board. Ideally, internal auditors should provide the board with an independent and objective assessment of the MFI's operations, improving the MFI's financial and social performance.

Larger boards can encourage members to be independent in their monitoring responsibilities, allowing for greater CEO independence. Yermack (1996) & Eisenberg et al. (1998) reported that a larger board was associated with lower firm performance, measured as Tobin's Q or ROA.

Bennedsen *et al.* (2008) confirmed a negative relationship in small family firms. Hartarska (2005) confirmed the results of the ROA regression for MFIs. Adams & Mehran (2003a) present conflicting evidence from banking firms in the United States, and de Andres & Vallelado (2008) agree based on Spanish data. Many MFIs are organizations non-profit organizations (NPOs). Handy (1995) proposes that board members in NPOs offer their reputation as collateral & Speckbacher (2008) argues that NPOs need larger boards because they lack owners with monetary incentives to monitor their investments. Referring to Hartarska (2005) findings, the Islamic MF research results with cash waqf expect a larger board (Board size) to reduce company performance.

Legal incorporation or type of ownership can play a role in MFI performance. Similar to ordinary banking (Hansmann 1996; Rasmussen 1988), MFI ownership differs significantly (Labie 2001; Mersland 2009). NPOs are often considered a weaker structure because MFIs lack owners with operating financial shares (Jansson & Westley 2004), this finding leads to lower financial performance than shareholder firms (SHF). Therefore, Ledgerwood & White (2006) and Fernando (2004) support the transformation of NPOs into SHF. However, NPOs are believed to be more effective in reaching poor customers. This finding implies that SHF should show better financial performance but reach fewer poor clients than NPOs. However, Mersland & Strøm (2008) found that the performance of SHF and NPO was equally good. Incentive issues between owners and managers may be more prominent in NPOs, but NPOs have the compensatory benefit of reducing adverse customer selection and avoiding moral hazards (Desrochers & Fischer 2002; Hansmann 1996; Mersland 2009). better able to enter the local information network. In comparison, Caprio & Vittas (1997) and Cull *et al.* (2006) proved by asserting that many SHFs are not run according to the shareholder value model because, generally, MFIs are committed to reaching the poor (Reille & Forster 2008). This study confirmed the findings of Valnek (1998), Crespi *et al.* (2004), and Mersland & Strøm (2008) that NPO (Islamic MF with Cash Waqaf) performs as well as SHF.

Stakeholders on the board arguably influence the MFI's governance, Hartarska (2005) finding that director employees are negatively related to financial performance and outreach. The research data found that no stakeholder group improved the company's performance or outreach (unpublished data). External governance mechanisms, such as product market competition and regulation, may be relevant for Islamic microfinance institutions (Islamic MFIs). Generally, the more intense the competition, the fewer owners will need internal governance mechanisms (Hart 1983; Schmidt 1997). However, Nickell (1996) argues that because increased competition can reduce costs, the negative effects of lower product prices may not be outweighed. Thus, the effect on performance is uncertain. Petersen & Rajan (1995) argue that banks and financial institutions earn rent from survivors in long-term relationships. When relationships are damaged by competition, banks/financial institutions stop lending to customers, which is risky and expensive. This can reduce outreach.

2 METHODS

This study uses a methodology with a qualitative approach with quantitative methods. Quantitative data is collected and constructed based on the theory that supports it. Data in the form of questionnaire answers were analyzed and integrated (Inquiry and sequentially concept). Data Analysis was done to adjust the research conducted with the research method used to gain a deeper understanding of the relationship and influence between cash waqf variables, Baitul Mal wa Tamwil, and Microfinance.

Table 1. Definition of operational variables and hypotheses.

Variables	Explanation	Hypothesis Governance	Hypothesis Outreach
CEO/Chairman Duality (CEO)	CEO and Chairman are the same person	+/–	–
Internal Board Auditor (AUDIT)	Islamic MFI has an internal auditor reporting to the board	+	-/+
Board Size (BORDSZ)	The number of directors	–	–
Shareholder Firm (SHF)	A shareholder firm	+	–
Individual Loan (LOAN)	Loans are made mainly to individuals	+	–
Competition (COMPET)	A self-constructed measure of the local level of competition	–	+
Bank regulated (REGUL)	The Islamic MFI is regulated by banking authorities	–/+	+
Urban Market (URBAN)	The market served is urban only	–	–
Islamic Microfinance Age (IMF)	Years of experience as an Islamic MFI		
Portofolio at Risk (RISK)	The fraction of the portfolio with more than 30 days in arrears		
Firm Size (SIZE)	The natural logarithm of assets		
Human Development Index (HDI)	A Composite country Index covering life expectancy, education, and income (GDP per capita)		

2.1 Measurement model results (Outer model)

This measurement model connects the indicator items as manifest variables that explain the latent variables as supporting concepts that meet the criteria of validity and reliability as a measuring tool; the results are as follows:

Table 2. Outer loading stage 2 results.

Code	Loading	Evaluations	Code	Loading	Evaluations
COMPET1	0,788	Valid	RISK1	0,724	Valid
COMPET2	0,772	Valid	RISK2	0,714	Valid
HDI2	0,748	Valid	SIZE2	0,777	Valid
IMF2	0,729	Valid	CWLS1	0,712	Valid
LOAN1	0,717	Valid	CWLS2	0,999	Valid
LOAN2	0,753	Valid			

Source: Processed Data (2022)

Table 2 shows The value generated by each indicator (manifest variable) to measure latent variables, such as CEO/chairman duality, internal board auditor, board size, shareholder firm, bank regulation, competition, individual loan, urban market, MFI age, portfolio at risk, firm size, and Human Development Index (HDI). This variable is divided into three dimensions of outreach. MFI-BMT manages cash waqf: First, internal dimensions, including CEO/chairman duality, internal board auditor, the board size, shareholder firm, and bank regulation. Second, the external dimension consists of a competition, individual loans, the urban market, and MFI age. The third is the dimension of control factors: portfolio at risk, firm size, and the Human Development Index (HDI).

2.2 *Structural model results (Inner model)*

Table 3. T-Statistic.

	Cash Waqaf Institusinal	T-Statistic	P-Value
Governance Islamic MFI –> Cash Waqaf Institusional	0,675	5,659	0,0000

Source: Processed data (2022)

Table 3 shows that the relationship between these two positive variables between Islamic Governance MFI and Institutional Cash Waqf has a significant level of 5.656 > 1.86 and a P-value of 0.0000 < 0.05; this proves the level of significance of this finding that The internal governance dimension shows that there is no relationship and there is no outreach between MFIs (BMT) on cash waqf management, where the correlation value of each variable is very low, namely: CEO/chair duality, internal board auditor, Board size, shareholder firm, and bank regulation. This happens because half of the MFIs (BMT) managing cash waqf in the sample of this study, the CEO or chairman of the MFI (BMT) has concurrent positions. These results cannot reveal whether the MFI (BMT) managing cash waqf would be better regulated by a CEO who is not chairman MFI (BMT). This finding was previously referenced by Brickly *et al.* (1997).

3 RESULT AND DISCUSSION

The results of hypothesis testing on Individual Loans at LKM-BMT with Cash Waqaf indicate that individual loans are a significant variable in this study; loans with business feasibility assessment (LOAN1) have a correlation of 71.7% and loans with collateral feasibility assessment (LOAN2) correlated by 75.3%. This shows that the sustainable financial performance of MFIs-BMT based on Cash Waqf can be achieved with individual and group loans. However, the proposed comparative efficiency in group lending is not confirmed.

The results of the external competition governance mechanism (COMPET) LKM-BMT with cash waqf management shows positive and significant results for competition between individual LKM-BMT (COMPET1) correlates with 78.8%. In industry competition, MFI has a correlation of 77.2% in the MFI-BMT study. Thus, it appears that the performance of MFI-BMT can increase with the increasing competition; this result supports the opinion of Nickell (1996), because the first entrants become more efficient when new MFIs enter the MFI-BMT industrial market.

The results on the Urban market variable show that the MFI-BMT urban market with cash waqf primarily concentrates its business only in rural areas. The result is very low, given the difficulty of reaching rural areas (Helms 2006). Currently, most MFIs only serve the rural market, and only a few serve both.

In the results of the Islamic Microfinance age (MFI age), this study shows that MFIs are generally young, although some institutions can trace their microfinance activities before the country's independence, when these microfinance institutions began to provide loans to small farmers, small traders, etc. However, MFIs generally have little time to build relationships with their clients. The results of the Portfolio at Risks show a positive and significant correlation between the credit risk profile (RISK1) of 72.4% and the interest rate risk profile of 71.4%. This result is counter-intuitive considering that the LKM-BMT is a sharia-based microfinance institution that must be analyzed further.

The results of the Human Development Index (HDI) show that the Human Development Index (HDI) in companies with human resources from various backgrounds, the inclusion of HDI can capture some of the differences in microfinance institutions. Firm size results show that the CEO/chairman can increase the number of customers; this is the priority of the firm size effect. The variable between CEO and Chairman duality interaction with firm size has a positive and significant correlation. Then by institutionalizing duality, MFI-BMT can pursue organizational goals by maximizing firm size, according to research by Berle & Means (1932).

Table 4. R-Square.

Variable	R Square	R Square Adjusted
Cash Waqf Institutional	0,455	0,449

Table 4 shows the simultaneous correlation of indicators of the LKM-BMT governance variable to the institutional cash waqf variable, which has a value of 45.5%. Moreover, the remaining 54.5% is another variable that is not a factor in this study. The overall conclusion is that several mechanisms of traditional internal and external corporate governance factors with controlling and outreach factors significantly affect the institutional or management of Cash Waqf.

4 CONCLUSION

Determinants of governance in microfinance institutions – BMT in cash waqf management are CEO/chairman duality, internal board auditor, board size, shareholder firm, individual loan, competition, regulated bank, urban market, Islamic microfinance Institutional age, portfolio at risk, firm size, and Human development index (HDI).

Outreach governance of LKM-BMT on cash waqf management is divided into three dimensions, namely: first, the internal dimensions are; CEO/chairman duality, internal board auditor, the board size, and shareholder firm. Second, the external dimensions are; individual loans, competition, bank regulation, urban market, Islamic microfinance Institutional age. Third, the dimensions of the control factor are; portfolio at risk, firm size, and Human development index (HDI).

LKM-BMT on cash waqf management in Indonesia in this study proves that outreach governance of LKM-BMT occurs only in external dimensions, including; individual loan, competition, bank regulated, urban market, Islamic microfinance Institutional age. This means that MFIs in Indonesia focus on individual loans, competition between MFIs and BMTs, are subject to banking regulations, are more developed in rural markets than urban markets, and this dimension of outreach governance depends on the age of MFIs-BMTs.

REFERENCES

Adams, R.B., Mehran, H., 2003b. Is corporate governance different for bank holding companies? *FRBNY Economic Policy Review*, 123–142.

Becht, M., Bolton, P., Röell, A., 2003. Corporate governance and control. In Constantinides, G., Harris, M., Stulz, R. (Eds.), *Handbook of the Economics of Finance 1A*. North Holland, Amsterdam.

Cizakca, Murat. Awqaf in history and its implications for modern islamic economies. *Islamic Economic Studies*, Volume 6, No. 1.(1998).

de Andres, P., Vallelado, E., 2008. Corporate governance in banking: The role of the board of directors. *Journal of Banking and Finance* 32 (12), 2570–2580.
Eisenberg, T., Sundgren, S., Wells, M.T., 1998. Larger board size and decreasing firm value in small firms. *Journal of Financial Economics* 48, 35–54.
Hardy, D., Holden, P., Prokopenko, V., 2003. Microfinance institutions and public policy. *Policy Reform* 6, 147.
Hartarska, V., 2005. Governance and performance of microfinance institutions in central and eastern Europe and the newly independent states. *World Development* 33 (10), 1627–1648.
Hartarska, V., Nadolnyak, D., 2007. Do regulated microfinance institutions achieve better sustainability and outreach? Cross-country evidence. *Applied Economics* 39 (10 12), 1207–1222.
Hermalin, B.E., Weisbach, M.S., 1991. The effect of board composition and direct incentives on firm performance. *Financial Management* 21 (4), 101–112.
Hermalin, B.E., Weisbach, M.S., 1998. Endogenously chosen boards of directors and their monitoring of the CEO. *American Economic Review* 88 (1), 96–118.
Hermalin, B.E., Weisbach, M.S., 2003. Boards of directors as an endogenously determined institution: A survey of the economic literature. *Economic Policy Review* 9 (1), 7–26.
Hsiao, T., 2003. *Analysis of panel data.* Cambridge University Press, Cambridge.
Human development report, 2006. *Beyond Scarcity: Power, Poverty and the Global Water Crisis.* UNDP/ Palgrave Macmillan: Gordonsville, USA.
Jansson, T.R.R., Westley, G., 2004. *Principles and Practices for Regulating and Supervising Microfinance.* Inter-American Development Bank, Washington, D.C.
Labie, M., 2001. Corporate governance in microfinance organizations: A long and winding road. *Management Decision* 39, 296–301.
Ledgerwood, J., White, V., 2006. *Transforming Microfinance Institutions.* The World Bank and The MicroFinance Network, Washington D.C.
McIntosh, C., Wydick, B., 2005. Competition and microfinance. *Journal of Development Economics* 78, 271–298.
Mersland, R., 2009. The cost of ownership in microfinance organizations. *World Development* 37 (2).
Morduch, J., 1999. The microfinance promise. *Journal of Economic Literature* 37 (4), 1569–1614.
Nickell, S.J., 1996. Competition and corporate performance. *Journal of Political Economy* 104 (4), 724–746.
Otero, M., Chu, M., 2002. Governance and ownership of microfinance institutions. In: Drake, D., Rhyne, E. (Eds.), *The Commercialization of Microfinance.* Kumarian Press, Bloomfield.
Robinson, M.S., 2001. *The Microfinance Revolution: Sustainable Finance for the Poor.* World Bank, New York.
Rock, R., Otero, M., Saltzman, S., 1998. *Microenterprise Best Practices.* Principles and Practices of Microfinance Governance. DAI/USAID, Washington. pp. 1–58.
Schreiner, M., 2002. Aspects of outreach: A framework for discussion of the social benefits of microfinance. *Journal of International Development* 14, 591–603.
Speckbacher, G., 2008. Nonprofit versus corporate governance: An economic approach. *Nonprofit Management and Leadership* 18 (3), 295–320.
Steinwand, D., 2000. *A Risk Management Framework for Microfinance Institutions.* GTZ, Financial Systems Development, Eschborn, Germany. pp. 1–70.
Van Greuning, H., Galardo, J., Randhawa, B., 1999. *A Framework for Regulating Microfinance Institutions.* World Bank Policy Research Working Paper No. 2061, The World Bank: Washington, D.C.
Yermack, D., 1996. Higher market valuation of companies with a small board of directors. *Journal of Financial Economics* 40, 185–212.

Religion, Education, Science and Technology towards a More Inclusive and Sustainable Future –
Rahiem (Ed.)

The concept of Dharurah: A review of its basic guidelines and application in shariah economic contracts

J.M. Muslimin* & F.M. Thamsin
UIN Syarif Hidayatullah Jakarta, Indonesia

W. Munawar
Djuanda University Bogor, Indonesia

ABSTRACT: This article aims to understand the linguistic meaning of the word emergency and to review the development of its concept in Islamic juridical philosophy, including its practice in shariah economic transactions. Through the qualitative library research method, the data were taken from some books, articles, journals, or magazines that discuss the meaning of emergency vocabulary in Islamic legal tradition. The data were analyzed using content analysis techniques and presented descriptively in the context of normative perspective. The study concluded that several linguistic terms have similar meanings to the concept of emergency, namely *mashaqqah* and *hājah*; These vocabularies were used in the similar context within Islamic economic legal discources.

Keywords: Dharurah, Hajah, Sharia Economic Contracts

1 INTRODUCTION

The willingness to walk on the rules has been considered as one of the proper ways to achieve success. However, sometimes that commitment faces challenges when dealing with sharp changing human conditions, especially those termed "emergency". It goes in Islamic economic transactions as well. In response to this, Posner (2006:1) says "A constitution (regulation) that will not bend will break". This expression emphasizes that regulatory changes are something that must be understood. That is due to responding abnormal conditions requires legal flexibility as a form of exception (Arsil *et al.* 2020:423–446). That does not mean ignoring commitment to the rules, because exceptions themselves have conditions and limitations. Posner's words are in line with the concept of never stop *ijtihād* which unites the practice of legal norms with the dynamic reality of social life (Hallaq 1994:29–65). In emergency cases, Islamic jurisprudence is known for the principle of *al-ḍarūrah tubiḥ al-maḥzurah* (the emergency cases may break the prohibition). This principle is explained conceptually by several contemporary scholars such as: Wahbah al-Zuḥayli (1985) and Muḥammad al-Jizānī (2007) with the aim that it might be properly used to solve problems, including in shariah economic transactions. However, the word "emergency" itself has the same meaning with other vocabularies, so the question arises what is the meaning of the concept of emergency and how does it differ from similar vocabularies and their applicative forms in shariah economics?

*Corresponding Author: muslimin@uinjkt.ac.id

DOI: 10.1201/9781003322054-44

2 METHODS

This research contains qualitative and normative library data, related to the concept of emergency in Islamic Legal Theory. The data were taken from books, articles, journals and magazines that discuss emergency and its concepts in Islamic legal discourses and its applications in some sharia key transaction practices. The data were gathered through linguistic parallelism of emergency concept *(al-dharurah)*. The gathered data were then seen as key concepts of legal underlying formula (concept of law) in solving the real transaction cases of sharia economic contracts. Hence it can be stated that the research was conducted by using content analysis techniques through in-depth understanding of Islamic Legal Philosophy and its application.

3 RESEARCH FINDINGS

3.1 *Emergency in legal linguistic vocabulary*

The word emergency in the original language (Arabic) is الضَرُورَة *(al-ḍarūrah)*. It is a derivative of the word الضَرَر *(al-ḍarar)* which means difficulty, damage or danger with the antonym النَفْعُ *(al-naf'u)* which means usefulness. Named by *al-ḍarar* is because it is the worst human condition that contains suffering, loss or pain that does not benefit the one who experiences it (Ibn Manẓūr 1999:46). Furthermore, this meaning is classified by Imam Fayrūzabādī (1998:428) into four uses, namely: a). Emergency to describe the condition of *al-ḍarar* (danger), b). Emergency describes the condition of *mashaqqah* (difficulty), c). Emergency describes the condition of *ḥājah* (need), and d). Emergency describes the condition of *iḍṭirār* (need to force something). Something that is a solution to overcome this condition is known as *al-ḍarūrī*, which is something that is urgently needed (necessary) (Qaldaji 1996:255). These meanings indicate that the use of the word emergency is quite broad. However, with the inclusion of the words *mashaqqah* and *ḥājah* in their linguistic meaning, it is necessary to review their meanings and differences.

As for the concept of emergency used in conventional legal studies, the word emergency is also known by several legal terms such as state of emergency or "extreme situation" to describe an unexpected situation that threatens a country, and cannot be prevented by the implementation of normal legal regulation, so that an extraordinary action must be taken immediately as a form of legal exception. This action is termed a "condition of necessity" or a "state of exception" (Zwitter 2012:95–111). In addition to the above terms, the term overmacht/force majeure is also used to describe someone who is forced to do an act due to an unavoidable external power. Eventhough, the concept force majeure is said to have originated from Rome which was later adopted by French law and English Common Law Doctrine of Frustration (ECLDF), but in Islam a term has been used to describe this condition, namely *al-quwwah al-qāhirah* (force of coercion). Classical Islamic jurist do not provide a specific definition for this term, they just mention things related to its linguistic meaning, so it can be concluded that the meaning of *al-quwwah al-qāhirah* is all disasters that are unexpected (sudden), cannot be anticipated (avoided), which have a major impact on the sustainability of a contract (business contract) so that it cannot be implemented as agreed, and ends in termination (Abdel Samad 2021:381–460).

3.2 *Emergency in Islamic jurisdiction*

The discourse on the concept of emergency stems from the legality of consuming forbidden food contained in five verses of the Qur'an, namely al-Baqarah verse 173, al-Maidah verse 3, al-An'am verse 145, al-Naḥl verse 115 and al-Maidah verse 145. An'am verse 119. These five verses form the basis of the birth of the principle of *al-ḍarūrah tubiḥ al-maḥzurah* (Khiṭāb

2008:168–170). This principle was then described conceptually by the scholars, starting with what was initiated by Abū Bakr al-Jaṣāṣ (d. 370 H), al-Sayūṭī al-Shāfi'ī (d. 911 H) with the meaning of the word emergency as worry about the dangers that befall themselves if they do not consume forbidden foods. This concept was later developed by al-Dardīrī al-Mālikī (d.1201H) with the assertion that the existence of the alarming danger does not have to occur at that time, but can be based on a strong suspicion that it will occur in the future (al-Ṣalābī 2002:116–120).

Al-Zuḥaylī (1985:67) views that the emergency condition in this contemporary era is not only limited to protecting the soul but also includes everything that can protect *al-uṣūl al-ḍarūriyyāt al-khamsah* (religion, soul, mind, lineage, property), and actions needed to become the solution is not only limited to the problem of consuming forbidden food, but also includes the act of committing prohibited acts or leaving mandatory orders or prohibiting permissible acts. Hence it can be seen that the first criterion that characterizes the concept of emergency is the feeling of worry over danger. The feeling of worry itself can only be known by oneself based on knowledge or feeling about something that threatens, either with certainty or just a prejudice, so it will be different for each person depending on his condition (al-Zaynī 1993:20). This allows an assumption that the existence of danger at that time is not absolute to be a measure, because if you consider al-Zarqa's view (2004:1004) that an emergency condition does not have to result in a danger of destruction if it does not violate the law, but is sufficient based on an estimate of something that will result in weakness or loss of self-ability.

This is what makes al-Zuḥaylī (1985:68–72) and al-Jizānī (2007:66) viewed that worry is not the only condition that must be met, so they developed several provisions in practicing this concept of Islamic juridical emergency. It is abstracted from their description with the following points: 1). There is a belief that the danger beyond the customary limit is occurring, or there is a strong expectation that the danger will occur in the future, 2). The level of danger is forcing to do forbidden acts, 3). Illegal acts are the only way to prevent this danger, 4). The permissibility of violating the law is only limited to the estimation of the danger being lost based on the size of the amount and the size of the time, 5). Eliminating a hazard should not carry a hazard equal to or greater than the hazard being eliminated, 6). Eliminating emergency conditions must not be in conflict with the basics of Islamic shari'ah, including safeguarding everyone's rights in the context of the benefit of *al-uṣūl al-ḍarūriyyāt al-khamsah*, 7). Consumption of illegal drugs must be based on recommendations from trusted health experts in terms of science and religion, 8). The determination of a wide-scale emergency is determined by the government authorities by ensuring that four things occur, namely: the occurrence of damage that exceeds the limit, the occurrence of extraordinary danger, the occurrence of extreme difficulties, and the state is threatened if it does not take global benefits.

3.3 *Emergency, mashaqqah and economic transactions*

The meaning of mashaqqah in terminology is not explained by the scholars in detail but only limited to a discussion of the principle of *al-mashaqqah tajlibu al-taysīr* (difficulty bringing relief), even so al-Zarqā (w.1420H) (2004:1001) assessing that not all *mashaqqah* causes the Shari'a to provide relief, because what is included as *mashaqqah* carriers of relief is *mashaqqah* that crosses customary boundaries. In this regard, al-'Izz Ibn 'Abd al-Salam (d.660H) (2000:13–14) has explained that the *mashaqqah* is divided into two parts. The first part is *mashaqqah mu'tādah* (the usual difficulties) namely *mashaqqah* that can be borne by the *mukallaf* without the need for waivers. This *mashaqqah* can be in the form of obeying the law such as carrying out the rules that apply in worship and mu'amalah, or in the form of punishment for violating legal deviations such as *hudūd, jināyāt, qiṣās, kafārah* and *ta'dhīr* (Hakeem *et al.* 2012:7–21). The second part is *mashaqqah gair al-mu'tādah*, namely difficulties that bring legal relief.

Based on the explanation, it can be seen that the relationship between emergency and *mashaqqah* in terms of language is in the similarity of meaning, namely difficulty. As for Islamic juridical, emergency is the effect of *mashaqqah fadiḥah* so that legal deviations that are allowed are doing things that are forbidden. So it can be understood that not all *mashaqqah* is an emergency, but every emergency is a *mashaqqah* and its implication in Islamic jurisprudence is *tajlib al-taysīr* (bringing relief) in the form of *tubīḥ al-maḥzūrah* (the difficulty justifies the breaking of prohibition). As for the relationship with mu'āmalah contracts, it can be seen in all transactions that cannot be avoided but can save property from threatening dangers, such as termination of the contract due to danger if it continues (force majeure), or contemporary electronic transactions such as the use of ATMs in waḍi'ah transactions.

3.4 *Emergency, ḥājah and mu'amalah transactions*

According to al-Shaṭybī (2004:222) *ḥājah* is something that is needed to overcome difficulties, besides that it can also be something needed to realize the spaciousness of the burdens faced. The fuqaha' often equate between *ḥājah* and *darūrah*, but according to al-Zarqā (2004:1005) the two are substantially different. To be able to practice the concept of *ḥājah*, Rashīd (2008:197–202) put forward several terms and criteria, namely: 1). Obliged to ask for a fatwa to scholars who are experts related to the law of *ḥājah*, 2). Trying as much as possible to leave the need for legal deviation, 3). The need for legal deviation is limited to the fulfillment of what is needed, 4). Legal deviations in *ḥājah* conditions are limited to the magnitude of *ḥājah* faced and can be of long duration, 5). The prohibition that is violated in the *ḥājah* condition is only the prohibition of *al-ḥaram ligairih.*

In addition to these criteria, ḥājah if he is general in nature can also occupy an emergency position so that the concept of emergency applies in the condition of *ḥājah*, as intended by the principle of *al-ḥājah qad tanzilu manzilah al-ḍarūrah 'āmah kānat aw khāṣah.* As for the meaning of *al-ḥājah al-'āmah* here is the need for legal exceptions to cover many people, and the nature of the law is *mustamirrah wa dāimah*, namely the law that continues for a long time and is not affected by the presence or absence of the condition of *ḥājah*. In economic transactions, this can be seen in the need to take advantage of other people's assets in *i'ārah, ijārah, qarāḍ* contracts, or other people's services such as *muzāra'ah, musāqah, syarikah, wakālah, hiwālah,* and *wadī'ah* or trust in debt like rahn, aman. In addition, it can be in the form of the need for transactions that contain *garar* such as greetings, *istisna,* and *ja'ālah.* While *al-ḥājah al-khāṣah* only limited to a person or a group of people and the law is *muáqqatah* (limited time) for those who need it, such as using silk cloth because of the need for testing or because of the need to overcome itching or because of the need to ward off fleas (Ismail 2019:23).

4 DISCUSSION

Based on the terms and conditions in the concept of emergency and the criteria for *ḥājah* and the meaning of *mashaqqah*, it can be understood that the difference in these words can be seen in the following four points, namely: 1). The level of *mashaqqah* in emergency is greater than *mashaqqah* in *ḥājah*, 2). The need for legal exceptions in emergency conditions is coercive because there is no other choice but to do so, while in *ḥājah* conditions legal exceptions are optional without coercion, 3). The rules that may be violated in an emergency are the prohibitions that are *al-ḥaram lizātihi*, which are forbidden because the substance and its actions cause harm based on the arguments of the Qur'an and Hadith, such as blood, wine, pigs, killing, adultery, etc., while what may be violated in the *ḥājah* condition are things that are prohibited not because of their substance but because of other factors as a form of *sadd al-dharī'ah*, because basically the act is permissible but it can cause harm, such as praying with stolen clothes can invalidate the prayer or seeing the opposite sex can lead to adultery. 4). The duration of the need for legal deviation in an emergency case is time-limited, namely as long as the emergency condition occurs, if the emergency condition ends then the legal exception also ends. As for the condition of *hājah*, the need for something that can overcome the difficulties faced can last a long time. This understanding also applies to transactions in the shariah economy.

5 CONCLUSIONS

Based on the above review, the linguistic meaning of the word "emergency" cannot be separated from the words *hājah* and *mashaqqah*. It's just that it has special characteristics that must be met, because an emergency is a moment of concern over the danger that threatens the safety of human *al-uṣūl al-ḍarūriyāt al-khamsah* (the safety of religion, soul (life and limbs), mind, offspring and treasure). This concern may be based on a belief that danger is occurring, or on a strong suspicion that danger is expected to occur in the future. And to eliminate the threat, the emergence of a compelling need to violate the rule of law as the only way to overcome that danger with the provision that the violation cannot be avoided so that it becomes an exception to the law. It's just that this exception is limited by the limitation of the disappearance of the worry and danger that threatens it. As for when dealing with *mashaqqah* which does not endanger the safety of *al-uṣūl al-ḍarūriyāt al-khamsah* but can cause difficulties in life, then if the benefit contained in the violation is more needed globally, it may becomes an alternative to overcome difficulties as long as it is needed. Meanwhile, if the needs are only personal or a group of people, the legal deviation only applies until the *mashaqqah* is lost. Therefore, practicing the concept of emergency must pay attention to the criteria and the possibility of avoiding violations of the law. This also applies to shariah economic transactions.

ACKNOWLEDGMENTS

We gratefully acknowledge the support from the committee of the 5th International Colloquium on Interdisciplinary Islamic Studies (ICIIS) Graduate School UIN Syarif Hidayatullah Jakarta Indonesia, of the year 2022.

REFERENCES

Abdel Samad, Hosni. 2021. 'The novel coronavirus pandemic between the theories of emergency conditions and force majeure and its impact on contractual obligations a comparative study in islamic jurisprudence'. *Journal Sharia and Law* 2021 (88): pp.381–460.

Arsil, Fitra, and Qurrata Ayuni. 2020. 'Model pengaturan kedaruratan dan pilihan kedaruratan Indonesia dalam menghadapi pandemi Covid-19'. *Jurnal Hukum & Pembangunan* 50 (2): 423–46. https://doi.org/10.21143/jhp.vol50.no2.2585.

Fayrūzābādī, Muhammad al-. 1998. *Al-Qamūs al-Muhīṭ*. Damaskus: Muassasah Ar-Risalah.

Hakeem, *et al.* 2012. 'The concept of punishment under sharia'. In *Policing Muslim Communities: Comparative International Context*, edited byFarrukh B. Hakeem, M.R. Haberfeld, and Arvind Verma, 7–21. New York, NY: Springer. https://doi.org/10.1007/978-1-4614-3552-5_2.

Hallaq, Wael B. 1994. 'From Fatwās to Furū': Growth and change in Islamic substantive law'. *Islamic Law and Society* 1 (1): 29–65. https://doi.org/10.2307/3399430.

Ibn Abd al-Salām, 'Izz ad-Dīn 'Abdul 'Azīz. 2000. *Al-Qawāid al-Kubrā al-Mausūm Bi Qawā'id al-Aḥkām Fi Iṣlāh al-Anām*. Edited byUtsmān Hammād. 1st ed. Vol. 2. 2 vols. Damaskus: Dār al-Qalam.

Ibn Manẓūr. 1999. *Lisān Al-'Arab*. Vol. 8. 17 vols. Bairut: Dār Iḥyā' al-Turāth al-'Arabī.

Ismail, Kūsi. 2019. 'Al-Ḍarūrah ash-Shariyyah Mafhūmuhā Asāsuhā Ḍawābituhā wa Namāzij min Taṭbīqātihā'. *at-Turaṣ* 8 (1): h. 19–39.
Jīzāni, Muhammad ibn Husayn al-. 2007. *Ḥaqīqah Al-Ḍarūrah Ash-Shar'iyyah Wa Taṭbīqātuhā al-Mu'āṣirah*. 1st ed. ar-Riyādh: Dār al-Minhāj.
Khiṭāb, Ḥasan as-Said. 2008. 'Qā'idah al-Ḍarūrāt Tubīḥ al-Maḥẓūrāt Wa Taṭbīqātuhā al-Mu'āṣirah Fī al-Fiqh al-Islāmi'. *Majallah Al-Ūṣūl Wa an-Nawāẓil* 2: 145–223.
Posner, Richard A. 2006. *Not a Suicide Pact: The Constitution in a Time of National Emergency*. Inalienable Rights Series. New York; Oxford: Oxford University Press.
Qaldaji, Muhamamd Ruwas. 1996. *Mu'jam Lugah al-Fuqahā'*. 1st ed. Bairut: Dār an-Nafāis.
Rashīd, Ahmad Ibn Abd Al-Raḥmān al-. 2008. *Al-Ḥajah Wa Atharuhā Fī al-Aḥkām, Dirāsah Naẓariyyah Taṭbīqiyyah*. al-Riyāḍ: Kunūz Ashbīliyā.
Ṣalābī, Usāmah Muḥammad Muḥammad al-. 2002. *Al-Rukhaṣ al-Shar'iyyah Aḥkāmuhā Wa Ḍawābituhā*. 1st ed. Iskandariah: Dār al-Īmān.
Shātybi, Abu Ishāq Ibrahim bin Musa al-. 2004. *Al-Muwāfaqāt Fi Uṣūl Al-Sharī'ah*. Edited by Abdullah Darāz. 1st ed. Bairut: Dār al-Kutub al-'lmiyah.
Zarqā, Muṣṭafā Aḥmad al-. 2004. *Al-Madkhal Fī al-Fiqh al-'Ām*. 2nd ed. Damaskus: Dār al-Qalam.
Zaynī, Maḥmūd Muḥammad Abd al-'Azīz al-. 1993. *Al-Ḍarūrah Fi al-Sharī'ah al-Islāmiyah Wa Al-Qanūn al-Waḍ'i*. Iskandariah: Muassah as-S\aqāfah al-Jāmi'iyyah.
Zuhaylī, Wahbah al-. 1985. *Naẓariyatu Al-Ḍarūrah al-Shar'iyyah Muqāranah Ma'a al-Qanūn al-Waḍ'i*. 4th ed. Bairut: Muassah Ar-Risālah.
Zwitter, Andrej. 2012. 'The rule of law in times of crisis: A legal theory on the State of Emergency in the Liberal Democracy'. *ARSP: Archiv Für Rechts- Und Sozialphilosophie / Archives for Philosophy of Law and Social Philosophy* 98 (1): 95–111.

Religion, Education, Science and Technology towards a More Inclusive and Sustainable Future – Rahiem (Ed.)

Islamic feminism in Indonesia: The case of Fiqh an-Nisa program P3M/Rahima

N. Hidayah*, S. Hidayati & K. Zada
UIN Syarif Hidayatullah Jakarta, Indonesia

ABSTRACT: This study aims to analyze progressive Muslim gender activism in responding to socio-political situations that affect women with a case study of the *Fiqh an-Nisa* program P3M/Rahima. The author will analyze the publication material and activities of the *Fiqh an-Nisa* program P3M/Rahima uses discourse analysis to study the relationship between gender, power, and ideology in Indonesia. Rahima contributed significantly to reforming Islamic discourse on gender by shifting the tendency of previous religious scripturalism that adhered to the ideology of gender inequality to become Islamic feminism which aspires to gender equality. Rahima actively disseminates her ideas about gender equality in Islam through programs to increase gender awareness and empower women, especially women in the Muslim community. However, such activism is influenced by the broader state power structure and local socio-political dynamics. Rahima was criticized for not only focusing on reforming Islamic discourse and implying that religion was the only contributing factor to the subordinated and marginalized position of Muslim women. In a narrowing backlash from conservative forces, it has to some extent, forced Rahima to deal with several negotiations about their progressive ideas and must exercise self-censorship and caution. Rahima's methods, paradigms, arguments, and implementation techniques for gender equality must be formulated systematically. Thus, it is hoped that there will be a democratic dialogue with various methods, primarily directed at groups that demand such explanations.

Keywords: Islam, Political Islam, Islamic Feminism, Fiqh an-Nisa program P3M, Rahima

1 INTRODUCTION

Experiences of some Muslim countries embarking on the Islamization programs of the state and society, such as in Iran and Pakistan, have borne lessons that such programs have resulted in adverse social-legal implications on women and minority groups. This is partly attributed to the perspective adopted as historical *sharia* that resonates some socio-political dissonances with contemporary Muslim contexts partly due to its worldview of women and non-Muslims as secondary subjects (Na'im 1990). This paper aims to analyze how Indonesian Muslim women has responded to the Islamization within the context of democratization in Indonesia since the 1990s to the post-reform era. It is argued that Muslim women activists in Indonesia have not been passive victims of this contestation but rather actively struggled within it and even further significantly contributed to reforming Islamic discourses on women's rights. However, such activism is influenced by the broader structure of state power and the local socio-religious-political.

Since the democratization momentum, these groups have been increasingly growing and actively speaking their voices and participating in the democratic political system. Ironically, often condemn it as a western import. They range from radical groups, such as FPI (Front Pembela Islam) and MMI (Majelis Mujahidin Indonesia), to conservative ones, such as DDII

*Corresponding Author: nurhidayah@uinjkt.ac.id

DOI: 10.1201/9781003322054-45

(Dewan Dakwah Islam Indonesia), HTI (Hizbut Tahrir Indonesia) and political parties such as PKS (Partai Keadilan Sejahtera) and PBB (Partai Bulan Bintang). Their agenda of Islamization of the state and society can be reflected in the attempt to bring back the Jakarta Charter (Hosen 2007) to the constitutional amendments to make the *shari'a* the ceremonial law for Muslims in Indonesia. Given the fact that such an attempt has resulted in vain, they refocus the Islamization program on the local governments as the decentralization policy has provided such an opportunity through the regional *shari'a* legislation or the so-called *shari'a*-based by-laws, particularly in some predominantly Muslim regions, such as Aceh, West Sumatra, Banten, West Java, and South Sulawesi. This has sparked controversies due to many contradictions of such by-laws with the upper laws, but many of them have primarily targeted women.

The fact that currently, political parties with Islamist ideologies build a coalition with incumbent ruling parties has put the state in an ambivalent position toward gender issues, reflected in its contradictory gender policies. Similar controversies have also sparked over the legislation on some controversial gender issues such as domestic violence, anti-pornography, the counter-legal draft to the compilation of Islamic law, and the draft for the amendment of the law on health that covers the issue of abortion (Fealy & White 2008). The state-backed up religious authority body of MUI (Majelis Ulama Indonesia/Council of Indonesian Ulama) has also shifted toward a more conservative trend during democratization resulting in some *fatwa* (edicts) that constrain Muslims' freedom of religion (Hosen 2004). Overall, while, on the one hand, democratization in Indonesia has successfully toppled the authoritarian New Order's state Ibuism gender ideology (Sears 1996), on the other hand, it has opened up another harsh battleground for women as they have to face more fierce struggles and challenges from both the state and non-state actors who have their gender ideologies to pursue and impose on women once they gain victory in the contest for power.

Under such circumstances, how do Muslim women respond to such cases? Are they just passive victims? Or do they actively respond to the reaction by protesting and reclaiming their rights, leading to their empowerment? This study will look at a case study of the Fiqh an-Nisa P3M (*Perhimpunan Pengembangan Pesantren dan Masyarakat*) program, which was later adopted by Rahima to analyze progressive Muslim gender activism in responding to socio-political situations that affect women. P3M was established with the initial goal of empowering the pesantren community by conducting training to foster critical awareness of religious ways in the dynamic current of development. P3M gave birth to a gender sensitivity approach in the Fiqh an-Nisa program in 1994, focusing on reinterpreting the understanding of Fiqh on women's reproductive rights. Meanwhile, Rahima is an NGO with membership affiliated to Nahdlatul Ulama (NU) with members who come from upper-middle class intellectuals who were educated at Islamic boarding schools or had attended college and have connections with feminists and human rights activists during the reform movement. The researcher chose Rahima as a research case study because Rahima is one of the pioneering NGOs in the struggle for women's rights and gender equality in Indonesia with an Islamic approach, primarily based on the *Kitab Kuning*, which is a pesantren tradition.

This study will analyze the publication material and activities of the Fiqh an-Nisa P3M/Rahima program using discourse analysis to study the relationship between gender, power, and ideology in Indonesia. The focus is on the strategies, struggles, challenges, and socio-political circumstances that shape the nature of progressive Muslim gender activism in Indonesia. This study examines the struggles of Muslim women in Indonesia more deeply in the context of the tension and contestation between political Islam promoted by Islamist groups and the process of modernization and democratization.

2 METHODS

The researchers will analyze the publication materials and activities of the *Fiqh an-Nisa* P3M and Rahima programs in Indonesia to study the relationship between gender, power, and ideology in Indonesia. Data analysis will use a qualitative descriptive approach. In this study, data were collected through observation and reading. Observations were made to observe, feel, and understand the Fiqh an-Nisa P3M/Rahima program through interviews with Muslim women activists in Rahima to obtain the information needed for research. In addition, the author also reads previous studies that are relevant to this study to support the research argument. The outline of this paper is: first, introduction; second, research methods; third, results and discussions that will discuss the general description, strategies, struggles, challenges, and socio-political conditions regarding the P3M and Rahima *Fiqh an-Nisa* programs in Indonesia; and finally, conclusion.

3 RESULT AND DISCUSSION

3.1 *Fiqh an-Nisa programme P3M (Perhimpunan pengembangan pesantren dan masyarakat) and Rahima*

From the 1980s until the time after the reform, several thinkers with NU backgrounds established Non-Governmental Organizations (NGOs), which concentrated on community empowerment based on pesantren. Some of these NGOs use feminism and gender as analytical tools to advocate for women's rights, starting from the reconstruction of the understanding of the Qur'an and Hadith. One NGO with a vision and mission to produce Islamic knowledge using a gender perspective is P3M (Perhimpunan Pembangunan Pesantren dan Masyarakat). P3M is an NGO affiliated with NU (Nahdhatul Ulama). P3M was founded in 1983, in line with the transformation of Indonesian Muslims from the tendency of formal political Islam to the socio-cultural development of Islam. Its emancipatory Islamic vision aims to transform NU's valuable traditionalist Islamic discourse into a progressive one, including on gender issues. Women's division was founded in 1993 with a focus on formulating progressive Muslim gender discourse and disseminating it through programs to increase gender awareness and empower women in the Muslim community, particularly in pesantren, which were previously known as Islamic patriarchal guidance sites. Its establishment is based on the assumption that Islamic boarding schools have cultural capital and potential as agents of social change (Ma'ruf *et al.* 2021).

Analysis of social and class injustice, which is P3M's primary concern, has begun to include gender analysis since the joining of staff, such as Lie-Marcos Natsir and Farha Ciciek with the existence of the *Fiqh an-Nisa* program at P3M. The *Fiqh an-Nisa* program emphasizes a contextualized approach to understanding verses and reflecting on the content of gender-biased religious texts (Hidayah 2013). This approach also considers several aspects of the theology of the Qur'an and hadith, classical books, and the development of national and international regulations, taking into account women's experiences. The pesantren feminist method, their typical model in reconstructing *Fiqh an-Nisa*, is to understand the Qur'an and al-Hadith by separating their teachings into two categories, namely universal principles (*qath'i*) and operational, technical provisions (*dzanni*). Universal principles are absolute because their position is *qath'i*, and all Muslims must submit, not know the place and time. Meanwhile, operational and technical provisions which have the meaning of *dzanni* are seen as binding substantially, not literally, because there are sociological and contextual dimensions. Therefore, literally (the letter of the law) can and may change based on universal principles and substantial values.

Issues regarding kinship, society, and community, including gender analysis, are categorized as zhanni texts, so interpreting this text is still possible. Women feminists add the principles of justice, equality, deliberation, mu'asyarah bi al-ma'ruf, and ta'awun as universal principles, which reinforce the teachings of the Qur'an about these principles and must be implemented in all things, including daily affairs days in family life. Rahima later adopted the *P3M Fiqh an-Nisa* program along with the reform and transformation period. Rahima is an organization founded in Jakarta in 2001, with membership affiliated with Nahdlatul Ulama (NU). In an interview conducted by the researcher on Farha Ciciek as one of the women activists, "she said that Rahima was built with various backgrounds, not only struggling with religious issues but also in advocacy, both national and international." With this transformation, the focus of the *Fiqh an-Nisa* program has also shifted not only to women's reproductive rights but also to gender issues in Islam. This shift gained momentum because, during this period, Islamic groups were actively struggling with their plans to Islamize the state and society.

3.2 *Islamic Feminism in Indonesia: The cases of Fiqh an-Nisa programme P3M/Rahima*

Assessment and reinterpretation, even at a certain level, deconstruction of gender-biased interpretations and understandings of Islam becomes very important (Luthfiyah 2015). Fiqh always has a plurality of opinions. There are no absolutes. This indicates the relativity and contextuality of thought. When viewed from a discourse perspective on gender, NU ulama can be traced back to Hasyim Asy'ari's thought that teaching women to write and read was permissible. By using the rule *al-hukmu yaduru ma'a illatihi wujudan wa adamant,* Hasyim Asy'ari broke the opinion forbidding teaching women to write and read. This study was also later developed on the problem of *khalwat* in the female learning process. Which is usually a male teacher; using the same rules, he explained that the law of learning was initially *makruh*, but if it was safe, then the law became *mubah*. This gender discourse continues to proliferate (Zulaiha & Busro 2020).

Several studies in Indonesia that discuss pesantren women and knowledge of gender and Islam, gender and politics, and gender and democracy have been successfully demonstrated by several researchers. Gender studies in the context of women's presidential discourse were discussed by (Doorn-Harder 2002) dan (Robinson 2004), polygamy in post-reform democratic culture was written by (Brenner 2006), and the critical Islamic feminist agency piety movement was discussed by (Rinaldo 2013). Other research aims to promote gender equality, which is Women, Business, and Law, by the World Bank. The results of this study indicate that, on average, women only have three-quarters of the legal rights granted to men. Therefore, observing the laws and regulations that limit women's economic rights is considered essential to serve as the basis for recommendations for legal reform toward gender equality (World Bank 2021).

This research discusses progressive Muslim gender activism in response to socio-political situations that affect women, especially the *Fiqh an-Nisa* program P3M/Rahima. Under the leadership of some activists of P3M, such as Masdar F. Mas'udi, Lies M. Natsir, Husein Muhammad, Syafiq Hasyim, and Farha Ciciek. The organization not only reformulated conventional *Fiqh* rulings about gender issues using the methodology of classical *ushul Fiqh* combined with gender analysis but also further disseminated such a new discourse. In the interview, Lies M. Natsir said that "Rahima's organization's focus on reforming Fiqh discourse might be responding to a broader critique of the works of the reformists Wadud, Hassan, and Barlas Muslims. They reinterpret the Qur'an using a women's perspective but ignore the influence culture that makes up Fiqh". Quoting Kecia Ali (Safi 2003), Doorn-Harder argues that the reality indicates that many women in Muslim countries continue to suffer injustices within their marriage because their husband's frame of reference about marriage adheres mainly to the *Fiqh*. When scholars who try to reread the Islamic texts on women ignore these *Fiqh* texts, she further argues that they ultimately failed to address the legal system that has reproduced Islamic misogynist teachings on women, which renders the reformist mode of interpretation often remaining vague and apologetic.

Farha Ciciek said the *Fiqh an-Nisa* program's focus has shifted to women's reproductive rights and gender issues in Islam after Rahima adopted its approach. The program was carried out as a step in transformative da'wah. Rahima's main program is the cadre of female ulama (*Perekrutan Ulama Perempuan/PUP*), which was initiated in 2005. One of the programs that Rahima carries out, also through *tadarus*, is a learning model for prospective cadres with discussions on themes such as gender perspective, social change, social analysis, and methodological discourses in Islam. Rahima has also actively disseminated its ideas on gender equality in Islam to grassroots levels through gender consciousness-raising and women empowerment programs, particularly for women in the Muslim community. There have been various responses among the program participants, ranging from those who resist to those who eclectically accept to those who find such a discourse liberating and empowering. While those who oppose might represent a slight minority, the majority responses fall between those who eclectically take to those who feel empowered. Such empowerment has been at the level of individuals where women can negotiate their relations with their husbands with the same texts that used to control them as docile bodies in the name of religious precepts on the requirements of wives' blind obedience to their husbands. At the community level, the organization targets strategic groups, particularly the local leaders of their communities.

Amidst this struggle and its achievements, Rahima has been criticized for not only its sole focus on reforming Islamic discourse as if implying religion was the sole contributing factor to Muslim women's subordinated and marginalized positions. Thus, the influence and transformation it has affected are perceived to be too limited in depth and breadth. Some scholars argue that an overreliance on scriptural sources tends to freeze Muslim women's social status (Jospeh 2007) and leave other concepts essential for women's empowerment, such as poverty elimination and welfare states, underdeveloped (Moghadam 2002). As such, its influence and the transformation it causes are considered too limitited in depth and breadth.

Among its target groups, the emancipatory vision of Rahima for effecting socio-cultural transformation tends to be used as jargon instead of systematic programs to liberate Muslim women from unequal social-cultural structures that have so far marginalized them. In the broader context of the women's movement, its focus on cultural advocacy at the expense of legal and structural advocacies has limited the organization's influence and left a vacuum in the national leadership on legal advocacy for Muslim women's rights. So far, such advocacy has been attempted by Muslim women individually, such as the figure of Musdah Mulia and Badriyah Fayumi, instead by organizations that have strong constituencies at the grassroots levels. Such a national vacuum can be potentially filled if progressive Muslim women's organizations, such as Rahima, reformulate their organizational mandates to face this challenge. Such a challenge has gained momentum in this *reform* period since Islamist groups have actively conducted public advocacy in all legal, structural, and cultural spheres.

Rahima's discourse on resisting the so-called formalization of the *shari'ah* has signified its challenge to patriarchal local government policies that discriminate against women and Islamist's agenda of Islamizing the state and society. Furthermore, these activists' involvement in the study forum for critical analysis of traditional Islamic textbooks resulted in a book criticizing the *'uqud al-lujjayn* (text used in many *pesantren* on wife-husband relations using traditionalist Islamic perspective) also represents its challenge to traditionalist *ulama* (Muslim scholars) who attempts to maintain their traditional religious authorities. However, these attempts have prompted conservative backlash from those forces whose interests have been challenged by these Islamic feminists. The media of Islamist groups has continuously waged a psychological war against Islamic feminists. The Majlis Ulama Indonesia (MUI), a state-backed up religious authority body 2005, issued a fatwa (edict) prohibiting secularism, pluralism, and liberalism. Such a fatwa has a further implication of limiting this progressive movement whose interpretation of Islam is based on reformist liberal traditions (Fealy & White 2008).

In a similar vein, conservative traditionalist Muslim scholars from NU also counter a book by progressive Muslims, signifying their attempts to exclude progressive interpretations from the Muslim public sphere and deny women access to their rights to interpret their religion using a female-inclusive perspective. Several maneuvers by Islamist groups also attack these Islamic feminists, ranging from subtle to harsh ones. One example is a website reporting Rahima's Seminar discussing the adverse impact of the Sharia Regional Legislation on women: The substance of the National Seminar "Women under the Sharia Regional Legislation (Lessons from Tasikmalaya, Garut, Cianjur, and Banten) was no more than straying doctrine and propaganda of women's NGO to the society, in this case, Rahima as an organizer." Rahima's silence on some of the controversial issues of the Counter Bill and homosexuality, as well as the fair use of the popular culture of shalawat, can be seen in the context of the ability of these progressive activists to continuously pursue their agenda amidst attacks and narrowing backlash from conservative forces. This backlash, to some extent, forced these activists to deal with some negotiations about their progressive ideas. Self-censorship and caution have been exercised. Rahima's methods, paradigms, arguments, and implementation techniques must be formulated systematically. Thus, it is hoped that there will be a democratic dialogue with various methods, primarily directed at groups who demand such explanations.

4 CONCLUSION

Rahima's gender discourse and activism have been shaped mainly by socio-political circumstances. Rahima contributed significantly to reforming Islamic discourse on gender by shifting the tendency of previous religious scripturalism that adhered to the ideology of gender inequality to become Islamic feminism which aspires to gender equality. Rahima actively disseminates her ideas about gender equality in Islam through programs to increase gender awareness and empower women, especially women in the Muslim community. However, such activism is influenced by the broader state power structure and local socio-political dynamics. Rahima was criticized for focusing on reforming Islamic discourse as if implying religion was the only contributing factor to the subordinated and marginalized position of Muslim women. The lack of broader structural change that progressive Muslim women's organizations impact is partly attributed to the organization's internal challenges, the country's ambivalent gender policies, and the complex and diverse realities of Muslim gender politics. Rahima needs to regenerate progressive thoughts on the issue of women's rights and gender equality along with advocacy efforts in the field by continuing to actively engage with the reality of women's lives at the practical level. Rahima's methods, paradigms, arguments, and implementation techniques in fighting gender equality must be formulated systematically. Thus, it is hoped that there will be a democratic dialogue with various methods, primarily directed at groups that demand such explanations. This research has limitation of studying only case study. It recommends further research to use multiple case studies in order to further reflect the progressive Muslim women's movement.

REFERENCES

Brenner, S., 2006. Democracy, polygamy, and women in post-reformasi Indonesia. *Social Analysis*, 50 (1), 164–170.

Doorn-Harder, P. van, 2002. The Indonesian Islamic debate on a woman president. *Sojourn: Journal of Social Issues in Southeast Asia*, 164–90.

Fealy, G. and White, S., 2008. *Expressing Islam: Religious Life and Politics in Indonesia*. Singapore: ISEAS.

Heryanto, A. and Mandal, S.K., 2003. *Challenging Authoritarianism in Southeast Asia: Comparing Indonesia and Malaysia*. New York: Routledge Curzon.

Hidayah, N., 2013. Pembaharuan Islam, Gerakan perempuan, dan masyarakat sipil di Indonesia: Studi Kasus Fiqh Nisa P3M dan rahima. *Jurnal Ahkam*, 8 (1), 75–90.

Hosen, N., 2004. Behind the scenes: Fatwa of MUI (1975–1998). *Journal of Islamic Studies*, 15 (2), 147–179.

Hosen, N., 2007. *Shari'a and Constitutional Reform in Indonesia*. Singapore: ISEAS.

Jospeh, S., 2007. *Encyclopedia of Women and Islamic Culture*. Leiden: Brill.

Luthfiyah, N., 2015. Feminisme Islam di Indonesia. *ESENSIA*, 16 (1).

Ma'ruf, A., Wilodati, and Aryanti, T., 2021. Kongres ulama perempuan indonesia dalam wacana merebut tafsir gender pasca reformasi: sebuah tinjauan genealogi. *Jurnal Musawa*, 20 (2), 127–145.

Moghadam, V.M., 2002. Islamic feminism and its discontents: Toward a resolution of the debate. *Signs*, 27 (4), 1158.

Na`im, A.A.A., 1990. *Toward an Islamic Reformation: Civil Liberties, Human Rights, and International Law*. Syracuse, n.Y: Syracuse Univeristy Press.

Rinaldo, R., 2013. *Mobilizing Piety: Islam and Feminism in Indonesia*. Oxford University Press.

Robinson, K., 2004. Islam, Gender, and politics in Indonesia. *Islamic Perspectives on the New Millennium (ISEAS Publishing)*, 183–196.

Safi, O., 2003. *Progressive Muslims: On Justice, Gender, and Pluralism*. Oxford: Oneworld.

Sears, L.J., 1996. *Fantasizing the Feminine in Indonesia*. Durham: Duke University Press.

World Bank, 2021. *Women, Business and the Law 2021*. Washington, DC: The World Bank.

Zulaiha, E. and Busro, B., 2020. Tradisi bahts al-masail Nahdatul Ulama (NU): Pematangan pemikiran fikih adil gender husein muhammad. *Jurnal Studi Gender dan Islam*, 19 (2), 205–218.

Religion, Education, Science and Technology towards a More Inclusive and Sustainable Future – Rahiem (Ed.)

Development of digital entrepreneurship programs in pesantren in Indonesia

Sarwenda*, H. Rahim, D. Rosyada, A. Zamhari & A. Salim
UIN Syarif Hidayatullah, Jakarta, Indonesia

ABSTRACT: *Pesantren*, as one of the oldest Islamic educational institutions in Indonesia, is known to be able to survive in various conditions and continues to improve itself to face the digital era. One of the *pesantren* programs that has continued to develop and has received support from many parties to the current government is *Pesantrenpreneur*. This study aims to show that digital entrepreneurship programs have been developed in the agribusiness *pesantren* in Indonesia. This study uses a qualitative method with a phenomenological approach, and the sample of this study is *Pesantren Al-Ittifaq* (Alif) in Bandung. The data was obtained from observations, interviews with 27 informants, and a number of questionnaires using open-ended questions. The results showed that: (1) *Pesantren* have started to abandon traditional farming methods and are using e-farming, such as smart greenhouses, to produce their agricultural products and students' learning practices; (2) Various online platforms and social media have also been used and developed by *pesantren* as digital marketing and entrepreneurship learning practices; and (3) Agribusiness *pesantren* are ready to face the 5.0 era by developing the Internet of Things in improving the quality of agricultural products. This study concluded that, although still in the process of developing and learning to adapt to digital technology, *pesantren* are able to produce quality agricultural products and can enter supermarket segments.

Keywords: digital entrepreneurship, digitalpreneur, pesantrepreneur, santripreneur, agribusiness Islamic boarding school

1 INTRODUCTION

Educational institutions in Indonesia are currently preparing to face the era of a 5.0 society (Nastiti & Abdu 2020). One of them is *pesantren*, the oldest Islamic educational institution (Kholili 2021). Based on the data from the Ministry of Religion of the Republic of Indonesia (2021), the number of *pesantren* currently registered is 27,722, with a total of 4,175,531 students. In the 21st century, *pesantren* has been using technological innovations in the field of entrepreneurship (Nurhattati 2020). Entrepreneurship is also an issue that continues to be developed in Islamic boarding schools to generate the potential of Islamic boarding schools in the economic field. However, *pesantren* still faced challenges in terms of the effort to adapt to digital usage, which is considered technologically stuttering (Kuswara nd). There are still many *pesantren* that are not open to using technology on the grounds of protecting students from the negative effects of technology (Ansori *et al.* 2022). It was evidenced by the data on the comparison chart of the economic potential of the Ministry of Religion of the Republic of Indonesia (2021), which revealed that there are still a few *pesantren* with potential in the agribusiness sector, which only amounts to 1479 *pesantren*, and the potential in the

*Corresponding Author: Sarwenda15@mhs.uinjkt.ac.id

DOI: 10.1201/9781003322054-46

technology sector is 366 *pesantren*. On the other hand, Kholil (2021) mentions that many *pesantren* have used technology and made it part of the learning process, either for learning media or for the institution's operational system. Indra (2020) said that educational institutions in Indonesia must be able to accommodate the various qualities of human resources needed in welcoming the 5.0 era, such as those "who have scientific insight, economics, entrepreneurial spirit, professional mentality, etc." This is in line with the function of *pesantren* as stated in Law no. 18 of 2019 concerning *pesantren*, Article 4. The scope of the function of the *pesantren* includes (a) Education; (b) *Da'wah*; and (c) community empowerment.

Additionally, there are previous studies regarding technology and digital entrepreneurship in education (Lynch 2021; Zhang 2021; Akhmetshin *et al.* 2019) and regarding digital entrepreneurship in *pesantren* (Ansori 2022; Hidayatulloh *et al.* 2019). Nevertheless, these studies did not discuss the development of technology in the field of agribusiness. Even though the use of digital entrepreneurship at Al-Ittifaq has also been previously studied by Basseri (2021), which revealed that Indonesia urgently needed the development of more agricultural-based *pesantren*, it did not discuss in detail the learning methods used in digital entrepreneurship in this *pesantren*.

Entrepreneurship learning is an attempt to make students learn or an activity to teach students by paying attention to cognitive aspects, affective aspects, and psychomotor aspects (Suprapto 2018), and also to train students in critical and creative thinking (Fayolle 2013). National entrepreneurship development coordination by providing access to a more structured coordination both in the world of education and technology to creative bodies is required to accommodate the results of *technopreneurship* (Marti'ah 2017). The development of technology in national educational institutions has been developed up to the university level. Nevertheless, in Islamic educational institutions such as *pesantren*, this is still very rarely encountered for various reasons and obstacles (Rohmatullah 2022). *Pesantren* still have to continue to learn in the field of digital mastery, coupled with the development of the entrepreneurial aspect, which also requires educators or educational staff who have qualified technological skills to carry out all of this (Kholili 2021).

Based on these circumstances, this paper is made to answer: How digital entrepreneurship programs are developed and taught in agribusiness *pesantren*, especially in the concept of modern agriculture in Indonesia? With the massive development of *pesantren* in Indonesia and the development of digital technology throughout the world, the public also expects improvements to the system and learning model in *pesantren* so that the advancement of technology will not leave behind students who study in *pesantren*. Hence, the purpose of this research is to provide an elaboration on the digital technology developed in Islamic boarding schools in terms of entrepreneurship learning related to producing quality goods. Therefore, this research is expected to be useful for academics as input for policymakers pertaining to Islamic educational institutions in Indonesia and to become a reference material and new knowledge for researchers and practitioners from various fields of study, especially for those who are interested in the development of Islamic boarding schools in facing the challenges of the times in the digital era.

2 METHODS

This study is a part of a larger research on the self-reliance and entrepreneurship of students in *pesantren*. This study applies a qualitative method with a phenomenological approach, but the approach in this context is not a research theory (Rosyada 2020). The object of this research was one *pesantren* in Indonesia, namely *Pesantren* Al-Ittifaq in Bandung-West Java, and the researcher was a participant-observer for this research (Creswell 2014). The data used in this study describes the development of digital entrepreneurship related to modern agriculture and the learning process at Al-Ittifaq. There are two data categories in this study, namely primary data and secondary data. The first primary data sources are obtained from: (1) the participant's moderate observation of the practice and reality of the phenomena that occur at Al-Ittifaq; (2) the second-primary data is obtained from the interviews with selected informants. The researchers used purposive sampling to select the informants who were being interviewed. The interviewees in this study are 27 people, with the following specifications: 3 *pesantren* leaders; 5 Alif learning center (ALEC) staff and teachers; 5 male and female students; and 4 alumni of Al-Ittifaq from *Salafiyah* and *Takhosus* programs. Ten questionnaires with an open-ended method for 10 informants are distributed to the sample. The secondary data sources are Al-Ittifaq's documents from various research results that have been done, books, and other sources related to the theme of this study.

3 RESULTS

Based on the analysis of the primary and secondary data sources, this research revealed that *Pesantren* Al-Ittifaq has implemented and taught e-farming using smart greenhouses. This fact is also seen in the observations conducted by the researchers. Moreover, in this *pesantren*, the methods that have been used in the learning process in the class are presentation, discussion, answer, and question for agribusiness and technology information (IT) lessons. Al-Ittifaq has been famous for using the 'learning by doing' approach in their agribusiness activities, either for their students in the *Takhasus* and *Salafiyyah* programs or participants in ALEC. In addition to successfully developing agribusiness programs using digital farming for themselves, Al-Ittifaq also has a business incubator named Alif Learning Center (I/Kopontren's Staff).

According to one of the staff members of ALEC, Al-Ittifaq already has seven greenhouses (SF/Kopontren's staff). The students and teachers assumed that using a greenhouse to grow various vegetables and fruits could produce high-quality or grade-A agricultural products. According to them, apart from avoiding plant pests, greenhouses can also be used in weather that is not good for plant growth (Alif students). This was further explained by one of the research informants, who explained that the *pesantren* use greenhouses as a place to learn to process their agricultural products, as stated in the following statement: *We go to the garden every day based on a predetermined group. Some go to the greenhouse to learn how to treat fruits and vegetables that have been planted there* (MR/Student). The teacher also explained the use of greenhouses for processing certain vegetables or fruits for quality agricultural products so that they can be sold in supermarkets, as stated by a teacher: *Our agricultural products have entered supermarkets such as Lotte Mart, Superindo, Hero, and several other supermarkets.* This finding is in line with Basseri (2021), which showed that Al-Ittifaq has already been using greenhouses, online platforms, and social media in its agribusiness programs to increase the yield of higher-quality food products.

In addition, the use of digital marketing methods, such as online platforms and social media has become part of the marketing method in the *pesantren*. Al-Ittifaq has used an online store platform in the form of a website and social media as their digital marketing media (Alifmart, nd). This was also supported by Al-Ittifaq's students' argument when they were taught directly the process of packaging agricultural products before being marketed to offline stores and online stores. MR said: *We are usually divided into groups every week; some are in the strawberry garden and some are in the greenhouse; some of them are in the greenhouse; and the others are in the packaging section in Kopontren* (MR/Student). According to the experience of students who have started to open their own businesses using online stores such as *Tokopedia*, he dared to start this business because he saw the success of sales that have been made by the Alif store. As he mentioned: *I have had a Tokopedia account for the past few months, selling fruit products, such as strawberries, in collaboration with one of the religious teachers who owns a strawberry garden* (A/Student). In the development of digital agriculture in Indonesian Islamic boarding schools, Al-Ittifaq is one of the *pesantren* that has become a pilot project by the government of the Republic of Indonesia (Lukihardianti & Hafil 2022). This *pesantren* also succeeded in becoming a role model for the development of *pesantren* for the economic empowerment of the community, which was one of the programs of the Governor of the West Java provincial government, Ridwan Kamil, through one *pesantren* one product. Al-Ittifaq has succeeded in using the Internet of Things (IoT) in its agricultural activities, as stated by Al-Ittifaq's alumnus, who already has a *pesantren* (AK/Alumni). This was also conveyed by the governor of West Java, Ridwan Kamil (Lukihardianti & Hafil 2022). The use of the IoT makes it easier for farmers. One of which is to monitor the temperature in the greenhouse, as said by another Alif alumnus: *I can set the temperature I want at any time through my cellphone without having to come directly to the garden* (HK/Alumni). This is in line with the research findings of Nurhattati (2020), which revealed that Al-Ittifaq has empowered and developed IoT-based e-farming, namely precision farming, unmanned aerial vehicle or drone farming, and smart greenhouses.

4 DISCUSSION

The development of the internet and technology has provided many benefits in the 21st century for the needs of agricultural development, for instance, detecting gardens that lack nutrients (Lynch 2021). Modern technology in greenhouse farming has provided many other benefits for climate change as well (Field 2021). According to Pandey (2015), more than 50 countries have developed the concept of greenhouses for various types of plants because they have already known their benefits. In Indonesia, the development of greenhouses to grow horticultural products is not only carried out by Al-Ittifaq in Bandung (Basseri 2021) but it has also been used in other *pesantren* because it has been proven that by using a greenhouse to grow various types of vegetables and fruits, there is no need to worry about bad weather (Helmida *et al.* 2021).

Unfortunately, there are still a few *pesantren* in Indonesia that are able to build this greenhouse because the price is very expensive, as stated by one of the alumni, Al-Ittifaq: *For the price of one greenhouse is around four hundred million rupiahs* (YS/Alumni). In line with this opinion, the teaching staff at ALEC also conveyed the same thing (I/*Kopontren*'s Staff). In this study, it was found that the *pesantren* and the students viewed the learning process to grow various commodities using e-farming, such as smart greenhouses, as beneficial for producing quality food products (D/Teacher) and training students to hone their hard skills and soft skills in processing vegetables and fruit using technology that is acceptable in domestic and international markets (Nurhattati 2020). Strickler (2020) said that greenhouse technology could be the future of agricultural development because the need for food is increasing but the land is increasingly limited in each country.

In addition, according to Kotler (Kotler & Armstrong 2018), digital marketing is one of the fastest growing in the world of marketing as a result of the development of the internet and various social media and advertising platforms. The use of digital marketing such as social media platforms needs qualified trading skills, or in Vieira's terms, it is called the strategy of tug of war and performance (Vieira *et al.* 2022). As stated by Kotler (Kotler & Armstrong 2018), to be able to attract customers or consumers in the digital world, you need to have special marketing skills. This fact was also acknowledged by one of the *Kopontren* Alif staff who is also a trainer at ALEC: *The obstacle that is often encountered by the trainees here is the lack of human resources in their Islamic boarding school; besides that, there are still many trainees who come from various regions saying that they don't get good support from their institutions, so even though the trainees already have the knowledge, if they don't get support from their institutions, they won't last long* (I/*Kopontren*'s staff).

Moreover, entrepreneurial technology development is a form of technological disruption (Zhang 2021; Agrawal & Ting 2022). One of the skills that is now also part of the challenge for entrepreneurs is to be able to combine technology and entrepreneurial activities, which is commonly called *technopreneurship* (Elijah 2017). This is needed to make it easier for entrepreneurs to develop various forms of entrepreneurship. Not only that, but the use of the internet in various business programs can increase productivity and produce excellent quality results (Nurhattati 2020).

5 CONCLUSION

From this research, it can be concluded that there are *pesantren* in Indonesia that are able to develop and use digital entrepreneurship for the economic stability of their institutions and the learning process as well. Al-Ittifaq in Bandung has proven that the typology of *salafiyah pesantren* is able to develop hi-tech and integrate it into its agribusiness program through integrated farming with technology and information (*infratani*). By utilizing supporting natural resources and skills in agriculture and technology as well as business, this *pesantren* is able to become a role model in the development of digital agriculture in Islamic educational institutions in Indonesia. This paper has limited discussions regarding the development of digital entrepreneurship in modern Islamic boarding schools, which have not been disclosed here. How it is developed in the learning process and the methods used, as well as the technology used for agribusiness development in modern Islamic boarding schools, need to be studied further to get the best development model. Researchers recommend comparing the development of digital entrepreneurship programs in Islamic educational institutions in Indonesia with other formal educational institutions. This needs to be done considering the diverse potential of *pesantren* in Indonesia. Developing a curriculum for this program is also needed because Islamic educational institutions in Indonesia still have shortcomings in these aspects, such as training programs to master digital entrepreneurship in the agricultural sector, resources needed, time required, cost of facilities, and technological tools.

REFERENCES

Agrawal, S. and Ting, H.-I. (2022) 'Digital learning and entrepreneurship Education: Changing paradigms', *International Journal of Social and Business Sciences*, 16(2), pp. 29–35.

Akhmetshin, E.M. et al. (2019) *'Innovative Technologies in Entrepreneurship Education: The Case of'*, 22(1), pp. 1–15.

Alifmart (no date). Available at: https://alifmart.online/ (Accessed: 15 July 2022).

Ansori et al. (2022) 'Digital innovation in pesantren education':, *Nazhruna: Jurnal Pendidikan Islam*, 5(2), pp. 645–661.

Basseri, G.R. (2021) *Al-Ittifaq Agricultural Pesantren (Traditional Islamic Boarding School): Empowering Youth in Agriculture*. Wageningen.

Creswell, J. (2014) *'Research Design Qualitative, Quantitative, and Mixed Method Approaches by John W. Creswell (z-lib.org).pdf'*, p. 285.

Elijah, O. (2017) *'Technopreneurship: A View of Technology'*, 17(7).

Fayolle, A. (2013) *Handbook of Research in Entrepreneurship Education*, Volume 1. https://doi.org/10.4337/9781847205377.

Field, A. (2021) *New Greenhouse Technology Takes on Climate Change to Feed the Planet*. Available at: https://www.theimpactivate.com/new-greenhouse-technology-takes-on-climate-change-to-feed-the-planet/ (Accessed: 18 July 2022).

Helmida, B.E., Khotmi, H. and Dkk (2021) 'Teknologi green house di pondok pesantren darul', *Alamtana: Jurnal Pengabdian Maysrakat*, 2(3).

Hidayatulloh, M.H. *et al.* (2019) 'Entrepreneurship education grows santri's entrepreneurial spirit (evidence from indonesia's islamic boarding school)', 2019, pp. 594–601. https://doi.org/10.18502/kss.v3i13.4233.

Indra, H. (2020) 'Challenges and response in Islamic education perspective in the digital media era', *Attarbiyah: Journal of Islamic Culture and Education*, 5(1), pp. 31–42. https://doi.org/10.18326/Attarbiyah.V5I1.31–42.

Kementrian Agama RI (2021) *Pangkalan Data Pondok Pesantren, Ditpdpontren*. Available at: https://ditpdpontren.kemenag.go.id/pdpp (Accessed: 8 October 2022).

Kholili, Y. (2021) 'Challenges for pesantren in the revolution era of society 5.0', *AMCA Journal of Religion and Society*, 1(1), pp. 8–12. https://doi.org/10.51773/ajrs.v1i1.33.

Kotler, P. and Armstrong, G. (2018) *Principles of Marketing 17th Global Edition, Pearson Education Limited*. (Accessed: 27 December 2020).

Kuswara, H. (no date) *Santri dan Kesenjangan Digital Tantangan Vs Peluang – PP Pergunu*. Available at: https://pergunu.or.id/santri-dan-kesenjangan-digital-tantangan-vs-peluang/ (Accessed: 8 October 2022).

Lukihardianti, A. and Hafil, M. (2022) *Ponpes Al-Ittifaq Jadi Percontohan Nasional Digitalisasi Pertanian | Ihram*. Available at: https://ihram.republika.co.id/berita/r95ft7430/ponpes-alittifaq-jadi-percontohan-nasional-digitalisasi-pertanian (Accessed: 18 July 2022).

Lynch, M. (2021) 'Technological Forecasting & Social Change Combining technology and entrepreneurial education through design thinking: Students' reflections on the learning process', *Technological Forecasting & Social Change*, 164(January 2018), p. 119689. https://doi.org/10.1016/j.techfore.2019.06.015.

Nastiti, F. E., & Abdu, A.R.N. (2020) 'Kajian: Kesiapan pendidikan Indonesia menghadapi era society 5.0. edcomtech jurnal kajian teknologi Pendidikan.', *Chronologia: Journal of History Education* [Preprint]. Available at: http://journal2.um.ac.id/index.php/edcomtech/article/view/9138/pdf (Accessed: 20 June 2022).

Nurhattati (2020) 'Pemberdayaan santri melalui e-farming pesantren berbasis internet of think: Studi Kasus di Ecopesantren Ittifaq Bandung', *HAYULA: Indonesian Journal of Mustidisciplinary Islamic Studies*, 4(2), pp. 171–188.

Pandey, S. and Pandey, A. (2015) 'Greenhouse technology', *International Journal of Research-Granthaalayah*, 3(9SE), pp. 1–3. https://doi.org/10.29121/granthaalayah.v3.i9se.2015.3176.

Rosyada, D. (2020) *Penelitian Kualitatif Untuk ILMU Pendidikan*. 1st edn. Edited by Murodi. Jakarta: Kencana.

Strickler, J. (2020) *High-Tech Greenhouses Could Be The Future Of Agriculture, Forbes*. Available at: https://www.forbes.com/sites/jordanstrickler/2020/08/28/high-tech-greenhouses-could-be-the-future-of-agriculture/ (Accessed: 18 July 2022).

Vieira, V.A. *et al.* (2022) 'Optimising digital marketing and social media strategy: from push to pull to performance', *Journal of Marketing Management*, 38(7–8), pp. 709–739. https://doi.org/10.1080/0267257X.2021.1996444.

Zhang, Z. (2021) *'The Impact of Digital Technologies on Entrepreneurship Education A Literature Review for Progress and Prospects'*, 543(Icssed), pp. 448–452.

Religion, Education, Science and Technology towards a More Inclusive and Sustainable Future – Rahiem (Ed.)

Biodiversity of freshwater microalgae of Jangkok River, Lombok, Indonesia: An analysis for sustainable future

N. Purwati* & E.T. Jayanti
UIN Mataram, Indonesia

ABSTRACT: Microalgae are the basis of life in aquatic ecosystems. As unicellular organisms, microalgae reproduce faster and can grow in extreme conditions. This study aims to reveal the diversity of freshwater microalgae found in the Jangkok River, Lombok, Indonesia. The research was conducted descriptively by determining the sampling station based on the different characteristics of the area through which the river flows. Sampling of microalgae was carried out in April 2022, three times for each station, with an interval of one week. The samples of microalgae were observed in the laboratory using a light microscope, while the identification process was carried out by referring to books and relevant journals. Data were analyzed descriptively by first determining the Shanon–Wiener (H') diversity index; 48 genera and 85 species were found. The diversity of microalgae is at a high level with an index (H') = 3.163. The genera with the most species were *Nitzschia*, *Scenedesmus*, *Gomphonema*, and *Oscillatoria*, while the most abundant species is *Synedra ulna*.

Keywords: biodiversity, freshwater microalgae, sustainable future

1 INTRODUCTION

The increasing human population and the rapid development of industrialization have a certain influence on environmental conditions. Among them is the release of pollutant substances that are organic and inorganic, such as heavy metals, resulting in various changes and serious damage to various living organisms on the earth. For a long time, heavy metal pollution has become a challenge for the future sustainability of the ecosystem and the wider environment. Several methods that have been implemented to control and overcome the problem of the entry of pollutant substances into the environment have not yet yielded maximum results. Utilization of living things through the mechanisms of green ecology and green technology is an alternative that can be done. Moreover, the use of this biological method is believed to have higher effectiveness and sustainability value than other methods such as physical and chemical methods (Kumar *et al.* 2021; Tripathy *et al.* 2021). The use of biological indicators in basic research that aims to determine the ecological status of an environment is no less important. In this case, including to determine the quality of waters that are in direct contact with all activities of daily human life, such as rivers.

Rivers are a source of fresh water with an ecosystem structure that is interconnected with other aquatic ecosystems, such as lentic, lotic, and wetland ecosystems. Overexploitation of the biota that makes up the river ecosystem can disrupt its balance and sustainability

*Corresponding Author: nining.purwati@uinmataram.ac.id

DOI: 10.1201/9781003322054-47

(Marganingrum *et al.* 2018). According to Venkatachalapathy and Karthikeyan (2015), rivers are water bodies that are most susceptible to pollution as a negative result of the use of rivers in daily life. The existence of bioindicators provides information on the chemical and physical conditions of water, including rivers (Bellinger & Sigee 2015; Khan 2016; Perić *et al.* 2018). In addition, bioindicators can also show the accumulated effects of various types of pollutants in an ecosystem and the estimated length of time these pollutants last in an environment (Parmar *et al.* 2016).

Microalgae can be used as bioindicators of water quality (Costa *et al.* 2016; Efroymson & Dale 2015; Lobo *et al.* 2016). Its cosmopolitan nature, from snowy areas to deserts, lakes, rivers, soil, and rocks, gives advantages to an ecosystem. This makes microalgae the basis of life in aquatic ecosystems. As unicellular organisms, microalgae reproduce faster and can grow in extreme conditions. Algal communities have a high sensitivity to changes in their habitat (Maznah & Makhlough 2014). Therefore, many species of algae are used as indicators of water quality (Sahoo & Seckbach 2015).

Based on this explanation, knowing the diversity of microalgae species in a river ecosystem is the first step to determining the quality of the waters in relation to the role of algae as bioindicators. This role is closely related to the ability of algae to remove various types of pollutants (Tripathy *et al.* 2021).

The Jangkok River is one of the largest rivers in Lombok Island and flows along areas ranging from mountainous areas to urban areas, making it an inseparable part of the living activities of the surrounding community. This river is used as a source of raw water to meet domestic, agricultural, and industrial needs. In addition, this river is a place for household waste and industrial waste.

2 METHODS

This study was exploratory qualitative research that aimed to determine the diversity of microalgae in the Jangkok River, Mataram, Indonesia. The research was carried out in April 2022 along the Jangkok River from upstream to downstream. The difference in environmental characteristics around the river becomes the basis for determining the observation station. There are six observation stations, with each station consisting of three sampling points.

Microalgae data collection begins by filtering 2 liters of water at each sampling point at each station using a plankton net so that 100 ml of sample water is obtained. The filter results are placed in a sample bottle, dripped with 4% formalin, and then labeled according to the station code, sampling point, and time of collection. Microalgae sampling was carried out three times for each station, with a span of 1 week for each sampling. The tools and materials used were a plankton net (20 μm), microscope, object and cover slide, coverslip, dropper, collection bottle, water samples, formalin (4%), and label paper.

Water samples that have been obtained are then observed in the laboratory to find microalgal specimens. The identification process for microalgae is carried out by referring to textbooks and scientific journals. The data were analyzed qualitatively by describing the microalgae species found and determining the diversity of microalgae based on the Shannon–Wiener diversity index (H').

3 RESULTS

Based on the identification results, 85 species of microalgae found were included in 4 divisions, namely *Bacillariophyta* (56 species), *Chlorophyta* (19 species), *Cyanophyta* (8 species), and *Euglenophyta* (2 species). The species found are presented in Table 1.

Table 1. Species of microalgae of Jangkok river.

Division	Species
Bacillariophyta	*Terpsinoe musica* Ehrenb, *Melosira varians* C. Agardh, *Diadesmis confervacea* Kütz, *Bacillaria paxillifera* (O. F. Müll.) Hendy, *Aulacoseira granulata* (Ehrenb.) Simonsen, *Cyclotella ocellata* Pant, *Lemnicola Hungarica, Craticula ambigua* (Ehrenb.), *Achnanthidium exiguum* (Grunow) Czarn, *Achnanthidium reimeri, Achnanthidium rosentocki, Pinnularia lundii* Hust, *Pinnularia viridis* (Nitzsch) Ehrenberg, *Pinnularia latera* Khammer, *Pinnularia acrosphaeria W. Smith, Rossithidium duthei, Nitzschia clausii* Hantzsch, *Nitzschia recta* Hantzsch, *Nitzschia angustata* (W. Sm.) Grunow, *Nitzschia amphibia* Grunow, *Nitzschia frustulum* (Kütz.) Grunow, *Nitzschia sigmoidea* (Nitzsch) W. Sm., *Nitzschia palea* (Kütz.) W. Sm., *Nitzschia dissipata* (Kütz.) Grunow, *Tryblionella calida, Trybionella debilis* (Arn.) Grunow in Cleve & Grunow, *Synedra ulna* (Nitzsch) Ehrenberg, *Stenopterobia sigmatella, Stauroneis anceps* Ehrenberg, *Sellaphora pupula* (Kützing) Mersechkowsky, *Sellaphora proxima, Planothidium apiculatum, Cymbella turgidula* Grunow, *Navicula salinarum* Grunow, *Navicula gregaria* Donkin, *Navicula lanceolata* (Agardh) Kützing, *Gomphonema truncatum* Ehrenberg, *Gomphonema sphaerophorum* Ehrenberg, *Gomphonema oliva-ceoides* Hust, *Gomphonema duplipunctatum, Gomphonema consector* M. H. Hohn & Hellerm, *Luticola geoppertina* (Bleisch in Rabenh.), *Luticola olegsakharovii, Luticola Mutica* (Kützing), *Gyrosigma acuminatum* (Kütz.) Rabenh, *Surrirela eleganish* Ehrenberg, *Surirella tenera* W. Greg, *Surirella biseriata* Bréb. & Godey, *Surrirela biseriata var bifrons* (Ehrenberg) Hust, *Achnanthes crenulata, Pseudostaurosira brevistriata* (Grunow in Van Heurck) D. M. Williams & Round, *Cymatopleura elliptica* (Bréb. & Godey) W. Sm, *Eucocconeis flexella* (Kütz.) Cleve, *Caloneis undulata* (W. Greg.) Krammer in Krammer & Lange-Bert, *Rhopalodia gibba* (Ehrenb.) O. Müll, *Eunotia arcus* Ehrenberg
Chlorophyta	*Monoraphidium contortum, Triploceras gracile* Bailey, *Haematococcus lacustris* (Girod-chantrans) Rostafinski, *Cosmarium botrytis* Borg, *Cosmarium granatum* Brebisson, *Cosmarium contractum* Kirchner, *Scenedesmus denticulatus* Lagerhiem, *Scenedesmus brasiliensis* Bohlin, *Scenedesmus quadricauda* var. *maximum* West and West, *Scenedesmus carinatus, Scenedesmus armatus* (Chodat) G. M. Smith, *Scenedesmus acuminatus* (Lagerheim) Chodat, *Ankistrodesmus falcatus* (Corda) Ralfs, *Coelastrum microporum, Coelastrum cambricum* Archer, *Coelastrum astroideum, Tetrastrum heteracanthum* (Nordstedt) Choda, *Pediastrum duplex* var *gracillimum, Pediastrum boryanum* (Turpin) Meneghini, *Pediastrum duplex* Meyen, *Dictyosphaer-ium granulatum*
Cyanophyta	*Spirulina sp, Glaucospira laxissima, Oscillatoria limosa* (Roth) C. Agardh, *Oscillatoria agardhii* Gomont, *Oscillatoria refringens, Oscillatoria subbrevis* Schmidle, *Microcystis aeruginosa* (Kutzing) Lemmermann, *Nostoc commune*
Euglenophyta	*Euglena sociabilis, Trachelomonas scabra*

Referring to the Shanon–Wiener diversity index calculation, it is known that the diversity index (H') of microalgae in the Jangkok River was 3.163, which means that the diversity of microalgae species in the river is in the category of high diversity level. The discovery of certain species with dominant numbers compared to other species as well as several potential microalgae genera as bioindicators such as *Nitzschia, Scenedesmus, Gomphonema*, and *Oscillatoria* indicate that although they have a high level of diversity, the quality of this river is in a lightly polluted state.

4 DISCUSSION

Changes in environmental conditions are mostly a result of technological advancements, human population growth, and increasingly diverse human activity. A significant environmental issue is climate change, which is exacerbated by the rising demand for food, clothing, land, and clean water because of an expanding human population. As a result of numerous changes and severe harm to the lives of various organisms on the earth, the entry of pollutants from organic, inorganic, and heavy metals into the environment has long been a threat to the sustainability of ecosystems and the environment. It is inevitable that various increasingly complicated environmental issues will arise.

In this sense, the rate at which environmental issues occur is not balanced with the human capacity to address them. This is in line with the statement of Widowati *et al.* (2021) that an increase in the type and level of environmental problems is not accompanied by an increase in measures to resolve them. Improper behavior and a lack of environmental responsibility are the root causes (Istiana *et al.* 2020). Knowledge and comprehension of the environment have an impact on a person's conduct toward it (Sukri *et al.* 2022). How effectively a person learns at school, especially in terms of his literacy skills toward the environment, has a direct impact on how well he knows, understands, and behaves toward his environment.

The entry of domestic waste, sanitation waste, agricultural waste (insecticides, pesticides, fungicides, and chemical fertilizers), as well as industrial factory waste, into the river causes it to become polluted. This condition will affect the community structure of living things in the river, including microalgae. Algal communities will change rapidly in response to changes in water quality (Sepriyaningsih & Harmoko 2020). Algae will change the composition or diversity of species, which was originally rich in species, to become a monotonous community. In this case, various human activities have caused an imbalance of ecosystems that risks the extinction of species and the loss of biodiversity. The first reaction is a quantitative change in the community, which is then followed by a qualitative change (Ale *et al.* 2019; Fetscher *et al.* 2014).

The findings in this study showed the same thing: one species of microalgae was found in more abundance (dominant) than other microalgae species. The microalgae found in more abundance are *Synedra ulna*. As a member of the Bacillariophyta division, *Synedra ulna* has good tolerance and adaptability to changing water conditions (Kim 2018), including changes caused by the entry of pollutants (Meeranayak *et al.* 2020). Changes in water conditions can occur because of the entry of pollutants into the waters, both in the form of organic materials and heavy metals. Algae have the ability to remove pollutants from a wide range of organic, inorganic, and radioactive materials. Therefore, the use of algae through the phytoremediation mechanism in controlling and solving environmental problems has enormous potential for a sustainable future.

Algae can remove pollutants such as CO_2, SO_2, and NO_2 and other particles dissolved in the air or water by utilizing them for growth or by accumulating these particles. As phytoplankton in aquatic ecosystems, algae unload CO_2 through the process of photosynthesis and then produce O_2, which is used by other organisms at higher trophic levels (Bourgougnon *et al.* 2021).

It was further stated that certain species of algae showed the ability to remove heavy metals, explosives, and petroleum contaminants from different water sources (Tripathy *et al.* 2021). The increasing use of chemical products and products made from plastic in modern society can intensively pollute water sources on the earth. As a result, the source of clean water for the community is decreasing. As a renewable agent for bioremediation, the ability of algae as a biofilter provides the possibility of easier access to clean water supply and sanitation (ElFar *et al.* 2021; Ullmann & Grimm 2021). Apart from that, algae also help to remove dissolved elements from the water, which lead to the growth of harmful bacteria and successive problems.

In connection with the role of algae in the environment and sustainability in the future, the provision of ecological understanding in the community through environmental education for sustainability must be a serious concern for all parties, especially educators, researchers, and policymakers. Besides preserving the environment and biodiversity, we also live a better life by synergizing with the natural environment.

Regarding the objectives of sustainable development, it is essential to empower people's perceptions, attitudes, and actions regarding the environment. One's actions and behaviors toward their environment can be influenced by their understanding of the diversity of species in nature and how they interact with the environments in which they are found. As with protecting rivers and the creatures that live in them, having a thorough understanding of the complexity of an ecosystem like a river ecology enables one to discover more practical means of using its resources in a sustainable manner (Bourgougnon *et al.* 2021).

5 CONCLUSION

This study was limited to the diversity of freshwater microalgae found in the Jangkok River, Lombok, Indonesia. We found 48 genera and 85 species of microalgae in the Jangkok River, Lombok, Indonesia. The diversity of microalgae in this river was included in the high category level with an index value (H') = 3.163. This indicates that the quality of the Jangkok River in Lombok, Indonesia is in a lightly polluted state. Given the important role of this river in people's lives, for sustainability in the future, more serious attention is needed to the condition of the river and the potential entry of pollutant substances into the river flow, both from domestic and industrial activities. Further research, such as the isolation of microalgae species that have been found and have potential as bioaccumulators of pollutants, especially heavy metals, and the development of environmental learning models for communities around rivers, is very important.

REFERENCES

Ale, O., Sawant, N., Gupta, P., and Bhot, M., 2019. Biodiversity of fresh water microalgae of Gauripada Lake, Bhagva Lake (Kala Talao), Masunda Lake (Talaopali) and Brahmala Lake of Thane District, Maharashtra. *International Journal of Emerging Technologies and Innovative Research*, 6 (5), 111–118.

Bellinger, E.G. and Sigee, D.C., 2015. *Freshwater Algae: Identification, Enumeration and Use as Bioindicators.* Second. Chichester: John Wiley & Sons, Ltd.

Bourgougnon, N., Burlot, A.S., and Jacquin, A.G., 2021. Algae for global sustainability? *In*: *Advances in Botanical Research*. Academic Press Inc., 145–212.

Costa, D., Burlando, P., and Priadi, C., 2016. The importance of integrated solutions to flooding and water quality problems in the tropical megacity of Jakarta. *Sustainable Cities and Society*, 20, 199–209.

Efroymson, R. and Dale, V., 2015. Environmental indicators for sustainable production of algal biofuels. *Ecological Indicators*, 49, 1–13.

ElFar, O.A., Chang, C.K., Leong, H.Y., Peter, A.P., Chew, K.W., and Show, P.L., 2021. Prospects of Industry 5.0 in algae: Customization of production and new advance technology for clean bioenergy generation. *Energy Conversion and Management: X*, 10.

Fetscher, A.E., Stancheva, R., Mazor, R.D., Ode, P.R., and Busse, L.B., 2014. Development and comparison of stream indices of biotic integrity using diatoms vs. non-diatom algae vs. a combination. *Journal of Applied Phycology*, 26 (1), 433–450.

Istiana, R., Sunardi, O., Herlani, F., Ichsan, I.Z., Rogayan Jr, D. v., Rahman, Md.M., Alamsyah, M., Marhento, G., Ali, A., and Arif, W.P., 2020. Environmentally responsible behavior and naturalist intelligence: Biology learning to support sustainability. *Biosfer: Jurnal Tadris Biologi*, 11 (2), 87–100.

Khan, A., 2016. Aquatic plant biodiversity: A biological indicator for the monitoring and assessment of water quality. *In: Plant Biodiversity: Monitoring, Assessment and Conservation.*

Kim, H.S., 2018. Diversity of phytoplankton species in cheonjin lake, northeastern south korea. *Journal of Ecology and Environment*, 42 (1), 1–19.

Kumar, V., Jaiswal, K.K., Verma, M., Vlaskin, M.S., Nanda, M., Chauhan, P.K., Singh, A., and Kim, H., 2021. Algae-based sustainable approach for simultaneous removal of micropollutants, and bacteria from urban wastewater and its real-time reuse for aquaculture. *Science of the Total Environment*, 774.

Lobo, E.A., Heinrich, C.G., Schuch, M., Wetzel, C.E., and Ector, L., 2016. Diatoms as Bioindicators in Rivers. *In: River Algae.* Switzerland, 245–271.

Marganingrum, D., Djuwansah, M.R., and Mulyono, A., 2018. Penilaian daya tampung sungai jangkok dan sungai ancar terhadap polutan organik. *Jurnal Teknologi Lingkungan*, 19 (1).
Maznah, W.O.W. and Makhlough, A., 2014. Water quality of tropical reservoir based on spatio-temporal variation in phytoplankton composition and physico-chemical analysis. *International Journal Environmental Science technology*, 12 (7), 2221–2232.
Meeranayak, U.F.J., Nadaf, R.D., Toragall, M.M., Nadaf, U., and Shivasharana, C.T., 2020. The Role of algae in sustainable environment: A review. *Journal of Algal Biomass Utilization*, 11 (2), 28–34.
Parmar, T.K., Rawtani, D., and Agrawal, Y.K., 2016. Bioindicators: The natural indicator of environmental pollution. *Frontiers in Life Science*, 9 (2), 110–118.
Perić, M.S., Kepčija, R.M., Miliša, M., Gottstein, S., Lajtner, J., Dragun, Z., Marijić, V.F., Krasnići, N., Ivanković, D., and Erk, M., 2018. Benthos-drift relationships as proxies for the detection of the most suitable bioindicator taxa in flowing waters – a pilot-study within a Mediterranean karst river. *Ecotoxicology and Environmental Safety*, 163 (June), 125–135.
Sahoo, D. and Seckbach, J., 2015. *The Algae World.*
Sepriyaningsih and Harmoko, 2020. Keanekaragaman mikroalga bacillariophyta di sungai mesat kota lubuk linggau. *Quangga: Jurnal Pendidikan dan Biologi*, 11 (2).
Sukri, A., Rizka, M.A., Purwanti, E., Ramdiah, S., and Lukitasari, M., 2022. Validating student's green character instrument using factor and rasch model. *European Journal of Educational Research*, 11 (2), 859–872.
Tripathy, A., More, R.D., Gupta, S., Samuel, J., Singh, J., and Prasad, R., 2021. Present and future prospect of Algae: A potential candidate for sustainable pollution mitigation. *The Open Biotechnology Journal*, 15 (1), 142–156.
Ullmann, J. and Grimm, D., 2021. Algae and their potential for a future bioeconomy, landless food production, and the socio-economic impact of an algae industry. *Organic Agriculture*, 11 (2).
Venkatachalapathy, R. and Karthikeyan, P., 2015. Diatom indices and water quality index of the cauvery river, India: Implications on the suitability of bio-indicators for environmental impact assessment. *Environmental Management of River Basin Ecosystems*, 707–727.
Widowati, C., Purwanto, A., and Akbar, Z., 2021. Problem-based learning integration in stem education to improve environmental literation. *International Journal of Multicultural and Multireligious Understanding*, 8 (7), 374.

Religion, Education, Science and Technology towards a More Inclusive and Sustainable Future – Rahiem (Ed.)

Diversity of pollinators in cucumber plantation on organic and conventional farming in Jambi

B. Kurniawan*, D. Putra, L. Anggriani, A.Q. Manurung, M. El Widah & Ramlah
UIN Sulthan Thaha Saifuddin, Jambi, Indonesia

ABSTRACT: Insect pollinators diversity is related to the type of farming system (organic or non-organic). This research was to compare diversity and abundance, number of species, and visitation rates per hour of insect pollinators at organic and conventional cucumber farms. The study was conducted in June 2020 on 300 cucumber plants on both organic and conventional land. The data analysis shows that organic land has an index in the medium category 1-3 (H' = 1.57), Simpson's dominance index (D = 0.40), and even Pielou index (E = 0.58). Conventional land has a low index (H' = 0.69), Simpson's dominance index (D = 0.72), and Pielou's evenness index (E = 0.33). Eleven species are found on the organic cucumber farm, namely *Xylocopa confusa, X. latipes, Apis cerana, A. dorsata, Heterotrigona itama, Hypolimnas bolina, H. Misippus, Euripus nyctelius, Euploea mulciber, Neptis hylas,* and *Delias hyparete. Euripus nyctelius* and *Euploea mulciber* were not found on the conventional farm.

Keywords: pollinator, cucumber, plantation, pesticide, biodiversity

1 INTRODUCTION

Along with human population growth and a shift towards vegetable consumption, the demand for fresh vegetables such as cucumbers is increasing (Motzke *et al.* 2016). Cucumber is a plant with male and female flowers found on the same plant, and the pollination process of cucumber flowers requires the help of pollinators. The pollinator plays a role in transferring pollen from the anther of male flowers to the stigma of female flowers for fertilization. Pollinating insects play a major role in the cycle of nutrient availability and plant fertilization (Kumar *et al.* 2018). Agricultural intensification for more than half a century has resulted in the loss of biodiversity due to the application of synthetic pesticides and fertilizers as well as the degradation of natural habitats (Goded *et al.* 2019).

Insect pollinators are sensitive to environmental changes caused by the application of pesticides and changes in temperature and humidity (Kumar *et al.* 2018). The active ingredient deltamethrin in insecticides has the potential to poison butterfly pollinators, which causes 100% mortality in pollinating insects such as *Apis cerana* and *A. mellifera* (Braak *et al.* 2018). Organic farming systems can increase and counteract the decline in biodiversity, especially pollinators, other invertebrates, and birds (Goded *et al.* 2019). Organic farming systems are more environmentally friendly because they produce less pollution (Meemken & Qaim 2018). Most of the studies analyzing the effects of organic agriculture on biodiversity have focused on species richness and species abundance (Goded *et al.* 2019).

Pollination studies on cucumbers that compare research designs using organic and conventional farming systems have not been widely reported. In this research, we want to see how the biodiversity on organic farming lands is changing, considering this organic farming system

*Corresponding Author: bayu.kurniawan@uinjambi.ac.id

DOI: 10.1201/9781003322054-48

is starting to be abandoned by the community. In this study, we analyzed how organic and conventional farming affected diversity and abundance, the number of species, and individuals against the visiting time of pollinators on organic and conventional cucumber farms.

2 METHODS

2.1 *Time and place of research*

The study was conducted in June 7–13, 2020, on organic and conventional agricultural land, Bagan Pete Village, Jambi City, at an altitude of 42 m above sea level with the coordinate point S 01°39.606 'E 103°33.592'.

2.2 *Planting conditions and experimental design*

This study was carried out on 300 plants, for each organic and conventional group, which were cultivated by standard cultivation methods. Cucumbers of organic groups were fertilized by manure without the application of insecticide, while the conventional group was fertilized by chemical fertilizer N-P-K 25 kg/375 m^2, followed by insecticide application (active ingredient deltamethrin 25 g / l). Insecticide application is applied once every 7 days in the morning at 08.00–09.00 WIB, and pesticide application has been done three times. Organic and conventional farming locations are 200 meters apart.

2.3 *Pollinator observations*

Observation of pollinator diversity was carried out in the vegetative phase of flowers using scan sampling, which was carried out from the age of the plant 15 days to 36 days, or for 21 consecutive days, but we only analyzed the data on the peak flowering period that occurred for 7 days. Pollinator observations were made in sunny weather with no rain. Environmental parameters measured include air temperature, air humidity, and light intensity every hour when observed every day. Observations were carried out for 45 minutes every hour at 07.00–12.00 am (Ruiz-Toledo *et al.* 2020). At the location of this study, there are 300 cucumber plants on organic farmland and 300 cucumber plants on conventional farmland, with each land area of 375 m2. Pollinator observations were made in sunny and not rainy weather. The environmental parameters measured included air temperature, humidity, and light intensity every hour, when observed every day.

2.4 *Collection and identification*

Insect collection is done by active and passive collection. The passive collection method used yellow pan traps (Hevia *et al.* 2016) and blue pan traps (Jeavons *et al.* 2020) conducted from 08.00 am to 12.00 am. The pan trap used is 15 cm in diameter and contains soapy water (half the volume of the trap pan). The height of the pan trap is placed 50 cm from the ground at a distance of 5 m each. The yellow pan trap and blue pan trap methods are very suitable for this type of research that compares pollinator communities in various types of research locations (Hevia *et al.* 2016) and active methods using insect nets by tracing the entire study area (Ruiz-Toledo *et al.* 2020).

All types of pollinating insects collected were stored in jars (coded with the date and time of capture) containing ethyl acetate, and each type of insect collected was recorded to determine the species of the pollinating insects (O'Brien & Arathi 2019).

Pollinators caught using insect nets and yellow and blue pan traps were preserved and identified to the genus level and, where possible, to the species level in the laboratory with the aid of a stereo microscope and using a book of determination (Wilson & Carril 2016).

The Shannon–Wiener diversity index (Qiu *et al.* 2018) and Simpson's abundance index (O'Brien & Arathi 2019) can effectively reflect the temporal and spatial changes of pollinator communities. The formula of Shannon–Wiener diversity index $H' = -\sum_{i=1}^{s}(\text{PilnPi})$, Simpson's abundance index $D = 1/\sum_{i=1}^{R}(Pi^2)$, and Pielou's evenness index € (Rusman *et al.* 2016).

Pearson correlation is used to determine whether there is a relationship between pollinator visits to agricultural land and environmental parameter variables. The diversity and abundance of pollinators were analyzed using the Shannon–Wiener index, Simpson's index (O'Brien & Arathi 2019) and Pielou's evenness index (E) (Rusman *et al.* 2016).

3 RESULTS AND DISCUSSION

3.1 *Diversity of insect pollinators in organic and conventional lands*

In this study, organic land has a greater number of species and individuals than conventional land. On organic land, 15 species were found (three orders of Hymenoptera, Lepidoptera, and Diptera), for a total of 771 individuals. On conventional land, 8 species were found (three orders of Hymenoptera, Lepidoptera, and Diptera) with a total of 250 individuals (Table 1). Hymenoptera is a group that has the highest abundance of individuals at the research location, namely 660 individuals with 6 species (organic land) and 238 individuals with 4 species (conventional land). Lepidoptera has an abundance of 99 individuals with 8 species (organic land) and 7 individuals with 3 species (conventional land) (Table 1).

Table 1. Species and number of individual insect pollinators on organic land cucumber plants.

Ordo Family Species	Number of Individual	
	Organic	Conventional
Hymenoptera		
Apidae		
Xylocopa confusa	15	2
X. latipes	12	–
Apis cerana	84	14
A. dorsata	33	11
Heterotrigona itama	474	211
Ceratina sp.	42	–
Lepidoptera		
Nymphalidae		
Hypolimnas bolina	16	3
H. misippus	14	2
Euripus nyctelius	13	–
Euploea mulciber	14	–
Neptis hylas	15	–
Pieridae		
Delias hyparete	13	–
Eurema hecabe	6	–
Appias olferna	8	2
Diptera		
Stratiomyidae		
Hermetia remittens	12	5
Total	771	250
Shannon–Wiener Index (H')	1.57	0.69
Simpson's Index (D)	0.4	0.72
Pielou's evenness Index €	0.58	0.33

Two types of agricultural land used in this study, organic land and conventional land, showed differences in information on the diversity index, number of species, and number of individuals in the dominant species found. Organic land has a higher species richness and abundance when compared with conventional land based on the diversity index, as well as the number of species and the number of individuals found. The effect of land use change (organic and conventional) on diversity can occur because it contributes to the overall understanding of the ecology, and the mechanisms driving the loss of biodiversity, and the impact these changes can have on ecosystems (Cadotte *et al.* 2011). In this study, organic land has a greater number of species and individuals than conventional land. On organic land, 15 species were found (3 orders of Hymenoptera, Lepidoptera, and Diptera), for a total of 771 individuals. On conventional land, 8 species were found (three orders of Hymenoptera, Lepidoptera, and Diptera) with a total of 250 individuals (Table 1). Hymenoptera is a group that has the highest abundance of individuals at the research location, namely 660 individuals with 6 species (organic land) and 238 individuals with 4 species (conventional land). Lepidoptera has an abundance of 99 individuals with 8 species (organic land) and 7 individuals with 3 species (conventional land). The results of the analysis on organic land have a pollinator diversity index 'H') indicating that organic land has a diversity index in the moderate category 1-3'(H' = 1.24), Simpson's dominance index (D = 0.49), Pielou's evenness index (E = 0.52). In organic land, the evenness of species with the number of individuals is relatively the same.

The results of the analysis on the conventional farm have a low index of low diversity, namely 'H' = 0.62), Simpson's dominance index (D = 0.74) shows that there are species that have high abundance when compared with the ratio of species present; Pielou's evenness index (E = 0.28) in the low category means that the individual wealth of each species has a significant difference.

Environmental parameters that have a relationship with visiting insect pollinators are temperature and humidity. The results of the Pearson correlation analysis showed that air temperature and humidity were positively correlated with visiting insect pollinators ($r = -0.89$, $P = 0.04$) ($r = 0.93$, $P = 0.21$), while light intensity was not significantly correlated ($P > 0.05$).

3.2 *Number of species and individuals with respect to time of visit to flower*

The results also showed that the number of individual pollinators during the 7 days of observation was relatively the same in organic land, ranging from 94–110 individuals. Conventional land has a smaller number of individuals when compared with organic land. At the time of application of the deltamethrin-type pesticide, the population decreased drastically, with 20 individuals found on the fourth day of observation. Before and after pesticide application, there was a relatively stable increase, but not as much as in organic land (Figure 1). The relationship between pollinators and plants is influenced by flower

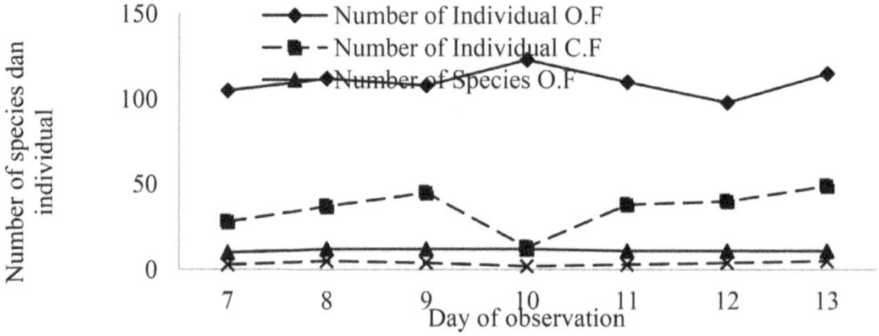

Figure 1. Comparison of the number of species and individual pollinators on organic and conventional agricultural lands.

morphology. The success of pollination by pollinators is also influenced by flower color. Flower color is related to the population of pollinating species (Reverté *et al.* 2016). Cucumber flowers produce a lot of nectar and pollen as an attractive pollinator, especially bees and beetles (carpenter bees).

In addition, the presence of pollinators is also important in increasing fruit sets in cucumbers. According to the results of the study by Hossain *et al.* (2018), pollinator activity is highest in the morning. The fruit set yield is 70% higher than without pollinator insects and the percentage of healthy fruit is 85%.

The results also showed that the number of individual pollinators during the 7 days of observation was relatively the same in organic land, ranging from 94–110 individuals. Conventional land has a smaller number of individuals when compared with organic land. At the time of application of the deltamethrin-type pesticide, the population decreased drastically with the number of 20 individuals found on the fourth day of observation. Before and after pesticide application, there was a relatively stable increase, but not as much as in organic land (Figure 1).

The activity of visiting the highest number of pollinator species was in the morning at 09.45 at an ambient temperature of 27°C with an air temperature of 80%, and the number of species found continued to decrease along with an increase in ambient temperature (29°C) and a decrease in air humidity (60%) until noon (Figure 2). This environmental factor also affects the presence of pollinators in flowers. Pollinator abundance has a positive relationship with relative humidity and ambient temperature (Munyuli 2012). This is in accordance with the results of research showing that the effect of temperature and relative humidity on the presence of pollinators at the study site (r = 0.999, P = 0.01), in contrast to the effect of light intensity (r = 0.103, P = –0.897) does not show a positive correlation. (Munyuli 2012) explained that the foraging behavior of bees also depends on temperature and light intensity.

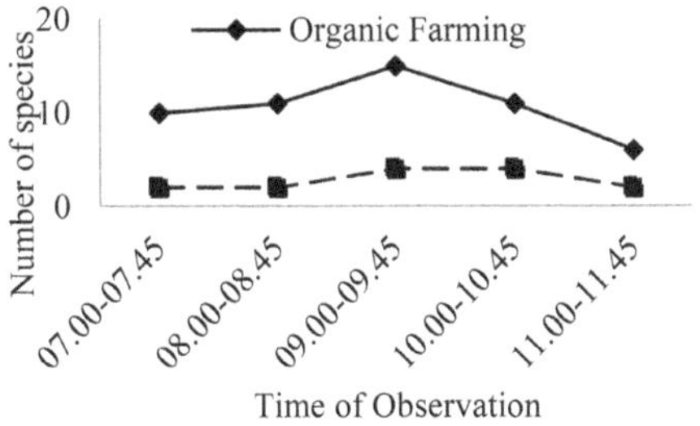

Figure 2. Comparison of the number of species visiting organic and conventional land.

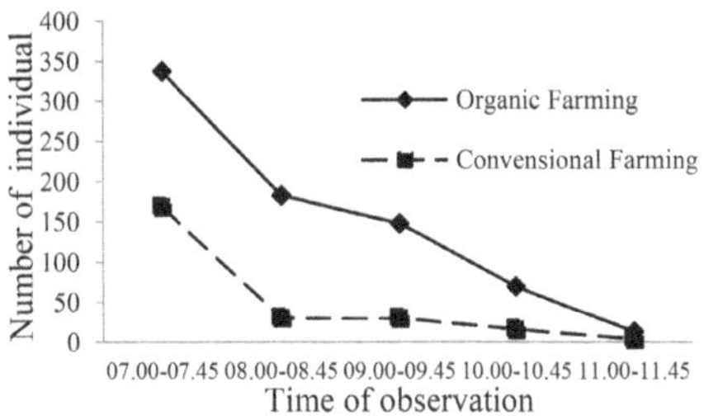

Figure 3. Comparison of time and the number of individual pollinators visiting flowers.

However, based on research (Munyuli 2012) species richness and abundance of pollinators have a negative correlation with the increase in mean annual temperature when comparing temperatures in previous years and temperatures in current years. This indicates the potential vulnerability of bee species in the future to face global climate change.

Pollinator visits to cucumber flowers with the highest species occurred at 09.00–10.45 on both organic and conventional land. At 09 am the organic land had the highest number of species, with 11 species and 9 species on conventional land. Along with the increase in ambient temperature, there is a decrease in the presence of pollinators. At 11.00–11.45 hours, only four species were found on organic land and two species on conventional land (Figure 2). In this study, there were two species that were not found on conventional land, namely *Euripus nyctelius* and *Euploea Mulciber*, for 7 consecutive days of observation (Table 1). The highest individual abundance occurred at 07.00, with as many as 317 on organic land and 259 on conventional land. With an increase in environmental temperature, the presence of pollinators is decreasing and the lowest is at 11.00–11.45, when only 13 individuals were found on the organic land and four individuals on conventional land (Figure 3).

4 CONCLUSION

Pollinating insects are very important in increasing cucumber production. The diversity of pollinators has differences in organic and conventional land. The highest diversity index was on the organic land with a medium category, and on the conventional land with a low category. Eleven species were found on the organic farm and nine species on the conventional farm. Efforts to conserve pollinators, use of organic land, and reduction in pesticide applications are needed in an effort to increase the success of the pollination process of food crops. Increasing the availability of plant resources in organic farming can provide benefits and richness to pollinator species.

REFERENCES

Braak, Nora, Rebecca Neve, Andrew K. Jones, Melanie Gibbs, and Casper J. Breuker. 2018. “The effects of insecticides on butterflies – a review.” *Environmental Pollution* 242: 507–18. https://doi.org/10.1016/j.envpol.2018.06.100.

Cadotte, Marc W., Kelly Carscadden, and Nicholas Mirotchnick. 2011. “Beyond species: Functional diversity and the maintenance of ecological processes and services.” *Journal of Applied Ecology* 48 (5): 1079–87. https://doi.org/10.1111/j.1365-2664.2011.02048.x.

Goded, Sandra, J. Ekroos, Joaquín G. Azcárate, José A. Guitián, and Henrik G. Smith. 2019. “Effects of organic farming on plant and butterfly functional diversity in mosaic landscapes.” *Agriculture, Ecosystems and Environment* 284 (October 2018). https://doi.org/10.1016/j.agee.2019.106600.

Hevia, Violeta, Jordi Bosch, Francisco M. Azcárate, Eva Fernández, Anselm Rodrigo, Helena Barril-Graells, and José A. González. 2016. “Bee diversity and abundance in a livestock drove road and its impact on pollination and seed set in adjacent sunflower fields.” *Agriculture, Ecosystems and Environment* 232: 336–44. https://doi.org/10.1016/j.agee.2016.08.021.

Hossain, M. S., F. Yeasmin, M. M. Rahman, S. Akhtarl and M. A. Hasnat. 2018. “Role of insect visits on cucumber (Cucumis Sativus L.) Yield.” *J. Biodivers. Conserv. Bioresour. Manag.* 4(2). https://doi.org/10.3329/jbcbm.v4i2.39854.

Jeavons, Emma, Joan van Baaren, and Cécile Le Lann. 2020. “Resource partitioning among a aollinator guild: A case study of monospecific flower crops under high honeybee pressure.” *Acta Oecologica* 104 (November 2019). https://doi.org/10.1016/j.actao.2020.103527.

Kumar, Sanjay, PC Joshi, Pashupati Nath, and Vinaya Kumar Singh. 2018. “Impacts of insecticides on pollinators of different food plants.” *Entomology, Ornithology & Herpetology: Current Research* 07 (02). https://doi.org/10.4172/2161-0983.1000211.

Motzke, Iris, Alexandra-maria Klein, Shahabuddin Saleh, and Thomas C Wanger. 2016. “Agriculture, ecosystems and environment habitat management on multiple spatial scales can enhance bee pollination and crop yield in tropical homegardens.” *“Agriculture, Ecosystems and Environment”* 223: 144–51. https://doi.org/10.1016/j.agee.2016.03.001.

Munyuli, M. B. Théodore. 2012. “Micro, local, landscape and regional drivers of bee biodiversity and pollination services delivery to coffee (Coffea Canephora) in Uganda.” *International Journal of Biodiversity Science, Ecosystem Services and Management* 8 (3): 190–203. https://doi.org/10.1080/21513732.2012.682361.

O’Brien, C., and H. S. Arathi. 2019. “Bee diversity and abundance on flowers of industrial hemp (Cannabis Sativa L.).” *Biomass and Bioenergy* 122 (January): 331–35. https://doi.org/10.1016/j.biombioe.2019.01.015.

Ruiz-Toledo, Jovani, Rémy Vandame, Patricia Penilla-Navarro, Jaime Gómez, and Daniel Sánchez. 2020. "Seasonal abundance and diversity of native bees in a patchy agricultural landscape in Southern Mexico." *Agriculture, Ecosystems and Environment* 292 (June 2019). https://doi.org/10.1016/j.agee.2019.106807.

Rusman, Ratih, Tri Atmowidi, and Djunijanti Peggie. 2016. "Butterflies (Lepidoptera: Papilionoidea) of Mount sago, west sumatra: Diversity and flower preference." *HAYATI Journal of Biosciences* 23 (3): 132–37. https://doi.org/10.1016/j.hjb.2016.12.001.

Stanley, Johnson, Khushboo Sah, S K Jain, J C Bhatt, and S N Sushil. 2015. "Chemosphere evaluation of pesticide toxicity at their field recommended doses to honeybees, Apis cerana and A. Mellifera through laboratory, semi-field and field studies." *Chemosphere* 119: 668–74. https://doi.org/10.1016/j.chemosphere.2014.07.039.

Vinod, Madhurima, Anjan Kumar Naik, Renuka Biradar, Prashant K Natikar, and R A Balikai. 2016. "Bee pollination under organic and conventional farming systems: A review." *Journal of Experimental Zoology India* 19 (2): 643–52.

Wilson, J.S, & Carril, O.M. 2016. *The Bees in Your Backyard: A Guide to North America's Bees*. New Jersey: Princeton University Press.

Religion, Education, Science and Technology towards a More Inclusive and Sustainable Future – Rahiem (Ed.)
© 2024 the Editor(s), ISBN: 978-1-032-56461-6

Author index

Abbas, P. 179
Acim, S.A. 172
Afiah, K.N. 235
Akasi, H.Y. 184, 213
Albantani, A.M. 196
Alfian 156
Ali, A.A. 103
Aliyas 218
Al Munir, M.I. 179
Anggriani, L. 285
Anwar, K. 90, 156
Ariani, R.D. 191
Arifin, A. 123
Arifullah, M. 161
Asmawi 61
Atika 108
Azizy, J. 61, 72, 240
Azlan, U. 161

Badarussyamsi, B. 50
Baharudin 108
Basuki, S. 223
Bintarawati, F. 44

Dasrizal 61
Del Castillo, F. 33
Dewi, F.K. 108
Diana, R.R. 149

El Widah, M. 285
El Widdah, M. 90
Ermawati, E. 50

Fahrurrozi 133
Fakhlina, R.J. 223
Fakhri, M. 133
Fatra, M. 97
Febriani, N.A. 72
Ferawati, R. 191
Fuadi, A. 172

Ginanto, D.E. 156

Habibullah, M. 179
Hadiyana, E. 133
Halim, A. 191
Hasan, H. 72, 123, 240
Helmi, M.I. 196
Helsing, C. 3
Henry, P.M. 24
Hidayah, N. 249, 267
Hidayati, S. 249, 267
Huda, M. 83, 113, 144

Idarianty, M.F. 90
Irfan, M.N. 44

Jahar, A.S. 254
Jamaludin 230
Jannah, S. 61
Jannah, S.R. 179
Jayanti, E.T. 279
Jehaut, R.M. 97

Kamarusdiana 249
Khairani, D. 144
Khoiri, A. 128, 235, 240
Kholiq, A.N. 44
Kultsum, U. 149
Kurniati, E. 207
Kurniawan, B. 285
Kusmana 235, 240

Latifa, R. 77
Lubis, M.R. 128
Lutfi, L. 44

Maani, B. 218
Mahida, N.F. 77
Maigahoaku, F.D. 97
Mailinar 161
Manurung, A.Q. 285
Masiyan 179
Masruroh, S.U. 144
Mohamed, H.I. 66
Muhtar, F. 172
Mujib, A. 149
Mulyadin, T. 156
Munawar, W. 261
Mursalin, A. 108
Musli 161
Muslimah, H.L. 149
Muslimin, J.M. 261
Mustika, D. 161
Musyaffa, A.A. 90
Muttaqin, S. 83, 123, 144

Naim, S. 37
Nata, A. 113
Ningrum, D.A. 235, 240
Noprival 156
Nukhbatunisa 128, 240

Octaberlina, L.R. 97
Omukaba, O.R. 167

Palejwala, I.Y. 128
Pihasniwati 149
Prasaja, A.S. 191
Prasetyowati, R.A. 254
Purwati, N. 279
Putra, D. 285
Putra, K. 156

Rahim, H. 83, 97, 273
Ramadhan, S. 196
Ramlah 285
Reksiana 113
Ribahan 172
Ridwan, M. 50
Rifai, A. 223
Rizal, A.N.S. 235
Rofiq, M.K. 44
Rosadi, K.I. 90
Rosyada, D. 83, 113, 273
Rosyid, M. 44
Rozelin, D. 161
Rusydy, M. 179

Saeful, A. 83, 144
Saepudin, D. 149
Salim, A. 273
Salsabila, A.J. 235
Sari, T.Y. 202
Sarwenda 273
Sihabussalam 61, 72
Sihombing, A.A. 97
Silalahi, K.P. 202
Soetomo, M. 191
Subagio, E.A. 139
Subchi, I. 196, 254
Sujoko, I. 123
Sukarno 118
Sya'roni 108
Syukri, A. 118
Syukri 139, 218

Tamam, B. 72
Thamsin, F.M. 261
Triana, W. 128

Ubaidah, H.H. 72
Usep, A.M. 56

Wasath, G. 61
Widjayanto, F.R. 37
Wulandari, T. 108

Yaqinah, S.N. 230
Yazid, R. 123
Yulianti, K. 156

Zada, K. 267
Zamhari, A. 113, 128, 196, 207, 273
Zander, L. 13
Zuhdi, M. 83
Zulkifli 223
Zulyadain 139